Oxidative Stress and Inflammation in Non-communicable Diseases – Molecular Mechanisms and Perspectives in Therapeutics

Advances in Experimental Medicine and Biology

Volume 824

For further volumes:
http://www.springer.com/series/5584

Jordi Camps
Editor

Oxidative Stress and Inflammation in Non-communicable Diseases – Molecular Mechanisms and Perspectives in Therapeutics

Springer

Editor
Jordi Camps
Unitat de Recerca Biomèdica, Hospital
Universitari de Sant Joan, Institut
d'Investigació Sanitària Pere Virgili
Universitat Rovira i Virgili
Reus, Spain

ISSN: 0065-2598 ISSN: 2214-8019 (electronic)
ISBN 978-3-319-07319-4 ISBN 978-3-319-07320-0 (eBook)
DOI 10.1007/978-3-319-07320-0
Springer Cham Heidelberg New York Dordrecht London

Library of Congress Control Number: 2014945107

Printed on acid-free paper

Springer is part of Springer Science+Business Media (www.springer.com)

Contents

1 Introduction: Oxidation and Inflammation, A Molecular Link Between Non-communicable Diseases

Jordi Camps and Anabel García-Heredia

Abstract

Non-communicable diseases are, by definition, those chronic diseases that are non-infectious and non-transmissible. The most common non-communicable diseases are obesity, diabetes, cancer, and cardiovascular, chronic respiratory and neurological diseases. Altogether, they are the commonest cause of death and disability in modern world. Recent investigations show that many of these diseases share common pathophysiological mechanisms and are, at least in part, different manifestations in different organs of similar molecular alterations. Mitochondrial alterations, oxidative stress and inflammation are inextricably linked and play major roles in the onset and development of non-communicable diseases. Therefore, it is conceivable that pharmacological or nutritional manipulation of oxidation and inflammation allows a significant decrease in the mortality and morbidity associated to these diseases.

Keywords

Inflammation • Metabolism • Non-communicable diseases • Oxidation

Alchemists, from Ancient India and China to the Middle Ages, developed great and fruitless efforts to find the elixir of eternal life. This is a mythical potion that, when drunk from a certain cup at a certain time, supposedly grants the drinker eternal life and/or eternal youth. Certainly, to prolong youth, health, and life span as much as possible is also the avowed goal of Medicine and biomedical research, although the procedures have changed considerably since the introduction of the scientific method. For many years, the different diseases have been considered as separate entities and, consequently, deserved different treatments. However, recent investigations show that many diseases share common pathophysiological mechanisms and are, at least in part, different manifestations in different organs of very similar molecular alterations. This allows a more holistic approach to the disease processes, considered as a response from the organism to the aggressions of the environment.

J. Camps (✉) • A. García-Heredia
Unitat de Recerca Biomèdica, Hospital Universitari de Sant Joan, Institut d'Investigació Sanitària Pere Virgili, Universitat Rovira i Virgili,
C. Sant Joan, s/n, 43201 Reus, Catalunya, Spain
e-mail: jcamps@grupsagessa.com

J. Camps (ed.), *Oxidative Stress and Inflammation in Non-communicable Diseases – Molecular Mechanisms and Perspectives in Therapeutics*, Advances in Experimental Medicine and Biology 824,
DOI 10.1007/978-3-319-07320-0_1, © Springer International Publishing Switzerland 2014

Whole genome sequencing, Epigenetics, and the different Omics technologies (Genomics, Transcriptomics, Proteomics, Metabolomics, Fluxomics) provide researchers with very powerful tools to study the human organism as a whole. Therefore, the question arises: If different diseases share common molecular mechanisms, is it possible to find common therapeutic agents? Or: Is it possible that finding a treatment for a specific disease would decrease the probability of development of several others? The issue of the elixir of eternal life, in a more modest and realistic version, is again put on the table.

Non-communicable diseases (NCD) are, by definition, those chronic diseases that are noninfectious and non-transmissible. The most common NCD are obesity, cardiovascular disease, diabetes, cancer, chronic respiratory diseases, and neurological diseases. Altogether, they have been the commonest cause of death and disability globally for at least the last three decades [1]. Even in sub-Saharan Africa, NCD contribute a third of the disability-adjusted life year burden [2]. Mitochondrial alterations, oxidative stress and inflammation underpin NCD and are molecular mechanisms playing major roles in their onset and development (Fig. 1.1) [3–8]. Oxidation and inflammation are inextricably linked. For example, chronic inflammation is associated with oxidation, anti-inflammatory cascades are linked to decreased oxidation, increased oxidative stress triggers inflammation, and a proper redox balance inhibits the inflammatory cellular response. Whether or not oxidative stress and inflammation represent the causes or the consequences of cellular pathology, they contribute significantly to the pathogenesis of NCD. The incidence of diseases involving oxidative stress, inflammation, and their related metabolic disturbances is rising, as are age-related diseases due to progressively aging populations. Interrelations between the mechanisms of oxidative stress and of inflammatory signaling and metabolism are, in the broad sense of energy transformation, being increasingly recognized as part of the problem in NCD [9]. Obesity and associated metabolic disturbances, for example, are important factors that underlie NCD and are the consequences of unhealthy diets and physical inactivity [10]. Obesity predisposes to diabetes mellitus, hypertension, atherosclerosis, dyslipidemia, cancer, and coronary heart disease [11]. Growing evidence links a low-grade, chronic inflammatory state to obesity and its coexisting conditions as well as to noncommunicable diseases [12–18]. This low-grade inflammatory state is aggravated by the recruitment of inflammatory cells, mainly macrophages, to adipose tissue. Inflammatory cell recruitment is likely due to the combined effects of the complex regulatory network of cells and mediators that are designed to resolve inflammatory responses. Anti-inflammatory drugs have shown to reverse insulin resistance and other related conditions that result from circulating cytokines that cause and maintain insulin resistance [19–24]. Therefore, it is likely that inflammation *per se* is a causal factor for NCD rather than an associated risk factor.

The objective of this book *Oxidative Stress and Inflammation in Non-communicable Diseases: Molecular Mechanisms and Perspectives in Therapeutics*, belonging to the *Advances in Experimental Medicine and Biology* series, is to review some current concepts indicating that oxidation and inflammation are key mechanisms linking the major NCD and, therefore, it is conceivable that their therapeutic manipulation allows the treatment of the most prevalent diseases in our societies.

We are going to pay special attention to some molecules that appear to play a crucial role in oxidative and inflammatory processes. The paraoxonases are a family of three enzymes that degrade lipid peroxides and play an antioxidant role in both circulation and cells [25]. Desaturases are the rate-limiting enzymes for the conversion of polyunsaturated fatty acids, and their alterations have been associated with several NCD [26]. The chemokine (C-C motif) ligand 2 is a chemokine, the suppression of which reduces the attraction of immune cells to the sites of inflammation and prevents atherosclerosis and obesity [27]. The advanced glycation end products are products of chronic renal diseases that have been associated with the higher risk for cardiovascular disease and diabetes in these patients [28].

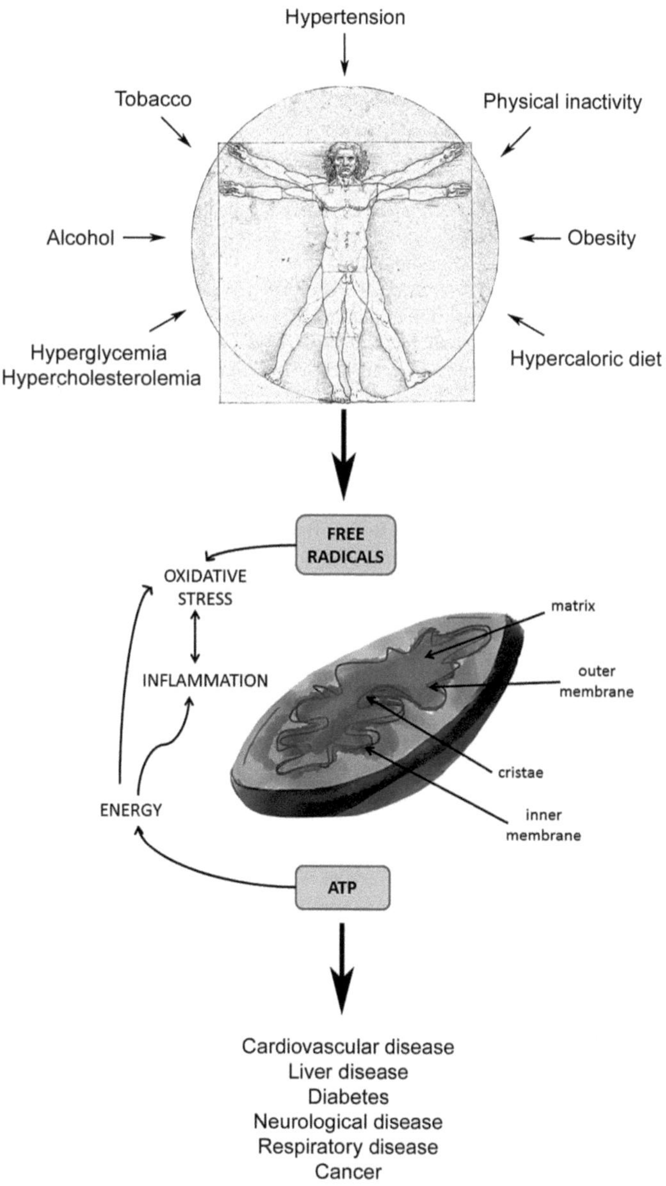

Fig. 1.1 Genetic and environmental factors, nutrition and lifestyle may induce a pro-oxidative and pro-inflammatory state, linked to alterations in mitochondrial structure and function. Many evidences strongly suggest that these processes underlie most non-communicable diseases

8-oxo-7,8-dihydro-2'-deoxyguanosine is a highly mutagenic altered nucleobase produced by oxidative stress that may be a link between obesity and cancer [29]. Other metabolic processes that are going to be described in the following chapters include the relationship between autophagy/mitophagy and inflammation, the role of adipokines, the polarisation of pro-inflammatory and anti-inflammatory macrophages, and the influence of the postprandial state on inflammation. Finally, the possibilities of therapeutic improvement by using synthetic high-density lipoproteins, apolipoprotein A-I mimetic peptides, or plant-derived polyphenols, are also going to be discussed.

References

1. Ebrahim S, Pearce N, Smeeth L, Casas JP, Jaffar S, Piot P. Tackling non-communicable diseases in low- and middle-income countries: is the evidence from high-income countries all we need? PLoS Med. 2013;10:e1001377.

2. Mensah GA, Mayosi BM. The 2011 United Nations high-level meeting on non-communicable diseases: the Africa agenda calls for a 5-by-5 approach. S Afr Med J. 2012;103:77–9.
3. Hoffmann JA, Kafatos FC, Janeway CA, Ezekowitz RA. Phylogenetic perspectives in innate immunity. Science. 1999;284:1313–8.
4. De Ferranti S, Mozaffarian S. The perfect storm: obesity, adipocyte dysfunction and metabolic consequences. Clin Chem. 2008;54:945–55.
5. Rull A, Camps J, Alonso-Villaverde C, Joven J. Insulin resistance, inflammation, and obesity: role of monocyte chemoattractant protein-1 (or CCL2) in the regulation of metabolism. Mediat Inflamm. 2010. doi:10.1155/2010/326580, pii: 326580.
6. Tarantino G, Savastano S, Colao A. Hepatic steatosis, low-grade chronic inflammation and hormone/growth factor/adipokine imbalance. World J Gastroenterol. 2010;16:4773–83.
7. Vinaixa M, Rodríguez MA, Rull A, Beltrán R, Bladé C, Brezmes J, et al. Metabolomic assessment of the effect of dietary cholesterol in the progressive development of fatty liver disease. J Proteome Res. 2010;9:2527–38.
8. Westermann B. Bioenergetic role of mitochondrial fusion and fission. Biochim Biophys Acta. 1817; 2012:1833–8.
9. Camps J, Rodríguez-Gallego E, García-Heredia A, Triguero I, Riera-Borrull M, Hernández-Aguilera A, et al. Adv Clin Chem. 2014;63:247–308.
10. Gaziano TA, Galea G, Reddy KS. Scaling up interventions for chronic disease prevention: the evidence. Lancet. 2007;370:1939–46.
11. Ferré N, Feliu A, García-Heredia A, et al. Impaired paraoxonase-1 status in obese children. Relationships with insulin resistance and metabolic syndrome. Clin Biochem. 2013;46:1830–6.
12. Kumar H, Lim HW, More SV, Kim BW, Koppula S, Kim IS, Choi DK. The role of free radicals in the aging brain and Parkinson's disease: convergence and parallelism. Int J Mol Sci. 2012;13:10478–504.
13. Osellame LD, Blacker TS, Duchen MR. Cellular and molecular mechanisms of mitochondrial function. Best Pract Res Clin Endocrinol Metab. 2012;26:711–23.
14. Arduíno DM, Esteves AR, Cardoso SM. Mitochondria drive autophagy pathology via microtubule disassembly: a new hypothesis for Parkinson disease. Autophagy. 2013;9:112–4.
15. Enache I, Charles AL, Bouitbir J, Favret F, Zoll J, Metzger D, et al. Skeletal muscle mitochondrial dysfunction precedes right ventricular impairment in experimental pulmonary hypertension. Mol Cell Biochem. 2013;373:161–70.
16. Medina-Gómez G. Mitochondria and endocrine function of adipose tissue. Best Pract Res Clin Endocrinol Metab. 2012;26:791–804.
17. Ouyang J, Wu M, Huang C, Cao L, Li G. Overexpression of oxidored-nitro domain containing protein 1 inhibits human nasopharyngeal carcinoma and cervical cancer cell proliferation and induces apoptosis: involvement of mitochondrial apoptotic pathways. Oncol Rep. 2013;29:79–86.
18. Pagano G, Castello G, Pallardó FV. Sjøgren's syndrome-associated oxidative stress and mitochondrial dysfunction: prospects for chemoprevention trials. Free Rad Res. 2013;47:71–3.
19. Masana L, Camprubi M, Sarda P, Sola R, Joven J, Turner PR. The Mediterranean-type diet: is there a need for further modification? Am J Clin Nutr. 1991;53:886–9.
20. Paul A, Calleja L, Camps J, Osada J, Vilella E, Ferré N, et al. The continuous administration of aspirin attenuates atherosclerosis in apolipoprotein E-deficient mice. Life Sci. 2000;68:457–65.
21. Rius B, López-Vicario C, González-Périz A, Morán-Salvador E, García-Alonso V, Clària J, Titos E, et al. Resolution of inflammation in obesity-induced liver disease. Front Immunol. 2012;3:257.
22. Rogge MM. The role of impaired mitochondrial lipid oxidation in obesity. Biol Res Nurs. 2009; 10:356–73.
23. Mirza MS. Obesity, visceral fat and NAFLD: querying the role of adipokines in the progression of nonalcoholic fatty liver disease. ISRN Gastroenterol. 2011;2011: 11 p. 592404.
24. Tous M, Ferré N, Rull A, Marsillach J, Coll B, Alonso-Villaverde C, et al. Dietary cholesterol and differential monocyte chemoattractant protein-1 gene expression in aorta and liver of apo E-deficient mice. Biochem Biophys Res Commun. 2006; 340:1078–84.
25. Camps J, Marsillach J, Joven J. The paraoxonases: role in human diseases and methodological difficulties in measurement. Crit Rev Clin Lab Sci. 2009;46:83–106.
26. Vanden Heuvel JP. Nutrigenomics and nutrigenetics of ω3 polyunsaturated fatty acids. Prog Mol Biol Transl Sci. 2012;108:75–112.
27. Rodríguez-Gallego E, Riera-Borrull M, Hernández-Aguilera A, Mariné-Casadó R, Rull A, Beltrán-Debón R, et al. Ubiquitous transgenic overexpression of C-C chemokine ligand 2: a model to assess the combined effect of high energy intake and continuous low-grade inflammation. Mediators Inflamm. 2013;2013:953841.
28. Ott C, Jacobs K, Haucke E, Navarrete Santos A, Grune T, Simm A. Role of advanced glycation end products in cellular signaling. Redox Biol. 2014;2:411–29.
29. Lowe FJ, Luettich K, Gregg EO. Lung cancer biomarkers for the assessment of modified risk tobacco products: an oxidative stress perspective. Biomarkers. 2013;18:183–95.

Oxidative Stress and DNA Damage in Obesity-Related Tumorigenesis

2

Concha Cerdá, Carlos Sánchez, Benjamín Climent, Antonio Vázquez, Antonio Iradi, Fátima El Amrani, Ana Bediaga, and Guillermo T. Sáez

Abstract

Reactive oxygen species induce oxidative modification of critical macromolecules. Oxygen derived free radicals may act as potential cytotoxic intermediates inducing inflammatory and degenerative processes, or as signal messengers for the regulation of gene expression. This dual effect mainly depends on the availability of free radicals in terms of concentration, as well as on the environmental characteristics in which they are produced. The formation of free radicals has been proposed to be the linking factor between certain metabolic disturbances and cancer. Circulating mononuclear cells of patients with high cholesterol levels, insulin resistance, metabolic syndrome or obesity present lower levels of antioxidant enzymes and increased concentrations of oxidative stress by-products such as isoprostanes or the DNA oxidized and highly mutagenic

C. Cerdá • A. Bediaga
Service of Clinical Analysis-CDB, Oxidative Stress Commission-SEQC, General University Hospital-CIBEROBN, University of Valencia, Valencia, Spain
e-mail: Concermi@gmail.com; anbecol@hotmail.com

C. Sánchez
Endocrinology and Nutrition Unit, General University Hospital, University of Valencia, Valencia, Spain
e-mail: Carlos.sanchez@uv.es

B. Climent
Service of Internal Medicine, General University Hospital, University of Valencia, Valencia, Spain
e-mail: Climent_ben@gva.es

A. Vázquez
Service of General and Digestive Surgery, General University Hospital, University of Valencia, Valencia, Spain
e-mail: Vprado.a@gmail.com

A. Iradi
Department of Physiology, Faculty of Medicine-CIBEROBN, University of Valencia, Valencia, Spain
e-mail: Antonio.iradi@uv.es

F. El Amrani
Department of Biochemistry and Molecular Biology, Faculty of Medicine, University of Valencia, Avda. Blasco Ibañez 15, 46010 Valencia, Spain
e-mail: Yolanda.ben@uv.es

G.T. Sáez (✉)
Service of Clinical Analysis-CDB, Oxidative Stress Commission-SEQC, General University Hospital-CIBEROBN, University of Valencia, Valencia, Spain

Department of Biochemistry and Molecular Biology, Faculty of Medicine, General University Hospital-CDB, University of Valencia, Oxidative Stress Commission-SEQC, Avda. Blasco Ibañez 15, 46010 Valencia, Spain
e-mail: Guillermo.saez@uv.es

J. Camps (ed.), *Oxidative Stress and Inflammation in Non-communicable Diseases – Molecular Mechanisms and Perspectives in Therapeutics*, Advances in Experimental Medicine and Biology 824, DOI 10.1007/978-3-319-07320-0_2, © Springer International Publishing Switzerland 2014

base 8-oxo-7,8-dihydro-2′-deoxyguanosine. Overweight or obese subjects also exhibit hormonal changes as a consequence of the increase of mass fat, and these hormonal alterations have been implicated in the alteration of different signal transduction mechanisms and in cell growth and differentiation. A significant correlation has been found between body mass index and cancer. The biological factors and molecular mechanisms implicated in obesity associated cancer susceptibility will be reviewed.

Keywords
Cancer • DNA damage • Free radicals • Obesity

2.1 Introduction

Oxidative stress (OS) is a major mechanism in the initiation and progression of different forms of non-communicable diseases, including cardiovascular disease, atherogenesis, neurodegeneration and cancer, all of which are associated with the aging process [1, 2].

A critical point of OS is the formation of reactive oxygen radicals (ROS), overwhelming the antioxidant defence systems of the cells and interacting with different macromolecules, which include carbohydrates, proteins, phospholipids and nucleic acids, that become structurally modified by the process of oxidation. This is not only a consequence but also the cause of tissue homeostatic disturbances and represents a plausible mechanism underlying the pathophysiology of a wide number of metabolic diseases. Of special importance is the oxidative modification of DNA which renders different by-products with an important mutagenic potential. This is the case of the damaged base 8-oxo-7,8-dihydro-2′-deoxiguanine (8-oxo-dG) which implication in tumour development has been extensively studied and demonstrated [3–8]. During the last decades, growing interest has been centered on the role of OS in metabolic disorders such as diabetes mellitus, metabolic syndrome, dyslipidaemia, insulin resistance and obesity. In addition to OS, high levels of damaged DNA products have been reported [9–14]. The increase of the mutagenic base 8-oxo-dG may represent a critical step in the link between metabolic disorders and cancer incidence based on the ability of this base to induce mismatch incorporation during DNA replication by DNA polymerase [15, 16]. This effect emphasises the importance of an efficient DNA repair mechanism [17].

As has been commonly accepted, overweight and obese patients present an increased incidence of tumour development. Although metabolic and hormonal alterations have been proposed as possible induction mechanisms, the role of ROS-induced OS and DNA damage cannot be discarded. The biological changes which take place in overweight or obese patients may have a profound implication in the carcinogenesis mechanism. An increase of fat mass due to adipogenesis results in an enhanced production of leptin which has been related to a special metabolic milieu favouring cell growth and inhibiting apoptotic signalling. In addition, the increase of the mutagenic base 8-oxo-dG observed in obese patients may accelerate the mutational rate of cells and or interfere with DNA repair mechanisms leading to an increase and accumulation of genetic events which characterize tumour development. The role of obesity in cancer incidence and mortality is based on large cohort studies and the evidence of a biological relationship including metabolic, endocrinological and inflammatory processes. There is also experimental evidence related with enhanced cell proliferation and decreased programmed cell death in experimental models of obesity, together with epidemiological studies showing a reduced incidence of cancer in women who have had sustained weight loss. Therefore, the link between obesity and cancer is thought to be causal [18, 19].

2.2 Obesity, A Growing Social Epidemics

To maintain a healthy weight by eating proper food in a society surrounded by large proportions of palatable, low cost and high energy food is an extraordinarily difficult task. The prevalence of obesity has increased all along the developed countries. In the US, 33.8 % of men and women aged >20 year were found to be obese in the 2008 Examination Survey. The World Health Organization (WHO) defines obese as having a body mass index (BMI) >30.0 kg/m^2. Overweight is defined as having a BMI between 25 and 29.9 kg/m^2 and this was found in an additional 34.2 % of the US population. WHO estimated that >1.6 billion people worldwide are overweight, about one quarter of whom are obese. There are some differences in rates of obesity among Americans with different ancestry origins, but an increased prevalence of obesity has been observed in every ethnic group. Different studies have been conducted to examine the prevalence of obesity in Europe. The most comprehensive data are from the WHO study termed MONICA. The final result was a prevalence of 10–20 % for men and 15–25 % for women. The prevalence was also higher in the Mediterranean countries and Eastern European countries than to that of North and West-central Europe. The prevalence of obesity is higher in women and increases with age and a low social and economic status. In men, the BMI was positively associated with age, alcohol consumption and the percentage of energy provided by fat. Obesity in infants and young people is especially dramatic, since it is the cause of future diseases. In Europe the highest prevalence of childhood obesity is found is Spain, along with Malta, Italy, UK and Greece [20–23].

The SESPAS (Spanish Society of Public Health and Health Administration) study reported in 2010 a prevalence of 35 % of excessive weight in children and adolescents in Spain. In the group between 8 and 13 years, excess weight exceeds 45 %, while for the group of 14–17 years, overweight is 25 %. This factor is, again, associated with the lowest social and educational strata [24].

Therefore, we are aware of a risk factor of increasing magnitude, and of great clinical importance, which should be monitored systematically and rigorously. Future can be very worrying, if remedy is not provided, as the WHO extrapolation of existing data suggests that by 2025 obesity levels could be 45–50 % in the US and Western Europe, 30–40 % in Australia, and over 20 % in Brazil [25].

2.3 Obesity and Cancer Incidence and Risk

Knowledge on the harmful effects of obesity on human health is as old as the first manifesto by the father of medicine, Hippocrates (460–370 BC). He warned about the dangers of too much food and too little exercise [26]. It was almost 2,000 years later when the first link between obesity and endometrial cancer was established by Robert Thomas [27].

Over the past century a significant improvement of our understanding on the interrelationships between overweight/obesity, energy balance and cancer risk as well as cancer recurrence and survival has been highlighted with a great amount of experimental data. Epidemiological studies show an elevated risk for cancer in people with a high BMI. Stratification of BMI allows to establish a relationship between normal weight, overweight and obesity, and the risk for cancer. Large cohort studies support a substantial relationship between overweight-associated metabolic changes and tumour outcome in overweight population [28–32].

Other factors such as the inflammatory milieu, the alteration of signal transduction pathways and transcription factors linked to cell proliferation and cell survival have also been implicated in the underlying mechanisms of obesity-induced carcinogenesis [19]. Therefore, the role of obesity in cancer incidence and cancer-induced mortality seems to be causal. As it was concluded from the Cancer Prevention Study II of the American Cancer Society (started in 1982 with a total of 900,000 adults), the death rate caused by all cancers in an extremely obese (BMI >40.0 kg/m^2) cohort was 52 % higher for men and 62 % higher for women than that of normal weight individuals. The relative risk for death from cancer in this group increased for both, men and women paralleling the BMI increase over the normal upper limit [18, 33]. Indeed, it has been estimated that about 20 % of all cancers are caused by excess weight [34] and the Million Women Study, which is considered the largest study of its kind on women, has shown that approximately a half of them can be attributed to obesity in postmenopausal women [35].

A meta-analysis of prospective observational studies with 282,000 incident cancer and a follow-up greater than 133 million person-years, has demonstrated that the obesity and cancer association is sex specific over a wide range of malignancies, and this remain mostly true for different geographic populations.

Epidemiological studies have also emphasized the role of obesity as a contributing factor in both increased incidence and mortality from different types of cancer. The higher rate of cancer deaths in obese population accounts for a decreased survival index which may be due to the enhancing effects of obesity on cancer potency and progression. The highest quartiles of BMI are associated with a higher tumour malignant phenotype, as has been shown in different cancer studies. In a study of 1,177 women with invasive ductal breast cancer, it was observed that those in the highest quartile of BMI developed tumours with a higher histological grade, mitotic cell count, and larger tumour size than those in the lower quartiles. In these patients, those in the highest BMI quartile expressed significantly higher levels of markers of proliferation and fast growth of tumour cells [36].

The association between increased BMI and cancer incidence is not equally extensive to all tumour types [37, 38]. Different meta-analyses have shown that there is a good association between BMI and endometrial, colorectal and postmenopausal breast cancer. In addition obesity has been also been shown to be associated with increased risk of oesophageal adenocarcinoma, renal cancer, gallbladder cancer in women, pancreatic cancer and ovarian cancer. Other cancer presentations such as thyroid cancer, malignant melanoma, non-Hodgkin's lymphoma, multiple myeloma and leukemia are less close associated with obesity while in the case of prostate cancer the reported data show that high BMI is associated with a higher risk of high-grade prostate cancer but with a lower risk of low-grade cancer [39].

2.4 The Biological Factors Linking Obesity and Cancer Susceptibility

Overweight or obese individuals present a number of differentiated metabolic and clinical characteristics when compared with the lean population. Regarding cancer incidence or susceptibility, there are a number of general aspects to be first considered. Excess body weight is usually a consequence of increase body fat accumulation with a more or less specific localisation. This abnormal situation is known to be associated with a number of different hormonal and metabolic changes which may compromise the whole organism or a specific tissue in a site specific manner. Therefore, conclusions on the mechanisms leading to cell degeneration in overweight or obese patients must be carefully considered.

Most of the types of cancer that are clearly associated with obesity are those related with endocrine or metabolic disturbances or affecting tissues with hormone activity. What have the scientific community learned and concluded from these data? The answer is a great amount of information that may be useful in the future for the better understanding and management of the patients with obesity associated cancer diseases. Different aspects must be reviewed in relation to the biological implications of obesity induced tumours.

Alterations in caloric intake or in the quality of diet may influence the risk of cancer development. An excess of calories and positive energy balance are critical factors in the induction of cell differentiation and progression leading to a tumour phenotype. On the contrary, a long-term caloric restriction decreases cancer incidence and extends longevity in different animal models [40, 41]. It has been also postulated that restriction of caloric intake reduces the formation of intracellular ROS and therefore prevents the oxidative damage to DNA which is considered an important step in cancer induction and progression [42].

Diet is an important factor in the induction or prevention of cancer. An increased intake of calories, alcohol beverages, and animal fats, and a reduced intake of vegetables and fruits are commonly associated as potentially harmful for health. On the contrary, low cancer risk has been associated with a higher intake of fruit and vegetables. High fibre cereal consumption has been related with reduced risk of colon cancer [43]. A negative energy balance has been also considered as a positive factor towards lowering cancer susceptibility.

Obesity also correlates with low serum levels of 25-hydroxy vitamin D (25-OH-D). The deficiency in this vitamin may be responsible for 20 % of the cancer risk linked to an excess of BMI [44]. In overweight patients, BMI positively correlates with circulating insulin levels, and insulin resistance is a common characteristic of obese persons. High levels of insulin are known to promote cell growth and this effect has been proposed as the mechanism of hyperinsulinaemia induced cancer development [45, 46]. It has been postulated that hyperinsulinaemia, by inhibiting the production of insulin-like growth factor-1 binding protein (IGFBP)-1 and IGFBP-2, would increase the availability of free IGF-1 which in turn may favour tumour development. Obese patients have high levels of circulating IGF-1 [47]. The IGF system represents a complex molecular machinery with different protein components interacting with two ligands IGF-1 and IGF-2. There are two principal receptors IGF-1R and IGF-2R and at least six high-affinity binding proteins IGFBP-1 to IGFBP-6 in addition to several binding protein proteases. IGF ligands bind to IGFBP-3 which is the main circulating binding protein together with an acid-labile subunit to form a very stable ternary complex [19]. Different epidemiological studies have shown a direct correlation between the serum levels of C-peptide and the development of some tumours [48–51]. Insulin activation of the insulin receptor (IR) triggers an intracellular signal transduction in both extracellular (ERK) and phospatidylinositol-3 kinase (PI-3 K) pathway, a mechanism which explains the mitogenic and antiapoptotic effect of the hormone at supraphysiological levels and probably in collaboration with IGF-I receptors [52]. It is has been show that, in obese patients, free IGF-1 does not respond to insulin administration, maintaining higher plasma concentrations compared with lean subjects [53]. Insulin and IGF-1 are believed to induce carcinogenesis through binding to the insulin receptor (IR) and IGF-1R. IGF-1 activates the phosphatidyl-Inositol 3-kinase (PI3K)-AKT system and the Ras/Raf/mitogen activated – protein-kinase (MAPK) system respectively with secondary inhibition of apoptosis and cell proliferation stimulation [54]. In addition to these carcinogenic effects, IGF-1 also exerts proangiogenic stimulation and induces tumour-related lymphangiogenesis through the induction of hypoxia-inducible factor-1a (HIF-1α) which together with vascular endothelial growth factor (VEGF) leads to neovascularisation and metastases of colon cancer cells [55]. Moreover, IGF-1 have been shown to inhibit tumour suppressor protein p53 by MDM2-dependent degradation in response to DNA damage [56] and, in so doing, to inhibit apoptosis and to induce metastasis [57].

The concept about the metabolic activity of adipose tissue has changed substantially over the years from an inert to an active tissue and even with attributed hormonal functions. There are more than 50 different types of adipokines identified and characterised which are synthesised in and released by adipose tissue. The most abundantly produced and studied in cancer development adipokines are leptin and adiponectin. Leptin is known to have marked effects on the synthesis of inflammatory mediators, such as interleukins IL6, IL1 and TNFα. These molecules activate the transcription factor NFκB and induce downstream effects through the mechanistic target of rapamycin (serine/threonine kinase) (mTOR) to initiate transcription. This sequence of molecular events is influenced by sex hormones such as oestradiol and testosterone, as well as by the vascular endothelial growth factor (VEGF). There is a notable molecular relationship between leptin and insulin. The expression of leptin is positively regulated by insulin to suppress appetite. Leptin acts as a pleiotrophic hormone with mitogenic effects for various cell types, including hematopoietic progenitor cells, normal and transformed epithelial cells, and vascular endothelial cells. It has also antiapoptotic and proangiogenic effects by itself and in synergy with VEGF [58, 59]. Long form leptin receptor (LRb) activates PI3 kinase, the mitogen-activated protein kinase (MAPK) and the signal transducer and activator transcription 3 (STAT3), which are essential mediators for cell survival, proliferation and differentiation. This receptor is present in nonmalignant and in cancer cell lines [60, 61].

Adiponectin is secreted by visceral adipose tissue. It is the most abundant adipokine secreted by mature adipocytes. Circulating adiponectin concentrations are mainly determined by genetic factors, nutrition and adiposity. With the increase of adiposity adiponectin concentrations are reduced. It has been proposed that adiponectin may be a biological link between obesity and increase cancer risk. Although adiponectin may influence cancer risk through its effects on insulin resistance, it has been also proposed that this cytokine acts on tumour cells directly. Indeed, several cancer cell types express adiponectin receptors that may mediate the effect of adiponectin on cellular proliferation [62]. Adiponectin and adiponectin receptors have been shown to play a role in the activation of the PPARγ pathway, which, in turn induces the transcription of different genes involved in the regulation of cell proliferation and differentiation. It has been proposed that functional reduction of PPARγ signalling, may lead to reduce levels of BRCA1 and the impairment of DNA repair mechanism [63]. Epidemiological studies show an inverse association between circulating adiponectin concentrations and the presence of some tumour types such as endometry [64], breast [65], prostate [66] and colon cancer [67].

Treatment with adiponectin has been shown to inhibit tumour progression although its precise mechanism is not fully understood. It has been proposed that adiponectin inactivates MAPK kinases 1 and 3 and ERK2 and ERK1, and the concomitant decrease of glucose uptake. On the other hand, the apoptotic effects induced by adiponectin can be related with the increase of p53 and Bax expression and the decrease of Bcl-2 [68]. In addition, the antiangiogenic effect of this cytokine is thought to occur through an apoptotic effect on vascular endothelial cells and the inhibition of cell migration [69]. Although extensive experimental and epidemiological data support the relationship between overweight and/or obesity and cancer, with solid established biological links, the precise involved mechanisms are not yet fully understand and require further identification of molecules and molecular interactions for better knowledge and interpretation.

2.5 Oxidative Stress in Obesity

Obesity and associated metabolic disturbances such as type 2 Diabetes (T2D) or the metabolic syndrome (MetS) are frequently associated with OS [12, 70]. Experimental studies in animal models reported that OS plays an important role in the pathophysiology of obesity. However, what remains to be resolved is whether OS is a cause or a consequence of body fat accumulation. Obesity has been reported to contribute to a pro-oxidant and pro-inflammatory status. Overweight or obese subjects are considered to be under a chronic OS state and this is a continuous potential danger for their health. The production of ROS in obese patients induces the oxidative modification of different macromolecules, the biological function of which becomes altered and can trigger different pathophysiological processes. All the disturbances associated with obesity are known to develop with a concomitant OS which has been demonstrated through the increase of OS by products in whole blood, circulating mononuclear cells or in the urine of affected subjects. OS may be linked to obesity and related complications through different mechanisms. High fat feed mice present an increased production of H_2O_2 within the heart, an effect which is rapidly accompanied by the expression of the antioxidant enzyme catalase [71]. Intracellular triglycerides inhibit adenosine nucleotide translocator (ANT) leading to the accumulation of ATP inside the mitochondria and the reduction of oxidative phosphorylation due to the decrease of ADP. This uncoupling effect results in the leakage of electrons and the partial reduction of molecular oxygen in form of superoxide ions ($O_2^{\cdot -}$) [72]. A similar effect has been mechanistically related to the increase of reduced intermediates ($NADH_2$ and $FADH_2$) produced by the glycolytic and tricarboxylic pathways leading to electron release at the mitochondrial complex III. The production of $O_2^{\cdot -}$ results in different and important carbohydrate metabolic changes and the activation of PKC with secondary formation of ROS and nitrous species. Advanced glycation end-products (AGEs) are also formed which may induce the production of ROS/RNS species by activating Nox and NFκB [73]. In obese patients, the intake of high energy nutrients is known to increase the expression of p47phox subunit of Nox2, the intracellular NFκB binding activity and the plasma concentrations of MMP-9, when compared with normal weight subjects. All these processes enhance the production of ROS [74]. Fat accumulation increases Nox activity and endoplasmic reticulum stress leading to ROS production [70]. In obese subjects, high levels of free fatty acids and carbohydrates induce the production of ROS which maintain reversible interactions with the transcription factor NFκB in the modulation of immunity, inflammation and cell survival [75]. Obesity is considered a mild inflammatory state where adipose tissue is infiltrated by macrophages thus leading to ROS production together with inflammatory cytokines which in turn produce and release more ROS to the environment and favours more macrophage recruitment to the adipose tissue in a vicious cycle [76] (Fig. 2.1). Different regulatory proteins have been recently implicated in the production of ROS-induced oxidative stress. Nuclear factor E2-related factor 2 (Nrf2) is a basic leucine zipper transcription factor encoded by the NFE2L2 gene in humans. Upon OS, Nrf2 enters into the cell nucleus to form a complex with Maf and Jun proteins, binds to the antioxidant response element (ARE) and induces the transcription of many antioxidant genes [77]. This molecular mechanism can be considered as a primary step in the cellular defence against the cytotoxic effects of OS. The production and reactivity of ROS under normal circumstances is controlled by different antioxidant molecules. The Nrf2 pathway also plays an important role in energy metabolism although with reported controversial effects. Nonetheless, the role of Nrf2 in adipose tissue regulation has been clearly established by using different experimental approaches showing a beneficial effect of the transcription factor in the prevention of body weight gain and the accumulation of white adipose tissue [78]. The levels and/or activities of antioxidants seem to be overwhelmed is obese patients and in those situations with impairment of the lipid profile. Alteration of oxidation defence mechanisms has been observed in overweight and obese people but without a direct correlation with BMI, or body fat [79]. In other studies, the activities of antioxidant enzymes superoxide dismutase, catalase and glutathione peroxidase have been found to be inversely related to BMI, both in obese children and adults [80]. The lower activity of antioxidants in obese patients may be related with their special characteristics, habits and increased requirements due to their special metabolic situations. The antioxidant enzyme paraoxonase-1 (PON1) has also been shown to be decreased in obese subjects [81] and the decline of the enzyme activity has been recently related with an increased OS and inflammatory state [82], insulin resistance and the metabolic syndrome (MetS) [83]. The multicentre prospective population study of diet and cancer in Europe (EPIC) showed that plasma

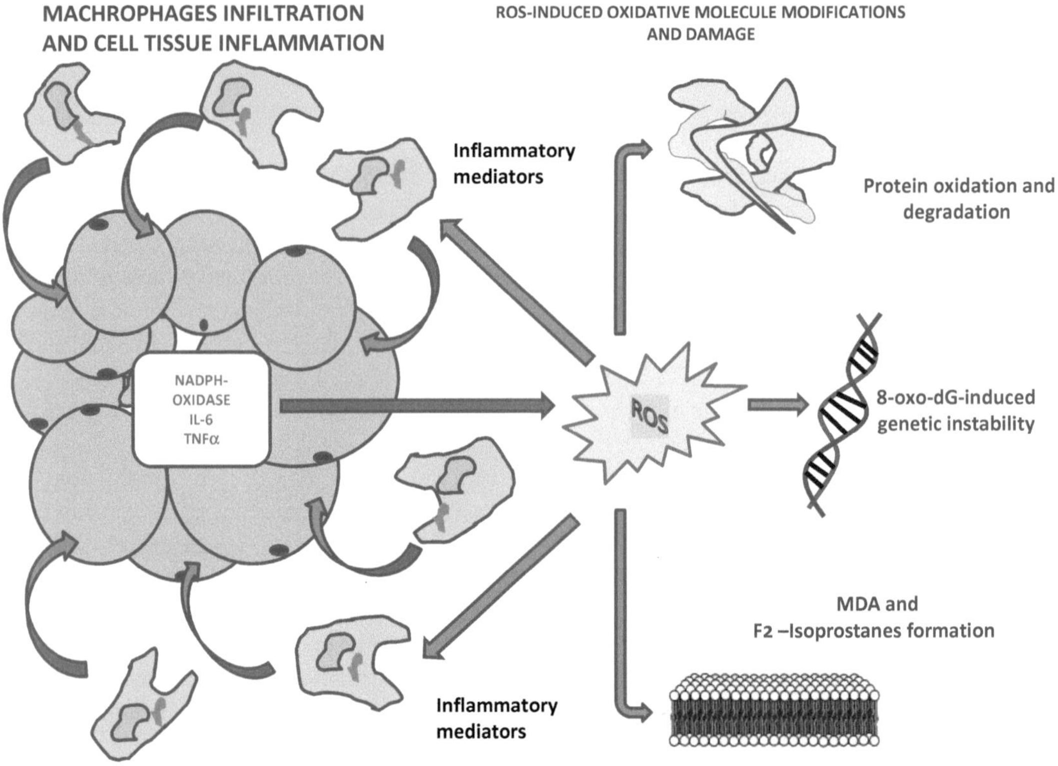

Fig. 2.1 Relationship between macrophage infiltration and reactive-oxygen species (ROS) in obesity. *8-oxo-dG* 8-oxo-7,8-dihydro-2′-deoxyguanosine; *IL-6* interleukin-6; *MDA* malondialdehyde; *TNFα* tumour necrosis factor alpha

vitamin C levels was inversely related to central fat distribution [84]. As well as in other pathological situations, OS in obesity can be monitored by the assay of most representative by-products. These are produced as a result of the oxidative modification of phospholipids, proteins and nucleic acids. Circulating oxidized low density lipoproteins and thiobarbituric acid-reactive substances are higher in obese patients compared with normal weight subjects [85]. In addition, BMI, total fat and waist circumference have been shown to be positive correlated with urinary F2-isoprostane levels and inversely correlated with PON1 [81]. The concentrations of malondialdehyde in circulating mononuclear cells [86, 87] and urinary F2-isoprostanes [88] are positive associated with circulating pro-inflammatory cytokines and clinical markers. It has been also proposed that F2-isoprostane levels may predict loss of total adiposity over time. A significant inverse correlation between urinary F2-isoprostanes and weight gain was demonstrated in two different follow-up studies [89, 90]. Morbid obese patients were examined for their levels of OS markers in circulating mononuclear cells and compared with those in a normal weight volunteers. The intracellular concentrations of malondialdehyde and of oxidised glutathione were found to be significantly higher in the morbid obese group than in the control group. Reduced glutathione was significantly reduced in the morbid obese patients with a marked increase of the oxidised/reduced glutahione ratio, indicating a high level of OS in these subjects. The comorbidities in the obese group did not influence the levels of OS markers. However after bariatric surgery, OS by-products progressively decreased, arriving, 1 year after intervention, to levels similar to those found in healthy people [86, 87].

2.6 DNA Damage in Obese Associated Metabolic Disturbances

Systemic alterations involved in obesity and related metabolic disturbances such as the metabolic syndrome are characterised by an increase of the circulating pro-inflammatory cytokines (TNFα, IL-1, IL-6 and others). These cytokines are known to induce the production of ROS which, due to their highly reactivity, modify the structure and function of lipids, proteins and nucleic acids [2]. Oxidative modification of DNA induces the formation of 8-oxo-dG, which increases genetic instability due to its mutagenic potential [5, 15, 16]. Indeed, the repair of oxidative lesions such as the formation of 8-oxo-dG has been proposed as part of the cellular response to OS. This has been observed by different experimental approaches. Oxidative DNA damage has been reported in mice with obesity and liver steatosis [91]. 8-oxoguanine DNA glycosylase (OGG1) deficient mice (Ogg1−/−) exposed to a high fat diet, had an increase of adiposity and hepatic steatosis. These animals also had higher insulin levels and impaired glucose tolerance upon a high fat feeding that that of their wild type counterparts [92]. Mice deficient in the antioxidant enzyme PON1, exhibit severe steatosis after feeding with high fat and high cholesterol diet. This derangement was associated with significant increases of the hepatic levels of lipid peroxides (8-isoprostanes and MDA), protein carbonyl concentrations and 8-oxo-dG together with a decrease in the activities of glycolysis and the urea and Krebs cycle [93]. Due to its mutagenic potential, 8-oxo-dG may be implicated in the pro-carcinogenic state which characterizes non alcoholic liver steatosis.

It seems to be an association between fat accumulation and DNA damage and DNA damage response (DDR). Overweight and obese patients have higher levels of DNA damage products in their white cells and in their urine [86, 87]. DNA damage has been also found in obesity associated metabolic alterations such as the MetS [12]. However, the role of DNA damage in the early stage of MetS has been recently challenged [94]. In some studies, higher levels of urine 8-oxo-dG have been found in the control group compared with the MetS group. This result has been explained by assuming a constant exposure to OS in the MetS group and an progressive adaptive response producing more antioxidant molecules. Similar results and explanation were reported in the cells of type 1 and 2 diabetic patients [95].

Finally, another aspect of great interest in relation with overweight and OS-induced DNA damage can be found in the mechanisms affecting decreased fertility. There has been a growing interest over the past few years in the impact of male nutrition on fertility. Infertility has been related with an increase of male overweight and obesity. Conventional semen parameter values seem to be altered in subjects with high BMI. Levels of sperm with high DNA damage were significantly higher in obese men than in normal-weight subjects [96, 97].

The metabolic interactions and molecular signalling underlying the carcinogenic mechanisms associated with overweight and obesity is an open research field in which many factors seems to be implicated and where the ROS-induced OS theory although consistent requires further efforts and support by the scientific community and the national and international health institutions.

Acknowledgements GTS thanks grants from Conselleria de Sanitat de la Generalitat de València and Instituto de Salud Carlos III: ACOM/2012/238; PI10/00802; PI13/01848; CIBEROBN 12/03/30016

References

1. Sies H, Cadenas E. Oxidative stress: damage to intact cells and organs. Philos Trans R Soc Lond B Biol Sci. 1985;311:617–31.
2. Halliwell B, Gutteridge JMC. Free radicals in biology and medicine. 4th ed. Oxford, UK: Oxford University Press; 2007.
3. Loft S, Danielson P, Löhr M, Jantzen K, Hemmingsen JG, Roursgaard M, et al. Urinary excretion of 8-oxo-7,8-dihydroguanine as biomarker of oxidative damage to DNA. Arch Biochem Biophys. 2012; 518:142–50.
4. Barregard L, Møller P, Henriksen T, Mistry V, Koppen G, Rossner Jr P, et al. Human and methodological

sources of variability in the measurement of urinary 8-oxo-7,8-dihydro-2′-deoxyguanosine. Antioxid Redox Signal. 2013;18:2377–91.
5. Oliva MR, Ripoll F, Muñiz P, Iradi A, Trullenque R, Valls V, et al. Genetic alterations and oxidative metabolism in sporadic colorectal tumors from a Spanish community. Mol Carcinogen. 1997;18:232–43.
6. Oltra AM, Carbonell F, Tormos C, Iradi A, Sáez GT. Antioxidant enzyme activities and the production of MDA and 8-oxo-dG in chronic lymphocytic leukemia. Free Rad Biol Med. 2001;30:1286–92.
7. Sánchez M, Torres JV, Tormos C, Iradi A, Muñiz P, Espinosa O, et al. Impairment of antioxidant enzymes, lipid peroxidation and 8-oxo-2′- deoxyguanosine in advanced epithelial ovarian carcinoma of a spanish community. Cancer Lett. 2006;233:28–35.
8. Collado R, Oliver I, Tormos C, Egea M, Miguel A, Cerdá C, et al. Early ROS-mediated DNA damage and oxidative stress biomarkers in monoclonal B lymphocytosis. Cancer Lett. 2012;317:144–9.
9. Abdilla N, Tormo MC, Fabia MJ, Chaves FJ, Sáez G, Redon J. Impact of the components of metabolic syndrome on oxidative stress and enzymatic antioxidant activity in essential hypertension. J Hum Hypertens. 2007;21:68–75.
10. Martinez-Hervas S, Fandos M, Real JT, Espinosa O, Chaves FJ, Sáez GT, et al. Insulin resistance and oxidative stress in familial combined hyperlipidemia. Atherosclerosis. 2008;199:384–9.
11. Fandos M, Corella D, Guillén M, Portloés O, Carrasco P, Iradi A, et al. Impact of cardiovascular risk factors on oxidative stress and DNA damage in a high risk Mediterranean population. Free Radic Res. 2009;43:1179–89.
12. Mitjavila MT, Fandos M, Salas-Salvadó J, Covas MI, Borrego S, Estruch R, et al. The Mediterranean diet improves the systemic lipid and DNA oxidative damage in metabolic syndrome individuals. A randomized, controlled trial. Clin Nutr. 2013;32:172–8.
13. López-Uriarte P, Nogués R, Saez G, Bulló M, Romeu M, Masana L. Effect of nut consumption on oxidative stress and the endothelial function in metabolic syndrome. Clin Nutr. 2010;29:373–80.
14. Real JT, Martínez-Hervás S, Tormos MC, Domenech E, Pallardó FV, Sáez-Tormo G, et al. Increased oxidative stress levels and normal antioxidant enzyme activity in circulating mononuclear cells from patients of familial hypercholesterolemia. Metabolism. 2010;59:293–8.
15. Kushino Y, Mori F, Kasai H, Inoue H, Iwai S, Miura K, et al. Misreading of DNA templates containing 8-hydroxy-deoxyguanosine at the modified base and at adjacent residues. Nature. 1987;327:77–9.
16. Shibutani S, Takeshita M, Grollman AP. Insertion of specific bases during DNA synthesis past the oxidation-damage base 8-oxo-Dg. Nature. 1991;349: 431–4.
17. Klungland A, Rosewell I, Hollenbach S, Larsen E, Daly G, Epe B, et al. Accumulation of premutagenic DNA lesions in mice defective in removal of oxidative base damage. Proc Natl Acad Sci U S A. 1999;96: 13300–5.
18. Lichtman MA. Obesity and the risk for a haematological malignancy: leukemia, lymphoma, or myeloma. Oncologist. 2010;15:1083–101.
19. Robert DL, Dive C, Renehan AF. Biological mechanisms linking obesity and cancer risk. New perspectives. Annu Rev Med. 2010;61:301–6.
20. Lobstein T, Frelut ML. Prevalence of overweight among children in Europe. Obes Rev. 2003;4: 195–200.
21. Caballero B. The global epidemic of obesity: an overview. Epidemiol Rev. 2007;29:1–5.
22. Wang Y, Lobstein T. Worldwide trends in childhood overweight and obesity. Int J Pediatr Obes. 2006;1: 11–25.
23. De Onis M, Onyango AW, Borghi E, Siyam A, Nishida C, Siekmann J. Development of a WHO growth reference for school-aged children and adolescents. Bull World Health Organ. 2007;85:660–7.
24. Sánchez-Cruz JJ, Jiménez-Moleón JJ, Fernández-Quesada F, Sánchez MJ. Prevalencia de obesidad infantil y juvenil en España en 2012. Rev Esp Cardiol. 2013;66:371–6.
25. Flegal KM, Carroll MD, Ogden CL, Curtin LR. Prevalence and trends in obesity among US adults. 1999–2008. JAMA. 2010;303:235–41.
26. Demark-Wahnefried W, Platz EA, Ligibel JA, Blair CK, Courneya KS, Meyerhardt JA, et al. The role of obesity in cancer survival and recurrence. Cancer Epidemiol Biomarkers Prev. 2012;21:1244–59.
27. Haslam D. Obesity: a medical history. Obes Rev. 2007;8 Suppl 1:31–6.
28. Renehan AG, Tyson M, Egger M, Heller F, Swahlen M. Body-mass index and incidence of cancer: a systematic review and meta-analysis of prospective observational studies. Lancet. 2008;371:569–78.
29. Cao Y, Ma J. Body mass index, prostate cancer-specific mortality, and biochemical recurrence: a systematic review and meta-analysis. Cancer Prev Res. 2011;4:486–501.
30. Ewertz M, Jensen MB, Grunnarsdottir KA, Højris I, Jakobsen EH, Nielsen D, et al. Effect of obesity on prognosis after early-stage breast cancer. J Clin Oncol. 2011;29:25–31.
31. Sinicrope FA, Foster NR, Sargent DJ. Obesity is an independent prognostic variable in colon cancer survivors. Clin Cancer Res. 2010;16:1884–93.
32. Protani M, Coory M, Marin JH. Effect of obesity on survival of women with breast cancer: systematic review and meta-analysis. Breast Cancer Res Treat. 2010;123:627–35.
33. Calle EE, Rodriguez C, Walker-Thurmond K, Thun MJ. Overweight, obesity and mortality from cancer in a prospectively studied cohort of U.S. adults. N Engl J Med. 2003;348:1625–38.
34. Wholin KY, Carson K, Colditz GA. Obesity and cancer. Oncologist. 2010;15:556–65.

35. Reeves GK, Pirie K, Beral V, Green J, Spencer E, Bull D. Cancer incidence and mortality in relation to body mass index in the million women study: cohort study. Brit Med J. 2007;335:1134–9.
36. Dalling JR, Malone KE, Doody DR, Johnson LG, Gralow JR, Porter PL. Relation of body mass index to tumor markers and survival among young women with invasive ductal breast carcinoma. Cancer. 2001;92:720–9.
37. De Pergola G, Silvestris F. Obesity as a major risk factor for cancer. J Obes. 2013;2013:291546.
38. Calle EE, Kaaks R. Overweight, obesity and cancer: epidemiological evidence and proposed mechanisms. Nat Rev Cancer. 2004;4:579–91.
39. Hsing AW, Sakoda LC, Chua Jr S. Obesity, metabolic syndrome, and prostate cancer. Am J Clin Nutr. 2007;86:843–57.
40. Spindler SR. Rapid and reversible induction of the longevity, anticancer and genomic effects of caloric restriction. Mech Ageing Dev. 2005;126:960–6.
41. Pallavi R, Giorgio M, Pelicci PG. Insights into the beneficial effect of caloric/dietary restriction for a healthy and prolonged life. Front Physiol. 2012;3:1–10.
42. Heydari AR, Unnikrishnan A, Lucente LV, Richardson A. Caloric restriction and genomic stability. Nucleic Acids Res. 2007;35:7485–96.
43. McMillan JR, Sattar N, Lean M, McArdle CS. Obesity and cancer. Brit Med J. 2006;333:1109–11.
44. Lagunova Z, Projnicu AC, Grant WB, Bruland O, Moan JE. Obesity and increased risk of cancer: dose decrease of serum 25-hydroxyvitamin D level with increasing body mass index explain some of the association? Mol Nutr Food Res. 2010;54:1127–33.
45. McKeown-Eyssen G. Epidemiology of colorectal cancer revisited: are serum triglycerides and/or plasma glucose associated with risk? Cancer Epidemiol Biomarkers Prev. 1994;3:687–95.
46. Giovannucci E. Insulin and colon cancer. Cancer Causes Control. 1995;6:164–79.
47. Frystyk J. Free insulin-like growth factors-measurements and relationships to growth hormone secretion and glucose homeostasis. Growth Horm IGF Res. 2004;14:337–75.
48. Becker S, Lossus L, Kaaks R. Obesity related hyperinsulinaemia and hyperglycaemia and cancer development. Arch Physiol Biochem. 2009;115:86–96.
49. Pisani P. Hyper-insulinaemia and cancer, meta-analyses of epidemiological studies. Arch Physiol Biochem. 2008;114:63–70.
50. Cust AE, Allen NE, Rinaldi S, Dossus L, Friedenreich C, Olsen A, et al. Serum levels of C-peptide, IGFBP-1 and IGFBP-2 and endometrial cancer risk: results from the European prospective investigation into cancer and nutrition. Int J Cancer. 2007;120:2656–64.
51. Verheus M, Peeters PH, Rinaldi S, Dossus L, Biessy C, Olsen A, et al. Serum C-peptide levels and breast cancer risk: results from the European prospective investigation into cancer and nutrition (EPIC). Int J Cancer. 2006;119:659–67.
52. Renehan AG, Frystyk J, Flyvbjerg A. Obesity and cancer risk: the role of the insulin-IGF axis. Trends Endocrinol Metab. 2006;17:328–36.
53. Richart W, Fernandez-Real JM. No decrease in free IGF-I with increasing insulin in obesity-related insulin resistance. Obes Res. 2001;9:631–6.
54. Pollak MN, Schernhammer ES, Hankinson SE. Insulin-like growth factor and neoplasia. Nat Rev Cancer. 2004;4:505–18.
55. Wu Y, Yakar S, Zhao L, Hennighausen L, LeRoith D. Circulating insulin-like growth factor-I levels regulate colon cancer growth and metastasis. Cancer Res. 2002;62:1030–5.
56. Heron-Milhavet L, LeRoith D. Insulin-like growth factor I induces MDM-dependent degradation of p53 via the p38 MAPK pathway in response to DNA damage. J Biol Chem. 2002;18:15600–6.
57. Canonici A, Steelant W, Rigot V, Khomitch-Baud A, Boutaghou-Cherid H, Bruyneel E. Insulin-like growth factor-I receptor, E-cadherin and alpha v integrin form a dynamic complex under the control of alpha-catenin. Int J Cancer. 2008;122: 572–82.
58. Bray GA. The underlying basis for obesity: relationship to cancer. J Nutr. 2002;132:3451s–5.
59. Rose DP, Komninou D, Stephenson GD. Obesity, adipocytokines, and insulin resistance in breast cancer. Obes Rev. 2004;5:153–65.
60. Hardwick JC, Van Den Brink GR, Offerhaus GJ, Van Deventer SJ, Peppelenbosch MP. Leptin is a growth factor for colonic epithelial cells. Gastroenterology. 2001;121:79–90.
61. Dieudonne MN, Machinal-Quelin F, Serazin-Leroy V, Leneveu MC, Pecquery R, Giudicelli Y. Leptin mediates a proliferative response in human MCF7 breast cancer cells. Biochem Biophys Res Commun. 2002;293:622–8.
62. Barb D, Williams CJ, Neuwirth AK, Mantzoros CS. Adiponectin in relation to malignancies: a review of existing basic research and clinical evidence. Am J Clin Nutr. 2007;86:s858–66.
63. Arcidiacono B, Iiritano S, Nocera A, Possidente K, Nevolo MT, Ventura V, et al. Insulin resistance and cancer risk: an overview of the pathogenetic mechanisms. Exp Diabetes Res. 2012;2012:789174.
64. Petridou E, Mantzoros C, Dessypris N, Koukoulomatis P, Addy C, Voulgaris Z, et al. Plasma adiponectin concentrations in relation to endometrial cancer: a case-control study in Greece. J Clin Endocrinol Metab. 2003;88:993–7.
65. Miyoshi Y, Funahashi T, Kihara S, Taguchi T, Tamaki Y, Matsuzawa Y, et al. Association of serum adiponectin levels with breast cancer risk. Clin Cancer Res. 2003;9:5699–704.
66. Goktas S, Yilmaz MI, Caglar K, Sonmez A, Kilic S, Bedir S, et al. Prostate cancer and adiponectin. Urology. 2005;65:1168–72.
67. Wei EK, Giovannucci E, Fuchs CS, Willett WC, Mantzoros CS. Low plasma adiponectin levels and

risk of colorectal cancer in men: a prospective study. J Natl Cancer Inst. 2005;97:1688–94.
68. Dieudonne MN, Bussiere M, Dos Santos E, Leneveu MC, Giudicelli Y, Pecquery R. Adiponectin mediates antiproliferative and apoptotic response in human MCF7 breast cancer cells. Boichem Biophys Res Commun. 2006;345:271–8.
69. Bråkenhielm E, Veitonmäki N, Cao R, Kihara S, Matsuzawa Y, Zhivotovsky B, et al. Adiponectin induced antiangiogenic and antitumor activity involve caspase-mediated endothelial cell apoptosis. Proc Natl Acad Sci U S A. 2004;101:2476–81.
70. Furukawa S, Fujita T, Shimabukuro M, Iwaki M, Yamada Y, Nakajima Y, et al. Increased oxidative-stress in obesity and its impact on metabolic syndrome. J Clin Invest. 2004;114:1752–61.
71. Rindler PM, Plafker SM, Szweda LI, Kinter M. High dietary fat selectively increases catalase expression within cardiac mitochondria. J Biol Chem. 2013;288:1979–90.
72. Dröse S, Brandt U. Molecular mechanisms of superoxide production by the mitochondrial respiratory chain. Adv Exp Med Biol. 2012;748:145–69.
73. Diaz-Marco MT, Moscat J. The atypical PKCs in inflammation: NFkB and beyond. Immunol Rev. 2012;246:154–67.
74. Patel C, Ghanim H, Ravishankar S, Sia CL, Viswanathan P, Mohanty P, et al. Prolonged reactive oxygen species generation and nuclear factor-kappaB activation after a high-fat, high-carbohydrate meal in the obese. J Clin Endocrinol Metabol. 2007;92: 4476–9.
75. Bubici C, Papa S, Dean K, Franzoso G. Mutual cross-talk between reactive oxygen species and nuclear factor-kappa B: molecular basis and biological significance. Oncogene. 2006;25:6731–48.
76. Surmi BK, Hasty AH. The role of chemokines in recruitment of immune cells to the artery wall and adipose tissue. Vasc Pharmacol. 2010;52:27–36.
77. Itoh K, Chiba T, Takahashi S, Ishii T, Igarashi K, Katoh Y, et al. An Nrf2/small Maf heterodimer mediates the induction of phase II detoxifying enzyme genes through antioxidant response elements. Biochem Biophys Res Commun. 1997;236: 313–22.
78. Shin S, Wakabayashi J, Yates MS, Wakabayashi N, Dolan PM, Aja S, et al. Role of Nrf2 in prevention of high-fat diet-induced obesity by synthetic triterpenoid CDDO-imidazolide. Eur J Pharmacol. 2009;620: 138–44.
79. Brown LA, Kerr CJ, Whiting P, Finer N, McEneny J, Ashton T. Oxidant stress in healthy normal-weight, overweight, and obese individuals. Obesity. 2009; 17:460–6.
80. Savini I, Catani MV, Evangelista D, Gasperi V, Avigliano L. Obesity-associated oxidative stress: strategies finalized to improve redox state. Int J Mol Sci. 2013;14:10947–538.
81. Ferretti G, Bacchetti T, Masciangelo S, Bicchiega V. HDL-paraoxonase and membrane lipid peroxidation: a comparison between healthy and obese subjects. Obesity. 2010;18:1079–84.
82. Krzystek-Korpacka M, Patryn E, Hotowy K, Czapińska E, Majda J, Kustrzeba-Wójcicka I, et al. Paraoxonase (PON)-1 activity in overweight and obese children and adolescents: association with obesity-related inflammation and oxidative stress. Adv Clin Exp Med. 2013;22:229–36.
83. Ferré N, Feliu A, García-Heredia A, Marsillach J, París N, Zaragoza-Jordana M, et al. Impaired paraoxonase-1 status in obese children. Relationships with insulin resistance and metabolic syndrome. Clin Biochem. 2013;46:1830–6.
84. Canoy D, Wareham N, Welch A, Bingham S, Luben R, Day N, Khaw KT. Plasma ascorbic acid concentrations and fat distribution in 19,068 British men and women in the European prospective investigation into cancer and nutrition Norfolk cohort study. Am J Clin Nutr. 2005;82:1203–9.
85. D'Archivio M, Annuzzi G, Varì R, Filesi C, Giacco R, Scazzocchio B, et al. Predominant role of obesity/ insulin resistance in oxidative development. Eur J Clin Invest. 2012;42:70–8.
86. De Tursi-Ríspoli L, Vázquez-Tarragón A, Vázquez-Prado A, Sáez-Tormo G, Alí-Mahmoud A, Gumbau-Puchol V. Estrés oxidativo; estudio comparativo entre un grupo de población normal y un grupo de población obesa mórbida. Nutr Hosp. 2013;28: 671–5.
87. De Tursi-Ríspoli L, Vázquez-Tarragón A, Vázquez-Prado A, Sáez-Tormo G, Mahmoud AL, Bruna-Esteban M, et al. Relationship of oxidative stress and weight loss achieved in morbid obese patients by means of bariatric surgery using the duodenal switch technique. Nutr Hosp. 2013;28:1085–92.
88. Dandona P, Aljada A, Chaudhuri A, Mohanty P, Garg R. Metabolic syndrome: a comprehensive perspective based on interactions between obesity, diabetes, and inflammation. Circulation. 2005;111: 1448–54.
89. Il'yasova D, Wang F, Spasojevic I, Base K, D'Agostino Jr RB, Wagenknecht LE. Urinary F2-isoprostanes, obesity, and weight gain in the IRAS cohort. Obesity (Silver Spring). 2012;20:1915–21.
90. Kanaya AM, Wassel CL, Stoddard PJ, Harris TB, Cummings SR, Kritchevsky SB, et al. F2-isoprostanes and adiposity in older adults. Obesity (Silver Spring). 2011;19:861–7.
91. Zhang H, Xie C, Spencer HJ, Zuo C, Higuchi M, Ranganathan G, et al. Obesity and hepatosteatosis in mice with enhanced oxidative DNA damage processing in mitochondria. Am J Pathol. 2011;178:1715–27.
92. Sampath H, Vartanian V, Rollins MR, Sakumi K, Nakabeppu Y, Lloyd RS. 8-Oxoguanine DNA glycosilase (OGG1) deficiency increases susceptibility to obese and metabolic dysfunction. PLoS One. 2012;7:e51697.

93. García-Heredia A, Kensicki E, Mohney RP, Rull A, Triguero I, Marsillach J, et al. Paraoxonase-1 deficiency is associated with severe liver steatosis in mice fed a high-fat high-cholesterol diet. A metabolomic approach. J Proteome Res. 2013;12:1946–55.
94. Milić M, Kišan M, Rogulj D, Radman M, Lovrenčić MV, Konjevoda P, et al. Level of primary DNA damage in the early stage of metabolic syndrome. Mutat Res. 2013;758:1–5.
95. Anderson D, Yu TW, Wright J, Ioannides C. An explanation of DNA strand breakage in the comet assay and antioxidant capacity in diabetic patients. Mutat Res. 1998;398:151–61.
96. Chavarro JE, Toth TL, Wright DL, Meeker JD, Hauser R. Body mass index in relation to semen quality, sperm DNA integrity, and serum reproductive hormone levels among men attending an infertility clinic. Fertil Steril. 2010;93:2222–31.
97. Dupont C, Faure C, Sermondade N, Boubaya M, Eustache F, Clément P, et al. Obesity leads to higher risk of sperm DNA damage in infertile patients. Asian J Androl. 2013;15:622–5.

High Density Lipoproteins and Ischemia Reperfusion Injury: The Therapeutic Potential of HDL to Modulate Cell Survival Pathways

3

Richard W. James and Miguel A. Frias

Abstract

The clinical importance of high density lipoproteins has grown in recent years with demonstrations of their impact on diverse pathological mechanisms implicated not only in vascular disease, but also in other physiological systems. This is related to the multiple functions associated with high-density lipoproteins (HDL), notably their ability to limit oxidant and inflammatory processes, which are common to different disease states. A second feature of particular clinical relevance is the possibility of synthesising a simplified form of HDL that exhibits some of the functions of the mature lipoprotein. The therapeutic potential of synthetic HDL is already under clinical scrutiny. To illustrate these points, the present chapter will discuss the role of HDL in limiting damage to the heart consequent to myocardial ischemia. It will review molecular survival pathways stimulated by HDL to combat oxidative stress and the potential of synthetic HDL to activate such pathways.

Keywords

High density lipoproteins • Ischemia reperfusion injury • Myocardial infarction • Oxidative stress

3.1 Introduction

Myocardial infarction is a frequently fatal, acute coronary syndrome arising from oxygen deprivation of heart tissue. Rapid re-oxygenation of ischemic tissues is the primary aim of remedial treatment. Paradoxically, the revascularisation procedure itself can further damage heart tissue due to oxidative stress, a well-established clinical complication termed ischemia reperfusion injury (IRI). Thus one major goal in cardiology is to limit the extent and reduce the clinical consequences of IRI.

High density lipoproteins (HDL) are a central component of algorithms currently used to assess risk of vascular disease. Their importance arises from their long-established

R.W. James (✉) • M.A. Frias
Department of Internal Medicine, Medical Faculty, University of Geneva,
Rue du Général- Dufour 24, Geneva 1211, Switzerland
e-mail: Richard.James@hcuge.ch; Miguel.Frias@unige.ch

J. Camps (ed.), *Oxidative Stress and Inflammation in Non-communicable Diseases – Molecular Mechanisms and Perspectives in Therapeutics*, Advances in Experimental Medicine and Biology 824,
DOI 10.1007/978-3-319-07320-0_3, © Springer International Publishing Switzerland 2014

association with atherosclerotic disease where raised concentrations of HDL cholesterol are associated with reduced risk. Attempts to explain the beneficial effects have focused on their ability to modulate blood cholesterol metabolism. HDL are suggested to facilitate cholesterol evacuation from the artery wall and its transfer to the bile for excretion. The process, referred to as reverse cholesterol transport, is thought to reduce risk by limiting availability of cholesterol for initiation and development of the atherosclerotic plaque [1].

Whilst the process presently remains central to explanations of the beneficial effects of HDL, its clinical importance has been called into question by the disappointing results of clinical trials aimed at reducing risk by pharmacologically increasing HDL-cholesterol [2]. It has cast the spotlight on growing evidence that HDL can influence vascular health by other mechanisms, not necessarily associated with cholesterol metabolism. These include anti-inflammatory, anti-oxidant, anti-apoptotic and anti-thrombotic functions [3], all of which are implicated in vascular diseases.

The present chapter will address one such novel function of HDL, its clinical implications and its potential as a therapeutic tool. It will review the impact of HDL on IRI and describe our present understanding of the molecular mechanisms by which the lipoprotein acts. It will also discuss how these observations could be exploited for therapeutic purposes.

3.2 Structure, Composition and Function of HDL

HDL constitute one of the three lipoprotein (lipid – protein) complexes that are present in fasting plasma. Their primary function is to enable transport of hydrophobic lipids (primarily triglycerides, esterified cholesterol) in the hydrophilic blood compartment. HDL are heterogeneous with respect to size, as well as protein and lipid composition [4].

The principal protein component of HDL is apolipoprotein AI (apoAI) which is present on all HDL particles. Together with apoAII, it constitutes >80 % of the total protein content of the lipoprotein. An extensive number of other peptides is also associated with HDL and the list has grown substantially with the advent of sophisticated proteomic techniques (Fig. 3.1) [5]. Whilst the historically recognised peptide components have functions primarily linked to lipid/HDL metabolism, many of those more recently identified do not. Their presence on HDL particles underlines the potential for the lipoprotein to influence areas of vascular physiology that are independent of lipid metabolism.

The principal lipid components of HDL are phospholipids, cholesterol and triglycerides, but there is a growing list of other lipids, albeit in minor quantities, associated with the complex [6]. Whilst this may reflect in part the role of lipoproteins to offer a hydrophobic refuge for lipids in blood, other HDL lipids appear to be metabolically active. A notable example is the sphingolipid group and particularly sphingosine-1-phosphate (S1P), which will be discussed in detail below.

There is an expanding list of HDL activities that can impact on vascular disease, although the depth of experimental evidence supporting these observations varies (Table 3.1). Moreover, the impact of HDL appears to extend beyond the vasculature, with effects on pancreatic beta-cells [7], muscle function [8] and the eye [9] having been reported. In this context, it has been suggested that specific HDL functions may be associated with discrete HDL particles. This could translate into the HDL compositional heterogeneity mentioned previously and allow a certain compartmentalisation of HDL activities.

The concept of discrete HDL particles defined by peptide composition was proposed over 50 years ago by Alaupovic and colleagues [10]. It is assuming greater relevance with the need to understand and explain the functional diversity of HDL. In the light of more recent studies, we must also add lipid composition as factor in HDL particle heterogeneity, as has been elegantly illustrated for S1P [11]. Unfortunately, there are virtually no studies

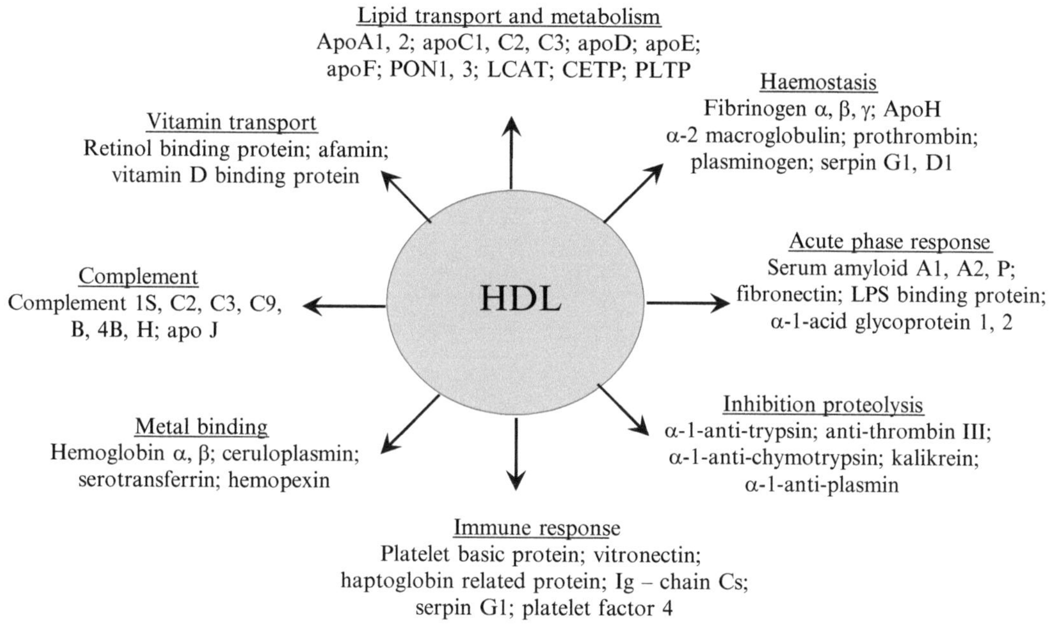

Fig. 3.1 Functionally related HDL-associated proteins identified by proteomic analyses

Table 3.1 Beneficial effects of HDL on the vascular system

Endothelial cell function
Vasorelaxation
Proliferation and migration
Precursor cell differentiation and maturation
Inflammatory reactions
Reduced expression of adhesion molecules
Regulates adaptive immunity (leukocytosis)
Reduces T-lymphocyte activation
Thrombosis
Inhibit platelet activation and aggregation
Reduces platelet production
Modulates coagulation pathway
Oxidative stress
Prevents oxidation of LDL
Limits ischemia reperfusion injury

that have been based, as yet, on HDL subfractions that are homogenous with respect to composition and/or function. It reflects technical difficulties in isolating such subfractions and is a challenge for future studies. Thus the data discussed below on the impact of HDL on IRI are derived from the use of total, non-fractionated HDL preparations.

3.3 HDL and IRI

It is unsurprising, given its clinical importance, that there is a vast literature on IRI. Interest has notably focused on the effects of ischemic pre- or post-conditioning, processes by which subjecting the heart to brief periods of ischemia reduces the clinical consequences of life-threatening, prolonged oxygen deprivation. The reader is referred to excellent reviews in this area [12, 13]. This chapter will focus on the impact of HDL on IRI, specifically with regard to cardiomyocytes.

The impact of HDL on the atherosclerotic process has largely dominated considerations of its favourable influence on myocardial infarction. However, even some 40 years ago, there were indications of a more direct impact of HDL on the myocardium. Several studies noted that low HDL cholesterol concentrations were a marker for less favourable recovery of ventricular function after infarction, irrespective of the severity of coronary atherosclerosis [14, 15]. Mochizuki et al. [16] were the first to investigate this observation using a rat model of the isolated, ex vivo heart. They reported a reduction in post-ischemic

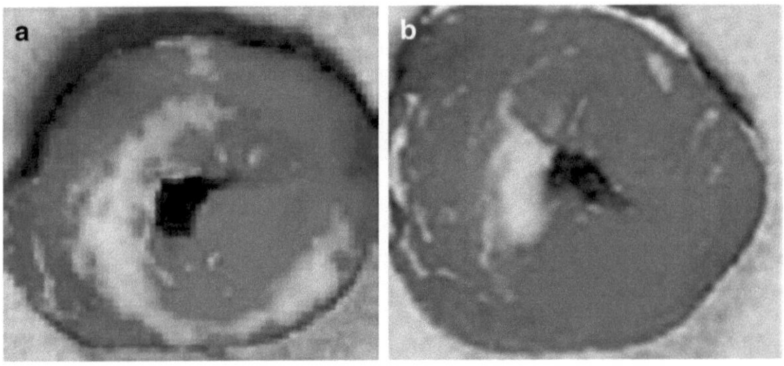

Fig. 3.2 Transverse sections of mouse hearts subjected to myocardial ischemia. Isolated, beating mouse hearts (ex vivo, Langendorff procedure) were subjected to ischemia (45 min without perfusion) then reperfused without (**a**) or with (**b**) added HDL. The whole heart was cut transversally into sections that were stained to identify infarcted areas (*white*)

arrhythmia following infusion of isolated HDL and invoked a possible impact on prostaglandin metabolism. Their report was not followed up until studies by Calabresi et al. [17, 18] confirmed the initial observations. They reported, also in an ex vivo model, that infusion of HDL improved several aspects of cardiac function. In their model, HDL were infused prior to inducing ischemia in the heart (pre-conditioning). An impact on prostaglandin metabolism, and, in particular, pro-inflammatory TNFα was observed. The authors suggested that HDL may bind and thus neutralise secreted TNFα preventing damage to the myocardium. In subsequent studies, Marchesi et al. [19] postulated that HDL may also act by reducing oxidative stress (see below).

Theilmeier et al. [20] extended these observations to an in vivo mouse model. They reported a 20–40 % reduction in the size of the infarct area (and a comparable reduction in cardiomyocyte apoptosis) in mice subjected to experimentally-induced coronary ischemia then reperfused for 24 h. As in the ex vivo studies, the model employed pre-treatment with HDL prior to inducing ischemia (Fig. 3.2). A strong inflammatory component was proposed based on neutrophil recruitment to the infarcted area. Theilmeier et al. [20] also demonstrated that the sphingolipid, sphingosine-1-phosphate (S1P), was as effective as HDL in their model. HDL are the principal source of S1P in plasma, suggesting that the effects of the lipoprotein may be mediated by associated S1P (see below for further discussion).

An important question is whether HDL act directly or indirectly on cardiomyocytes. There is an abundant literature showing that endothelial cells are particularly receptive to HDL, which can, by influencing nitric oxide metabolism, modulate various aspects of vascular physiology [21]. Indeed, Theilmeier et al. [20] suggest that in their model the protective effects of HDL were mediated by nitric oxide. In vitro, cardiomyocyte culture models have been used to address this question. Paralleling studies with the in vivo model, Theilmeier et al. [20] showed that the lipoprotein reduced apoptotic cell death in a rat neonatal cardiomyocytes, although it was not based on an ischemia-induced apoptosis model. Our studies also demonstrate a direct, protective effect of HDL on doxorubicin-induced oxidative stress in rat neonatal cardiomyocytes [22, 23]. We have recently confirmed the observations in an ischemic, in vitro model (hypoxia-reoxygenation) more closely resembling the ex vivo and in vivo experimental settings (Frias et al. submitted). Cell death was reduced by some 60–70 % in the doxorubicin model, and by 55 % in the hypoxia model. We have also investigated the role of S1P (see below). Our observations are corroborated by Tao et al. [24]. They reported a 25 % increase in HDL-mediated cell survival using adult rat cardiomyocytes in a hypoxia-reoxygenation model. A role for S1P was likewise observed.

The ex vivo and in vivo studies outlined above have employed a pre-treatment procedure with HDL to analyse their effects on IRI. Virtually no studies have analysed the impact of HDL on the myocardium when administered at the time of reperfusion, which more accurately reflects the clinical context for treatment of acute myocardial infarction. Our recent studies have addressed this question using in vivo and ex vivo models. In both cases, HDL proved to be as effective when administered in a post-ischemic context, that is at the time of reperfusion. Infarct size was reduced 35 % (ex vivo) and 50 % (in vivo) respectively (Frias et al. submitted). These observations are of particular relevance to a potential therapeutic application of HDL in IRI (see below).

3.3.1 Signalling Pathways

There is an extensive literature on signalling pathways implicated in the survival of cardiomyocytes subjected to IRI [25, 26]. A role for HDL has, however, only been identified in the last decade, and thus data on signalling pathways relaying the influence of HDL are more sparse. This overview will focus on our current understanding of the pathways linked to the impact of HDL.

An evident candidate to link HDL with intracellular signalling pathways is the scavenger receptor type BI (SRBI), an extensively characterised receptor for the lipoprotein. It is present on a wide variety of cell types, including cardiomyocytes. HDL activates several signalling pathways via the receptor [27]. However, surprisingly little attention has been paid to a possible role for SRBI in protection against IRI. Interest has centred primarily on S1P, which has a widespread impact on signalling in the vascular system [28]. HDL is a major source (50–70 %) of S1P in serum [11].

In their earlier studies, Calabresi et al. [17] proposed that HDL-mediated cardio-protection involved down regulation of TNFα expression and up-regulation of cardio-protective prostaglandins. There was no investigation, however, of signalling pathways that could effect these changes. As described above, Theilmeier et al. [20] confirmed the impact of HDL but also firmly fixed attention on the S1P component of HDL. S1P alone was sufficient to reproduce the protective effects of HDL. In particular, they attributed a major role to the S1P receptor type 3 (S1P3R) as S1P3R deficient mice could not be protected against IRI by S1P or HDL. An equally important role for nitric oxide was also observed as inhibition of NO synthase abrogated the impact of HDL. The study did not analyse the involvement of signalling pathways. However, Egom et al. [29] working with S1P alone in myocyte cultures and the ex vivo, whole heart model, have demonstrated a potential pathway. They propose activation via PI3K of Akt/Pak 1 (p21 activated kinase) leading to increased NO synthase activity and NO production, although only the role of Pak 1 activation was confirmed (using cardiac-specific Pak 1 knockout mice). The authors also postulated that IRI-induced production of TNFα by myocytes may also play a key role by stimulating myocyte production of S1P. Early studies by Lecour et al. [26, 30] had already provided evidence for a role of TNFα in protection against IRI. Our recent use of TNFα knockout models confirm the importance of the cytokine for HDL-mediated protection against IRI [31]. There would appear to be a contradiction between the results outlined above and those outlined earlier from Calabresi et al. [17]. As discussed by Lecour and James [26], an explanation may reside in the relative concentrations of TNFα, where the cytokine may have deleterious effects if the concentration rises excessively. Indeed, Calabresi et al. [17] proposed that one role of HDL was to absorb excess TNFα.

Our own studies [22, 23, 31] have analysed signalling pathways participating in HDL-mediated survival pathways. Using cardiomyocyte cultures and the ex vivo whole heart model, we have identified pivotal roles for ERK1/2 and STAT3 as well as TNFα as indicated above. These studies have entailed the use of specific agonists and antagonists, as well as cardiospecific STAT3 knockout mice. Inhibition of PI3K and p38 MAPK did not affect protection by HDL or S1P. Our observations also underlined the role of S1P receptors, and specifically identified the importance of the S1PR2 subtype using pharmacological activation or inhibition of the receptor subtypes. Comparable results were obtained using either HDL isolated from human serum or purified S1P. Our studies also clearly demonstrated the central importance of the S1P component of HDL to cardio-protection: synthetic HDL lacking S1P had a greatly reduced protective capacity (discussed below). Further refinements of survival pathways mediated by HDL/S1P were provide by Tao et al. [24], although they have also added to the complexity of HDL mediated signalling pathways in cardiomyocytes. Using an in vitro hypoxia model of IRI, they observed that HDL could protect against cell death by pathways involving S1PR1-MEK-ERK1/2 and S1PR3-PI3K-Akt (with GSKβ inactivation).

With so little data available, it is not, as yet, possible to achieve a consensus of the pathways involved. The variability in signalling pathways indicated above may reflect several factors including different IRI protocols and different models. It is also possible that given the importance of limiting damage to the heart, several survival pathways are available, as described for ischemic pre-conditioning [32]. One should also be wary of interpreting data obtained with S1P alone as being equivalent to the impact of S1P associated to HDL. As discussed by Sattler and Levkau [11], there are valid reasons for suggesting that the two sources of S1P are not functionally identical. HDL associated S1P is the form prevalent in human serum.

3.4 Therapeutic Potential of HDL

Studies of HDL as a therapeutic are greatly facilitated by the relative ease with which a synthetic form of the lipoprotein can be manufactured. It also circumvents potential safety problems arising from the use of HDL isolated from human serum. A basic form of HDL can be prepared from phospholipids and apoAI, the structural peptide of the lipoprotein. Termed reconstituted HDL (reHDL), it exhibits many of the functionalities of native HDL [33].

Synthetic HDL (reHDL) is a clinically acceptable therapeutic formulation of the lipoprotein that has already been investigated in several small scale studies in man. Nissen et al. [34] analysed the impact in coronary patients of weekly infusions of reHDL on the properties of atherosclerotic plaques. They reported a significant regression in plaque volume after 5 weeks, suggesting that it reflected the reverse cholesterol transport function of HDL. Shaw et al. [35] also observed a favourable impact on plaque morphology 1 week after a single infusion of reHDL. Other studies in man have shown that reHDL can improve endothelium function in hypercholesterolemic men [36] and reduce inflammatory reactions in type 2 diabetic patients [37].

Such clinical applications are addressing chronic features of vascular disease. It would presumably entail repeated infusions of reHDL perhaps over a prolonged period of time, which may limit its use. In contrast, exploiting reHDL in acute coronary syndromes, such as limiting IRI during cardiac reperfusion appears a more attractive and feasible therapeutic application. Surprisingly, investigations of this use of synthetic HDL are very limited, whilst virtually no studies have been undertaken with a post-ischemic model of HDL treatment.

Studies in experimental models of ischemia (ex vivo (rat) [17] and an in vivo (rabbit) [38]) showed that reHDL was capable of reducing the infarct size. Both involved treatment with the lipoprotein prior to initiating ischemia. Moreover, Calabresi et al. [17] suggested that post-ischemia treatment with reHDL was less effective in preventing IRI. Our own studies, both in vitro and ex vivo [23, 31], clearly indicate that HDL used at the time of reperfusion efficiently protects the cardiomyocytes from IRI, reducing the size of the infarct area in the ex vivo model. Moreover, our recent studies (Frias et al., submitted) have confirmed that post-ischemic treatment is equally effective in reducing infarct size with an in vivo mouse model, although such studies are presently limited to native HDL. Confirmation of the suitability of using reHDL as a therapeutic arises from our studies showing that native and reHDL activate similar signalling pathways, suggesting that reHDL is mimicking the activity of the native lipoprotein (Frias et al., submitted).

Finally, a notable advantage of reHDL is that modulating its composition can improve or add to its functionalities. We have underlined this point in our studies demonstrating the importance of S1P to protection against IRI afforded by HDL. reHDL lacking S1P is unable to, or weakly protects against IRI in vitro and ex vivo [23, 31] (the studies referred to above used reHDL containing S1P).

3.5 Conclusions

There is convincing evidence that HDL can reduce the consequences of oxidative stress in cardiac tissues subjected to ischemia. The signalling pathways involved are being identified, but we have, as yet, few indications of the target genes for these pathways. Available data suggest that synthetic HDL can be as effective as native HDL in reducing oxidative stress. The feasibility of using reHDL in man has already been demonstrated in other small scale, clinical studies. Treatment of myocardial infarction, which is a major clinical problem, offers a propitious clinical context for analysing and exploiting the therapeutic potential of reHDL. Future studies should focus on extending observations in animal models to studies in man.

Acknowledgements RWJ gratefully acknowledges support received from the Swiss National Research Foundation, the Novartis Consumer Health Foundation and Unitec, University of Geneva. MF has received support from Fondation Gustave et Simone Prévot, Wolfermann Naegli Stiftung, Jubiläumsstiftung and Fondation pour la luttre contre le cancer et investigations médico-biologiques.

RWJ and MAF are members of the COST Action BM0904.

References

1. Barter PJ, Rye KA. Molecular mechanisms of reverse cholesterol transport. Curr Opin Lipidol. 1996;7:82–7.
2. Schwartz GG, Olsson AG, Abt M, Ballantyne CM, Barter PJ, Brumm J, et al. Effects of dalcetrapib in patients with a recent acute coronary syndrome. N Engl J Med. 2012;367:2089–99.
3. Rohrer L, Hersberger M, von Eckardstein A. High density lipoproteins in the intersection of diabetes mellitus, inflammation and cardiovascular disease. Curr Opin Lipidol. 2004;15:269–78.
4. Annema W, von Eckardstein A. High-density lipoproteins. Multifunctional but vulnerable protections from atherosclerosis. Circ J. 2013;77:2432–48.
5. Shah AS, Tan L, Long JL, Davidson WS. Proteomic diversity of high density lipoproteins: our emerging understanding of its importance in lipid transport and beyond. J Lipid Res. 2013;54:2575–85.
6. Kontush A, Lhomme M, Chapman MJ. Unraveling the complexities of the HDL lipidome. J Lipid Res. 2013;54:2950–63.
7. Petremand J, Puyal J, Chatton JY, Duprez J, Allagnat F, Frias M, et al. HDLs protect pancreatic beta-cells against ER stress by restoring protein folding and trafficking. Diabetes. 2012;61:1100–11.
8. Lehti M, Donelan E, Abplanalp W, Al-Massadi O, Habegger KM, Weber J, et al. High-density lipoprotein maintains skeletal muscle function by modulating cellular respiration in mice. Circulation. 2013;128:2364–71.
9. Ishida BY, Duncan KG, Bailey KR, Kane JP, Schwartz DM. High density lipoprotein mediated lipid efflux from retinal pigment epithelial cells in culture. Br J Ophthalmol. 2006;90:616–20.
10. Alaupovic P. Significance of apolipoproteins for structure, function, and classification of plasma lipoproteins. Methods Enzymol. 1996;263:32–60.
11. Sattler K, Levkau B. Sphingosine-1-phosphate as a mediator of high-density lipoprotein effects in cardiovascular protection. Cardiovasc Res. 2009;82:201–11.
12. Hausenloy DJ. Signalling pathways in ischaemic postconditioning. Thromb Haemost. 2009;101:626–34.
13. Hausenloy DJ, Yellon DM. Preconditioning and postconditioning: underlying mechanisms and clinical application. Atherosclerosis. 2009;204:334–41.
14. Berge KG, Canner PL, Hainline Jr A. High-density lipoprotein cholesterol and prognosis after myocardial infarction. Circulation. 1982;66:1176–8.
15. Kempen HJ, van Gent CM, Buytenhek R, Buis B. Association of cholesterol concentrations in low-density lipoprotein, high-density lipoprotein, and high-density lipoprotein subfractions, and of apolipoproteins AI and AII, with coronary stenosis and left ventricular function. J Lab Clin Med. 1987;109:19–26.
16. Mochizuki S, Okumura M, Tanaka F, Sato T, Kagami A, Tada N, et al. Ischemia-reperfusion arrhythmias and lipids: effect of human high- and low-density

lipoproteins on reperfusion arrhythmias. Cardiovasc Drugs Ther. 1991;5 Suppl 2:269–76.
17. Calabresi L, Rossoni G, Gomaraschi M, Sisto F, Berti F, Franceschini G. High-density lipoproteins protect isolated rat hearts from ischemia-reperfusion injury by reducing cardiac tumor necrosis factor-alpha content and enhancing prostaglandin release. Circ Res. 2003;92:330–7.
18. Rossoni G, Gomaraschi M, Berti F, Sirtori CR, Franceschini G, Calabresi L. Synthetic high-density lipoproteins exert cardioprotective effects in myocardial ischemia/reperfusion injury. J Pharmacol Exp Ther. 2004;308:79–84.
19. Marchesi M, Booth EA, Rossoni G, Garcia RA, Hill KR, Sirtori CR, et al. Apolipoprotein A-IMilano/POPC complex attenuates post-ischemic ventricular dysfunction in the isolated rabbit heart. Atherosclerosis. 2008;197:572–8.
20. Theilmeier G, Schmidt C, Herrmann J, Keul P, Schafers M, Herrgott I, et al. High-density lipoproteins and their constituent, sphingosine-1-phosphate, directly protect the heart against ischemia/reperfusion injury in vivo via the S1P3 lysophospholipid receptor. Circulation. 2006;114:1403–9.
21. Andrews KL, Moore XL, Chin-Dusting JP. Anti-atherogenic effects of high-density lipoprotein on nitric oxide synthesis in the endothelium. Clin Exp Pharmacol Physiol. 2010;37:736–42.
22. Frias MA, James RW, Gerber-Wicht C, Lang U. Native and reconstituted HDL activate Stat3 in ventricular cardiomyocytes via ERK1/2: role of sphingosine-1-phosphate. Cardiovasc Res. 2009;82:313–23.
23. Frias MA, Lang U, Gerber-Wicht C, James RW. Native and reconstituted HDL protect cardiomyocytes from doxorubicin-induced apoptosis. Cardiovasc Res. 2010;85:118–26.
24. Tao R, Hoover HE, Honbo N, Kalinowski M, Alano CC, Karliner JS, et al. High-density lipoprotein determines adult mouse cardiomyocyte fate after hypoxia-reoxygenation through lipoprotein-associated sphingosine 1-phosphate. Am J Physiol Heart Circ Physiol. 2010;298:H1022–8.
25. Hausenloy DJ, Yellon DM. Myocardial ischemia-reperfusion injury: a neglected therapeutic target. J Clin Invest. 2013;123:92–100.
26. Lecour S, James RW. When are pro-inflammatory cytokines SAFE in heart failure? Eur Heart J. 2011; 32:680–5.
27. Mineo C, Shaul PW. Regulation of signal transduction by HDL. J Lipid Res. 2013;54:2315–24.
28. Alewijnse AE, Peters SL. Sphingolipid signalling in the cardiovascular system: good, bad or both? Eur J Pharmacol. 2008;585:292–302.
29. Egom EE, Mohamed TM, Mamas MA, Shi Y, Liu W, Chirico D, et al. Activation of Pak1/Akt/eNOS signaling following sphingosine-1-phosphate release as part of a mechanism protecting cardiomyocytes against ischemic cell injury. Am J Physiol Heart Circ Physiol. 2011;301:H1487–95.
30. Lecour S, Smith RM, Woodward B, Opie LH, Rochette L, Sack MN. Identification of a novel role for sphingolipid signaling in TNF alpha and ischemic preconditioning mediated cardioprotection. J Mol Cell Cardiol. 2002;34:509–18.
31. Frias MA, Pedretti S, Hacking D, Somers S, Lacerda L, Opie LH, et al. HDL protects against ischemia reperfusion injury by preserving mitochondrial integrity. Atherosclerosis. 2013;228:110–6.
32. Heusch G, Boengler K, Schulz R. Cardioprotection: nitric oxide, protein kinases, and mitochondria. Circulation. 2008;118:1915–9.
33. Cho KH. Biomedicinal implications of high-density lipoprotein: its composition, structure, functions, and clinical applications. BMB Rep. 2009;42:393–400.
34. Nissen SE, Tsunoda T, Tuzcu EM, Schoenhagen P, Cooper CJ, Yasin M, et al. Effect of recombinant ApoA-I Milano on coronary atherosclerosis in patients with acute coronary syndromes: a randomized controlled trial. JAMA. 2003;290:2292–300.
35. Shaw JA, Bobik A, Murphy A, Kanellakis P, Blombery P, Mukhamedova N, et al. Infusion of reconstituted high-density lipoprotein leads to acute changes in human atherosclerotic plaque. Circ Res. 2008;103:1084–91.
36. Spieker LE, Sudano I, Hurlimann D, Lerch PG, Lang MG, Binggeli C, et al. High-density lipoprotein restores endothelial function in hypercholesterolemic men. Circulation. 2002;105:1399–402.
37. Patel S, Drew BG, Nakhla S, Duffy SJ, Murphy AJ, Barter PJ, et al. Reconstituted high-density lipoprotein increases plasma high-density lipoprotein anti-inflammatory properties and cholesterol efflux capacity in patients with type 2 diabetes. J Am Coll Cardiol. 2009;53:962–71.
38. Marchesi M, Booth EA, Davis T, Bisgaier CL, Lucchesi BR. Apolipoprotein A-IMilano and 1-palmitoyl-2-oleoyl phosphatidylcholine complex (ETC-216) protects the in vivo rabbit heart from regional ischemia-reperfusion injury. J Pharmacol Exp Ther. 2004;311:1023–31.

The DING Family of Phosphate Binding Proteins in Inflammatory Diseases

4

Daniel Gonzalez, Mikael Elias, and Eric Chabrière

Abstract

Human paraoxonase 1 (hPON-1) is a protein that has been studied in relation to its antioxidant and anti-atherosclerotic properties. Despite extensive studies, the molecular mechanisms responsible for its functional properties remain unclear. During the last decade, a new partner of hPON-1 has been identified. Hidden for a long time because of a similar molecular weight with hPON-1, this protein, termed human phosphate-binding protein (HPBP), may contribute to the biological functions of hPON-1. Belonging to the DING protein, a sub-family of phosphate binding proteins (PBP or pstS), HPBP stabilizes hPON-1 and might prevent calcification of arteries in case of advance atherosclerosis. The role of other DING proteins in some calcification processes (*i.e.* nephrolithiasis) and the identification of HPBP in the atheroma plaque support this hypothesis. Nevertheless, the relevance of hPON-1/HPBP as well as the molecular determinants in atherosclerosis remains to be elucidated.

Keywords

DING proteins • HPBP • Inflammation • Phosphate-binding proteins

4.1 The Serendipitous Discovery of HPBP

Human paraoxonase-1 (hPON-1) is largely studied due to its relationship with human atherosclerosis. Nevertheless, the molecular determinants responsible of its anti-atherosclerotic properties remain unclear. While performing structural studies on hPON-1, Morales *et al.* serendipitously discovered a hPON-1 associated protein [1]. Astonishingly, from a supposedly "pure" sample of hPON-1,

D. Gonzalez • E. Chabrière (✉)
URMITE UMR CNRS-IRD 6236, IFR48, Faculté de Médecine et de Pharmacie, Université de la Méditerranée, 27, Boulevard Jean Moulin, CS 30064, Marseille Cedex 13385, France
e-mail: gonzalez.daniel.mrs@gmail.com; eric.chabriere@univ-amu.fr

M. Elias
Biological Chemistry, Weizmann Institute of Science, Rehovot, Israel
e-mail: mikael.elias@gmx.fr

J. Camps (ed.), *Oxidative Stress and Inflammation in Non-communicable Diseases – Molecular Mechanisms and Perspectives in Therapeutics*, Advances in Experimental Medicine and Biology 824, DOI 10.1007/978-3-319-07320-0_4, © Springer International Publishing Switzerland 2014

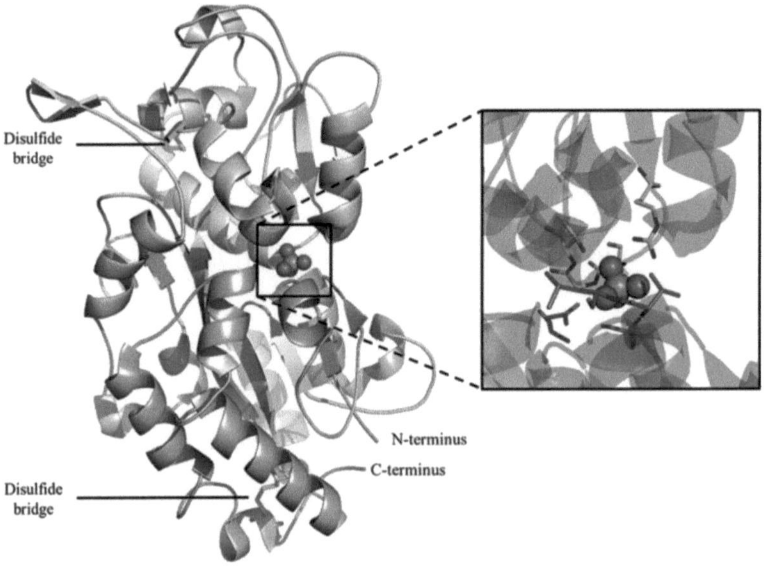

Fig. 4.1 **Structure of HPBP (pdbid: 2V3Q).** HPBP presents two globular domains colorized in *green* (*upper domain*) and *blue* (*bottom domain*). The disulfide bridges have been represented as sticks. The phosphate has been represented as spheres. At the interface of both domains, there is the phosphate binding cleft. Residues involved in phosphate binding have been represented as sticks and colorized using the same color-code as the domain

they crystallized a new protein, absent from genomic databases. This protein, termed human phosphate binding protein (HPBP), has been for a long time hidden behind hPON-1 glycosylated isoforms on SDS-PAGE because they have similar molecular weight (*i.e.* 38 kDa) [1]. Since HPBP was absent from genomic databases, the sequence of the protein has been determined using crystallographic data, coupled with mass spectrometry [2]. The obtained sequence and structure shows that HPBP belongs to the phosphate binding proteins (PBP) and more precisely to a sub-family called "DING proteins".

The structure of HPBP is composed of two globular domains linked together with a flexible hinge (Fig. 4.1). The phosphate anion binding site is located at the interface of both domains. Due to its ability to fix phosphate (with submicromolar affinity), HPBP constitutes the first phosphate transporter identified from human tissues, being homologous to the prokaryotic periplasmic PBP associated with the ABC transporter system (*i.e.* PstS) [3]. Noteworthy, some differences are noticed between HPBP's and PstS' structure, such as the presence of two disulfide bridges and four protuberant loops (Fig. 4.1). Nevertheless, the physiological role of these differences remains to be elucidated.

4.2 Properties of DING Proteins

DING proteins constitute a poorly characterized family of proteins, having usually 38–40 kDa, which were named for their eponymous N-terminus extremity (D-I-N-G-G-G) [1]. The first DING proteins were identified in animals and in some plants at the end of the 1990s. The genomic era has greatly increased the number of new DING protein sequences, especially in bacteria. Moreover, it allowed the first identifications of DING protein's genes in prokaryotes (and mainly from genus *Pseudomonas*). Today, more than 50 DING proteins (and genes) have been identified in the prokaryote kingdom [4, 5].

The situation is different in eukaryotes: although DING-coding sequences were amplified from eukaryotic genomic sequences; no open reading frame nor locus has yet been identified on published eukaryotic genomes [5]. This critical absence of eukaryotic genetic information constitutes a major barrier which has drastically hampered their studies. As a consequence, eukaryotic DING proteins were serendipitously identified due to their implication in pathological processes and, in particular, in inflammation processes [6].

In humans, six different DING proteins have been identified; HPBP, synovial stimulatory Protein (SSP), X-DING-CD4, crystal adhesion inhibitor (CAI), genistein-binding protein and a DING protein from GSS database. In the next part, we will focus on DING proteins which have been clearly involved in inflammatory diseases [1, 7–11].

4.2.1 Implications of DING Proteins in Inflammatory Processes

4.2.1.1 Rheumatoid Arthritis

DING proteins were discovered because of their involvement with rheumatoid arthritis (RA), an inflammatory auto-immune disease causing a deformation or destruction of articulations. In the aim to unravel the poorly understood origin of RA, the composition of rheumatoid fluids were widely studied, leading to the identification of the synovial stimulatory protein (SSP) [7]. This protein has been partially sequenced, revealing the characteristically N-terminus "D-I-N-G-G-G" and leading to the first isolation of a DING protein in a human disease. SSP has been shown to interact with human sera containing rheumatoid factor (*i.e.* auto-antibody) and to possess stimulatory capacity of T cell proliferation. Both abilities suggest that SSP is involved in the RA inflammatory process [7].

4.2.1.2 Nephrolithiasis

The implication of DING proteins in inflammatory processes is not only limited to SSP, and numerous examples show that this family of proteins is tightly correlated with inflammation [9, 12]. Nephrolithiasis, also known as kidney stone disease is a common derangement with an increasing prevalence in the population (up to 5 %) [13]. This illness is caused by the aggregation of calcium oxalate or calcium phosphate with proteins and/or bacteria [14]. Growth and migration of renal calculi into the lower urinary tract cause severe pain and in some cases complications and inflammation leading to death [14]. However, the exact mechanism of renal calculi formation remains unclear even if many theories have been proposed (saturation of urine, nanobacteria) [13, 15–17].

The crystal adhesion inhibitor (CAI) is a DING protein studied in relation to renal calculi formation. CAI is a 39 kDa protein which has been isolated in monkey renal epithelial cells [10, 18]. It has been demonstrated that CAI inhibits renal calculi growth by significantly reducing the binding of calcium oxalate monohydrate (COM) onto crystal [10]. Indeed, CAI sequesters/covers crystal surface avoiding its growth. This fixation capability may be due to CAI affinity towards phosphate (like other DING) since renal calculi crystals are often composed of calcium phosphate [13, 14]. Even whether the implication CAI in inhibition of renal calculi growth is clear, other proteins seem to be more crucial (for instance the Tamm-Horsfall Protein) [18].

4.2.1.3 Atherosclerosis

Atherosclerosis represents the leading cause of mortality/morbidity in western countries [19]. The molecular mechanisms involved in the formation of the atheroma plaque are complex, and a key phenomenon is the accumulation of oxidized low density lipoprotein particles (oxLDL) in the inner walls of arteries [20, 21]. The atherosclerotic plaque, once unhooked, may lead to ischemic stroke/heart attack [22, 23]. The accumulation of oxLDL in the atheroma plaque is counteracted by high density lipoproteins (HDL) that modify the balance oxLDL/LDL, slowing down the formation of atherosclerotic plaque [24]. Particularly, it has been found that hPON-1, a HDL-associated protein, plays a protective role by degrading peroxidized lipids and reducing the level of oxLDL in human plasma [25–28]. Nevertheless, the exact the molecular determinants remain unclear.

Recent studies showed that HPBP, a DING protein, is associated with hPON-1 [1, 29], and

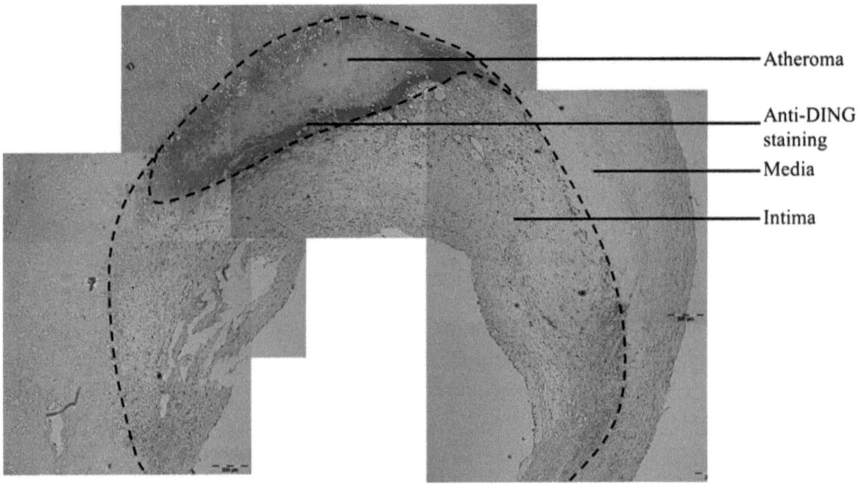

Fig. 4.2 Immunohistochemical staining of DING proteins in an atheroma. The endothelium is sub-divided in two parts, the intima and the media. The separation between the two layers is represented as a *dashed line*. The atheroma has been limited with *dashed line*

stabilizes the active conformation of this enzyme [29–31]. The association hPON-1/HPBP may thus influence the related anti-atherosclerotic properties of hPON1. Moreover, since DING proteins belong to the phosphate binding protein superfamily, it has been hypothesized that HPBP may have a role in preventing phosphate salt formation, specifically with calcium. HPBP may thus be implicated in calcification processes which are widely common in advance atherosclerosis. This hypothesis is consistent with the co-localization of DING proteins and hPON1 within the atheroma plaque.

Even if the biochemical mode of action of HPBP is yet unclear, the relationships between inflammation and other DING proteins closely related to HPBP have been recently investigated. DING proteins have been related to the phosphorylation of several proteins in the mitogen-activated protein (MAP) kinase pathway, including extracellular signal-regulated kinases (ERK) 1 and 2 and c-Jun N-terminal kinases (JNK) [32]. This impacts their downstream targets like the signal transducer and activator of transcription (STAT) 3, the cyclic AMP response element-binding (CREB), c-Jun and more importantly NF-κB [32]. Moreover, it has been shown that some DING proteins interact with the p50 sub-unit of NF-κB [33], a major pathway in inflammatory response. Indeed, the production of numerous pro-inflammatory cytokines has been shown to be NF-κB-dependent in atherosclerosis [34]. The modulation of the association of NF-κB subunits may thus lead to the modulation of inflammatory response in atherosclerosis process. Furthermore, DING proteins interact with the CCAAT/enhancer-binding protein (C/EBP) β binding site altering its nuclear localization. This was demonstrated by DNA band-shift assay, that showed that the presence of DING proteins reduces significantly the ability of C/EBP β to bind to DNA [35]. C/EBP β is important in the regulation of genes involved in immune and inflammatory responses and has been shown to bind to regulatory regions of several acute-phase and cytokine genes, and is critical for normal macrophage functioning, and in the modulation of inflammatory cytokine such as interleukins-1 and -6 [35, 36]. All these molecular mechanistic insights, and the presence of DING proteins *in situ* (*i.e.* in atheroma, Fig. 4.2) suggests that DING proteins may play a role in the pathophysiology of atherosclerosis.

4.3 Conclusion

Results from recent research show that DING proteins are probably involved in inflammatory process. In the case of human atherosclerosis, the association between HPBP and hPON1 may be relevant since it contributes to stabilize this antioxidant and anti-inflammatory enzyme. However, numerous questions remain and future structural/mechanistic studies on HPBP/hPON-1 complexes may unravel new potential functions of DING proteins in atherosclerosis.

References

1. Morales R, Berna A, Carpentier P, Contreras-Martel C, Renault F, Nicodeme M, et al. Serendipitous discovery and X-ray structure of a human phosphate binding apolipoprotein. Structure. 2006;14:601–9.
2. Diemer H, Elias M, Renault F, Rochu D, Contreras-Martel C, Schaeffer C, et al. Tandem use of X-ray crystallography and mass spectrometry to obtain ab initio the complete and exact amino acids sequence of HPBP, a human 38-kDa apolipoprotein. Proteins. 2008;71:1708–20.
3. Luecke H, Quiocho FA. High specificity of a phosphate transport protein determined by hydrogen bonds. Nature. 1990;347:402–6.
4. Berna A, Scott K, Chabriere E, Bernier F. The DING family of proteins: ubiquitous in eukaryotes, but where are the genes? Bioessays. 2009;31:570–80.
5. Bernier F. DING proteins: numerous functions, elusive genes, a potential for health. Cell Mol Life Sci. 2013;70:3045–56.
6. Berna A, Bernier F, Chabriere E, Elias M, Scott K, Suh A. For whom the bell tolls? DING proteins in health and disease. Cell Mol Life Sci. 2009;66:2205–18.
7. Adams L, Davey S, Scott K. The DING protein: an autocrine growth-stimulatory protein related to the human synovial stimulatory protein. Biochim Biophys Acta. 2002;1586:254–64.
8. Belenky M, Prasain J, Kim H, Barnes S. DING, a genistein target in human breast cancer: a protein without a gene. J Nutr. 2003;133:2497S–501.
9. Ivanova A, Shilpi RY, Sachdeva R, Li G, Simm M. Native X-DING-CD4 protein secreted by HIV-1 resistant CD4+ T cells blocks activity of IL-8 promoter in human endothelial cells infected with enteric bacteria. Innate Immun. 2012;18:571–9.
10. Kumar V, Yu S, Farell G, Toback F, Lieske J. Renal epithelial cells constitutively produce a protein that blocks adhesion of crystals to their surface. Am J Physiol Renal Physiol. 2004;287:F373–83.
11. Lesner A, Shilpi R, Ivanova A, Gawinowicz MA, Lesniak J, Nikolov D, et al. Identification of X-DING-CD4, a new member of human DING protein family that is secreted by HIV-1 resistant CD4(+) T cells and has anti-viral activity. Biochem Biophys Res Commun. 2009;389:284–9.
12. Mukerjee R, Deshmane SL, Darbinian N, Czernik M, Khalili K, Amini S, et al. St. John's Wort protein, p27SJ, regulates the MCP-1 promoter. Mol Immunol. 2008;45:4028–35.
13. Hall PM. Nephrolithiasis: treatment, causes, and prevention. Cleve Clin J Med. 2009;76:583–91.
14. Worcester EM, Coe FL. Nephrolithiasis. Prim Care. 2008;35:369–91.
15. Drancourt M, Jacomo V, Lepidi H, Lechevallier E, Grisoni V, Coulange C, et al. Attempted isolation of Nanobacterium sp. microorganisms from upper urinary tract stones. J Clin Microbiol. 2003;41:368–72.
16. Kumar V, Farell G, Yu S, Harrington S, Fitzpatrick L, Rzewuska E, et al. Cell biology of pathologic renal calcification: contribution of crystal transcytosis, cell-mediated calcification, and nanoparticles. J Investig Med. 2006;54:412–24.
17. Khullar M, Sharma SK, Singh SK, Bajwa P, Sheikh FA, Relan V, et al. Morphological and immunological characteristics of nanobacteria from human renal stones of a north Indian population. Urol Res. 2004;32:190–5.
18. Kumar V, Pena De La Vega L, Farell G, Lieske JC. Urinary macromolecular inhibition of crystal adhesion to renal epithelial cells is impaired in male stone formers. Kidney Int. 2005;68:1784–92.
19. Go AS, Mozaffarian D, Roger VL, Benjamin EJ, Berry JD, Borden WB, et al. Heart disease and stroke statistics-2013 update: a report from the American heart association. Circulation. 2013;127:e6–245.
20. Steinbrecher UP, Parthasarathy S, Leake DS, Witztum JL, Steinberg D. Modification of low density lipoprotein by endothelial cells involves lipid peroxidation and degradation of low density lipoprotein phospholipids. Proc Natl Acad Sci U S A. 1984;81:3883–7.
21. Yoshida H, Kisugi R. Mechanisms of LDL oxidation. Clin Chim Acta. 2010;411:1875–82.
22. Bui QT, Prempeh M, Wilensky RL. Atherosclerotic plaque development. Int J Biochem Cell Biol. 2009;41:2109–13.
23. Lusis AJ. Atherosclerosis. Nature. 2000;407:233–41.
24. Assmann G, Gotto Jr AM. HDL cholesterol and protective factors in atherosclerosis. Circulation. 2004;109:III8–14.
25. Shih DM, Gu L, Xia YR, Navab M, Li WF, Hama S, et al. Mice lacking serum paraoxonase are susceptible to organophosphate toxicity and atherosclerosis. Nature. 1998;394:284–7.
26. Tward A, Xia YR, Wang XP, Shi YS, Park C, Castellani L, et al. Decreased atherosclerotic lesion formation in human serum paraoxonase transgenic mice. Circulation. 2002;106:484–90.
27. Mackness B, Mackness MI, Arrol S, Turkie W, Durrington PN. Effect of the human serum paraoxonase

55 and 192 genetic polymorphisms on the protection by high density lipoprotein against low density lipoprotein oxidative modification. FEBS Lett. 1998;423:57–60.
28. Gupta N, Gill K, Singh S. Paraoxonases: structure, gene polymorphism & role in coronary artery disease. Indian J Med Res. 2009;130:361–8.
29. Renault F, Chabriere E, Andrieu JP, Dublet B, Masson P, Rochu D. Tandem purification of two HDL-associated partner proteins in human plasma, paraoxonase (PON1) and phosphate binding protein (HPBP) using hydroxyapatite chromatography. J Chromatogr B Analyt Technol Biomed Life Sci. 2006;836:15–21.
30. Rochu D, Chabriere E, Renault F, Elias M, Clery-Barraud C, Masson P. Stabilization of the active form(s) of human paraoxonase by human phosphate-binding protein. Biochem Soc Trans. 2007;35:1616–20.
31. Rochu D, Renault F, Clery-Barraud C, Chabriere E, Masson P. Stability of highly purified human paraoxonase (PON1): association with human phosphate binding protein (HPBP) is essential for preserving its active conformation(s). Biochim Biophys Acta. 2007;1774:874–83.
32. Bookland MJ, Darbinian N, Weaver M, Amini S, Khalili K. Growth inhibition of malignant glioblastoma by DING protein. J Neurooncol. 2012;107:247–56.
33. Lesner A, Li Y, Nitkiewicz J, Li G, Kartvelishvili A, Kartvelishvili M, et al. A soluble factor secreted by an HIV-1-resistant cell line blocks transcription through inactivating the DNA-binding capacity of the NF-kappa B p65/p50 dimer. J Immunol. 2005;175:2548–54.
34. Monaco C, Andreakos E, Kiriakidis S, Mauri C, Bicknell C, Foxwell B, et al. Canonical pathway of nuclear factor kappa B activation selectively regulates proinflammatory and prothrombotic responses in human atherosclerosis. Proc Natl Acad Sci U S A. 2004;101:5634–9.
35. Darbinian-Sarkissian N, Darbinyan A, Otte J, Radhakrishnan S, Sawaya BE, Arzumanyan A, et al. p27(SJ), a novel protein in St John's Wort, that suppresses expression of HIV-1 genome. Gene Ther. 2006;13:288–95.
36. Akira S, Isshiki H, Sugita T, Tanabe O, Kinoshita S, Nishio Y, et al. A nuclear factor for IL-6 expression (NF-IL6) is a member of a C/EBP family. EMBO J. 1990;9:1897–906.

Inflammation, Infection, Cancer and All That...The Role of Paraoxonases

5

Asokan Devarajan, Diana Shih, and Srinivasa T. Reddy

Abstract

The paraoxonase (PON) gene family consists of three members, PON1, PON2 and PON3. All PON proteins possess antioxidant properties and lipo-lactonase activities, and are implicated in the pathogenesis of several inflammatory diseases including atherosclerosis, Alzheimer's, Parkinson's, diabetes and cancer. Despite the role of PON proteins in critical cellular functions and associated pathologies, the physiological substrates and molecular mechanisms by which PON proteins function as anti-inflammatory proteins remain largely unknown. PON1 is found exclusively extracellular and associated solely with high-density lipoprotein (HDL) particles in the circulation, and, in part, confers the anti-oxidant and anti-inflammatory properties associated with HDL. Recent studies demonstrated that the intracellular PON proteins; PON2 and PON3 (i) are associated with mitochondria and mitochondria-associated membranes, (ii) modulate mitochondria-dependent superoxide production, and (iii) prevent apoptosis. Overexpression of PON2 and PON3 genes protected (i) mitochondria from antimycin or oligomycin mediated mitochondrial dysfunction and (ii) ER stress and ER stress mediated mitochondrial dysfunction. These studies illustrate that the anti-inflammatory effects of PON2 and PON3

A. Devarajan
Department of Obstetrics and Gynecology, David Geffen School of Medicine at UCLA, 90095 Los Angeles, CA, USA
e-mail: ADevarajan@mednet.ucla.edu

D. Shih
Department of Medicine, David Geffen School of Medicine at UCLA, 90095 Los Angeles, CA, USA
e-mail: dshih@mednet.ucla.edu

S.T. Reddy (✉)
Department of Obstetrics and Gynecology

Department of Molecular and Medical Pharmacology

Division of Cardiology, Department of Medicine, David Geffen School of Medicine at UCLA, 10833 Le Conte Avenue, Box 951679, 90095-1679, 43–144 CHS, Los Angeles, CA 167917, USA
e-mail: sreddy@mednet.ucla.edu

J. Camps (ed.), *Oxidative Stress and Inflammation in Non-communicable Diseases – Molecular Mechanisms and Perspectives in Therapeutics*, Advances in Experimental Medicine and Biology 824, DOI 10.1007/978-3-319-07320-0_5, © Springer International Publishing Switzerland 2014

may, in part, be mediated by their role in mitochondrial and associated organelle function. Since oxidative stress as a result of mitochondrial dysfunction is implicated in the development of inflammatory diseases including atherosclerosis and cancer, these recent studies on PON2 and PON3 proteins may provide a mechanism for the scores of epidemiological studies that show a link between PON genes and numerous inflammatory diseases. Understanding such mechanisms will provide novel routes of intervention in the treatment of diseases associated with pro-inflammatory oxidative stress.

Keywords
Antioxidants • Atherosclerosis • Cancer • Endoplasmic reticulum stress • Inflammation • Paraoxonase 1 • Paraoxonase 2 • Paraoxonase 3 • Quorum sensing

5.1 Introduction

The biology of oxidative stress underlies the molecular mechanisms responsible for the development of a number of inflammatory and infectious diseases, including atherosclerosis, diabetes mellitus, systemic lupus erythematosus, rheumatoid arthritis, cancer, and several ailments associated with age. A number of enzymes have evolved with pro-oxidant or anti-oxidant functions for maintaining oxidative homeostasis in cells and tissues. Understanding the function of such enzymes will pave way for the discovery of novel therapeutic agents in the fight against inflammatory diseases including atherosclerosis.

The paraoxonase (PON) gene family consists of three members, PON1, PON2, and PON3. Epidemiological studies suggest that expression of all three PON genes negatively correlates with a number of inflammatory diseases including atherosclerosis [1, 2]. PON genes are located on the long arm of chromosome 7 in human and chromosome 6 in mice [3]. PON2 appears to be the oldest member of the family, followed by PON3 and PON1, which most likely resulted from gene duplication [4]. PON1 was not only the first one of the family to be identified in a screen for plasma hydrolases of paraoxon, the active metabolite of insecticide parathion [5, 6], but also the predominant member of the PON family in the circulation. Most of the early investigations were centered on toxicological research on PON1. Following the discovery that PON1 is associated with high-density lipoproteins (HDL) in the circulation and that PON1 plays an important role in the protective antioxidant and anti-inflammatory effects of HDL [6], a new area of research emerged for PON1 in both its disease association and physiological function. In contrast to PON1, both PON2 and PON3 are predominantly localized to intracellular compartments (although small amounts of hPON3 is also associated with HDL) and modulate cellular oxidative stress generated both by intracellular mechanisms and in response to extracellular stimuli [1] (Fig. 5.1).

Battacharya et al. provided evidence for a mechanistic link between genetic determinants and activity of PON1 with systemic oxidative stress and prospective cardiovascular risk, indicating a potential mechanism for the atheroprotective

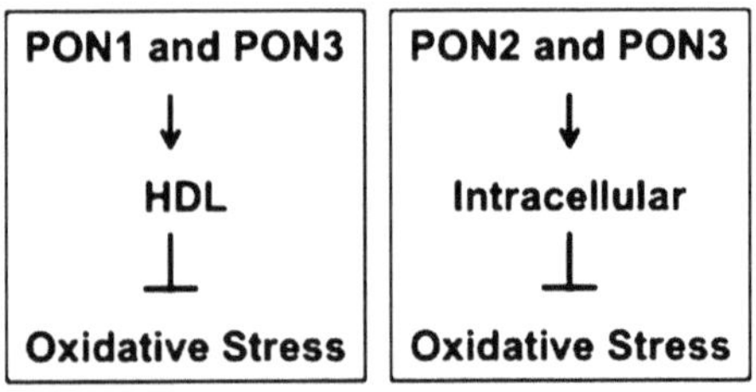

Fig. 5.1 PON genes are implicated in the mechanisms of oxidative stress

function of PON1 [7]. Stevens et al. showed that recombinant human PON1 protects against organophosphate poisoning [8]. Stoltz et al. demonstrated that PON proteins can interfere with quorum sensing *in vivo* and suggested a potential role for PON proteins as regulators of normal bacterial florae, a link between infection/inflammation and cardiovascular disease [9]. Witte et al. [10] demonstrated that PON2 provides apoptosis resistance and stabilizes tumor cells, and Schweikert et al. [11] showed that PON3 is upregulated in cancer tissue and prevents cell death. Each of the studies described above [7–11] make the compelling argument for a role for paraoxonases in inflammation, toxicology, infection, and cancer.

5.2 Inflammation and Atherosclerosis

A definitive proof for the antiatherogenic role for PON genes came first from PON1 deficient and transgenic mice, which are susceptible to organophosphate toxicity and atherosclerosis [12–14] and more recently from PON2 deficient [15], which are also susceptible to atherosclerosis. In contrast, overexpression of human PON3 [16] as well as the human paraoxonase gene cluster rendered mice resistant to atherosclerosis [17]. Thus, all three members of the PON gene family are antiatherogenic in mouse models of atherosclerosis (Fig. 5.2).

5.2.1 PON1

PON1 protects LDL against oxidation and preserves function of HDL [18, 19]. PON1 null mice by gene targeting [12, 13] and transgenic

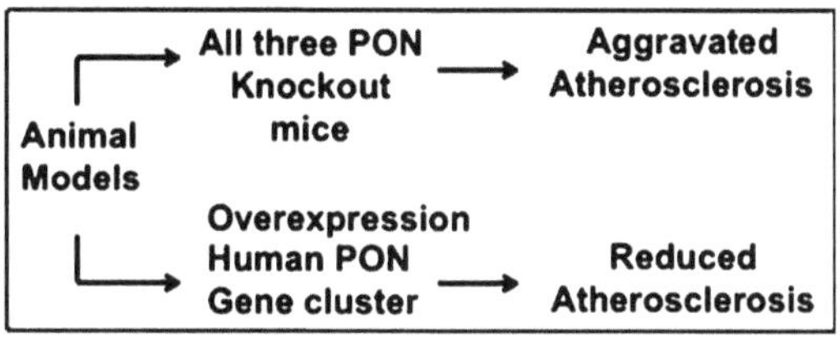

Fig. 5.2 PON genes are anti-atherogenic in mouse models of atherosclerosis

mice [14] corroborate the hypothesis that PON1 protects against atherogenesis and is an important contributor to HDL's antioxidant capacity. The *in vivo* studies combined with the antioxidant and antiatherogenic nature of PON1 underscore the potential of PON1 as a therapeutic agent to prevent atheroma [20, 21]. The association of PON1 with HDL and its functional consequences support a causal relationship between PON1 and cardiovascular and its associated inflammatory pathologies [22–26].

Multiple *in vitro*, animal and human studies have demonstrated anti-inflammatory and anti-oxidative functions of PON1. For example, PON1 deficient mice have been shown to be more susceptible to lipoprotein oxidation, inflammation, atherosclerosis [12, 13, 27], and hepatic steatosis [28], whereas PON1 transgenic mice over-expressing human PON1 are more resistant to inflammation and atherosclerosis [14]. PON1 has been shown to prevent LDL oxidation *in vitro* [18, 19] and decreased levels of PON1 are associated with increased risk for cardiovascular disease in humans [7, 29–32].

Recent detailed biochemical studies have further elucidated how PON1 exerts its anti-inflammatory and anti-oxidative functions [23, 33]. Besler et al. [33] demonstrated a role of HDL-associated PON1 activity in maintaining the endothelial atheroprotective effects of HDL, namely HDL-mediated stimulation of endothelial NO production. The study showed PON1 prevented the formation of the lipid peroxidation product malondialdehyde (MDA) in HDL. Increased MDA in HDL leads to activation of endothelial lectin-like oxidized LDL receptor 1 (LOX-1), triggering endothelial PKCβII activation, which in turn inhibited eNOS-activating pathways and eNOS-dependent NO production. The authors showed that HDL from PON1 deficient mice failed to stimulate NO production in mouse aortic endothelial cells. Subsequent supplementation of HDL from PON1 deficient mice with purified PON1 partially improved the capacity of HDL to stimulate endothelial NO production. Huang et al. [23] demonstrated that under inflammatory condition, myeloperoxidase (MPO), PON1, and HDL bind to one another, forming a ternary complex, wherein PON1 partially inhibits MPO activity, while MPO inactivates PON1. MPO is a leukocyte-derived heme protein that promotes protein and lipid oxidation [34, 35]. During inflammation, MPO binds to HDL and increases oxidant stress and promotes atherosclerosis. MPO-generated oxidant caused site-specific oxidative modification of certain tyrosine and methionine residues of PON1, leading to reduced PON1 activity [23]. Conversely, PON1 binds and partially inhibits MPO activity. Therefore, recent evidence suggests that PON1 may exert its anti-inflammatory, anti-oxidative functions, in part, by the prevention of MDA formation leading to HDL-mediated eNOS activation in endothelial cells, and inhibition of MPO activity of inflammatory HDL. PON1 has been and continues to be a target/candidate for developing therapeutic interventions for both inflammatory diseases and toxicology applications.

5.2.2 PON2

PON2 deficiency impairs respiratory complex activity and mitochondrial oxidative stress in liver, peritoneal macrophages, and aorta. PON2 protects against atherogenesis *in vivo* by modulating lipoprotein oxidation through the reduction of intracellular oxidative stress [36]. The principal source of cellular free radicals and oxidative stress, ROS generated by mitochondria play a fundamental role for many of the signaling pathways contributing to cardiovascular pathologies [37]. Devarajan et al. [15] hypothesized that PON2 deficiency may be influencing mitochondrial oxidative status. The authors evaluated the activities of mitochondrial ETC complexes from the livers of PON2 deficient and control C57BL/6 J mice administered an atherogenic diet, and results revealed that complex I and complex III activities were more than 50 % lower in PON2 deficient mice than in controls on a corresponding diet [15]. Moreover, the mitochondrial superoxide levels were significantly increased in PON2 deficient mice fed an atherogenic diet, and the ATP levels were reciprocally decreased when compared to control mice [15]. PON2 deficient mice showed a significantly lower level of basal mitochondrial oxygen consumption than the control peritoneal macrophage. Mitochondrial superoxide levels in peritoneal macrophages from PON2 deficient mice were significantly higher ($p<0.05$) and ATP levels were significantly lower compared with control mice [15]. Furthermore, superoxide levels (using Mitosox) were significantly higher in the supernatants of whole aorta lysates of PON2 deficient mice relative to controls. PON2 deficient mice backcrossed onto the hyperlipidemic apoE deficient background develop significantly larger atherosclerotic lesions in the aorta and higher levels of macrophage immunoreactivity in the aortic sections compared to their apoE deficient controls [15]. These results demonstrated that the anti-atherogenic activity of PON2 might be linked with mitochondrial function.

5.2.2.1 PON2 Is Associated with Mitochondria, ER, and Plasma Membrane

To determine whether the changes in mitochondrial oxidative stress are due to a direct or indirect effect of PON2 on mitochondrial function, Devarajan et al. [15] isolated and analyzed mitochondria from HeLa cells for the presence of PON2. PON2 protein is present in percoll-purified mitochondria from HeLa cells. To further determine the precise submitochondrial localization of PON2, inner and outer mitochondrial membrane preparations from the livers of C57BL/6 J mice were utilized to demonstrate that PON2 is associated with the inner mitochondrial membrane (IMM). Moreover, individual ETC complex pull-down experiments showed that PON2 is associated with complex III. Western blot analyses for COX IV, an IMM associated protein, and VDAC, an outer mitochondrial membrane-associated protein, showed negligible cross-contamination of the two preparations in these experiments. Similar results were observed with mitochondria isolated from mouse heart tissue (unpublished STR). Altenhöfer et al. [38] demonstrated that PON2 reduced superoxide release from the inner mitochondrial membrane, irrespective whether resulting from complex I or complex III of the electron transport chain by modulating quinones.

5.2.2.2 PON2 Overexpression Protects Against Mitochondrial Dysfunction

HeLa cells overexpressing human PON2 under the control of a tetracycline-inducible promoter [39] were treated with antimycin, a compound known to release ubisemiquinone from the ETC, thereby generating mitochondrial superoxide [40], or with oligomycin, a compound known to inhibit ATP synthesis [41]. HeLa cells overexpressing PON2 had significantly lower superoxide and significantly higher ATP levels than control cells [15].

5.2.2.3 PON2 Protects Against Endoplasmic Reticulum Stress

Horke S et al. [42] were the first to show that PON2 decreases endoplasmic reticulum stress (ER)-induced caspase activation. PON2 was found associated with the nuclear membrane and endoplasmic reticulum and induced at both the promoter and protein levels by endoplasmic reticulum stress pathway unfolded protein response [42]. The authors concluded that PON2 is an endogenous defense mechanism against vascular oxidative stress and unfolded protein response-induced cell death. Horke et al. [43] demonstrated that PON2 protects against ER stress mediated cell death by modulating calcium homeostasis. Devarajan et al. [44] reported that macrophage PON2 regulates calcium homeostasis and cell survival under ER stress conditions and is sufficient to prevent the development of aggravated atherosclerosis in PON2 and apoE double-deficient mice on a Western diet, suggesting that macrophage PON2 modulates mechanisms that link ER stress, mitochondrial dysfunction and the development of atherosclerosis. Taken together, these studies above [10, 15, 36, 38, 39, 41–44] suggest that PON2 plays an important protective anti-oxidant role in the development of inflammatory diseases.

5.2.3 PON3

Of the three members of the PON family, PON3 appears to be the least studied to date. Although these two proteins; PON2 and PON3 appear to be similar in function and their cell-type association [1, 45, 46] recent studies suggest that both intracellular localization and role in inflammatory diseases are likely to be distinct from each other [11, 16].

5.2.3.1 PON3 Protects Against the Development of Diabetes and Atherosclerosis in Mice

Reddy et al. [46]. were the first to demonstrate that PON3 prevents the oxidation of low-density lipoprotein *in vitro*. To test the role of PON3 in atherosclerosis and related traits, Shih et al. [16] generated two independent lines of human PON3 transgenic (Tg) mice on the C57BL/6 J (B6) and showed that atherosclerotic lesion areas were significantly smaller in both lines of male PON3 Tg mice as compared with the male non-Tg littermates on C57B6 background fed an atherogenic diet. When bred onto the low-density lipoprotein receptor knockout mouse background, the male PON3 Tg mice also exhibited decreased atherosclerotic lesion areas. In addition, decreased adiposity and lower circulating leptin levels were observed in both lines of male PON3 Tg mice as compared with the male non-Tg mice. Shih et al. demonstrated for the first time that elevated PON3 expression significantly decreases atherosclerotic lesion formation and adiposity in male mice.

Interestingly, isolated mitochondria from the PON3 deficient livers exhibited impaired mitochondrial function as compared to the wild type mitochondria [11], suggesting that PON3 deficiency, similar to PON2 deficiency, affects mitochondrial function.

5.3 Infection and Quorum Quenching

All PON proteins possess lipo-lactonase activity [47] and hydrolyze acyl- homoserine lactones (AHLs) [48], which mediate bacterial quorum-sensing (QS) signals. The mechanism of QS has been extensively characterized in several pathogenic bacteria including *P. aeruginosa*, which causes catastrophic infections in immunocompromised hosts, such as individuals with cystic fibrosis, cancer and severe burns [49]. About a decade ago, Greenberg's laboratory was the first to show that PON proteins possess AHL-inactivation activity [50] laying the foundation for a role for PON proteins in infection and quorum quenching.

Stoltz et al. [51] investigated the role of PON1, PON2, and PON3 in airway epithelial cell inactivation of N-(3-oxododecanoyl)-l-homoserine lactone (3OC12-HSL), a quorum-sensing molecule produced by gram-negative microbial pathogens such as *P. aeruginosa* (PAO1). Lysates of tracheal epithelial cells from PON2, but not PON1 or PON3, deficient mice had impaired 3OC12-HSL inactivation compared with wild-type mice. Overexpression of PON2 enhanced 3OC12-HSL degradation by human airway epithelial cell lysates.

Devarajan et al. [53] examined the susceptibility of PON2 deficient mice towards *P. aeruginosa* and demonstrated that both intact cells and membrane-enriched protein lysates obtained from PON2 deficient macrophages reveal a marked impairment in their ability to hydrolyze 3OC12-HSL. A decrease in bacterial clearance was noted in the spleen, lungs, and liver of PON2 deficient mice by 2.5, 5.7, and 14.8 fold, respectively, following administration of 1.6×10^{7}CFU of PAO1. In an ex vivo model, macrophages of PON2 deficient mice had significantly reduced phagocytosis function compared to control macrophages following PAO1 infection. These results suggested that PON2 regulates innate immune defense in PAO1 infection model [52].

Kim JB et al. [53] investigated the common and distinct pro-inflammatory pathways activated by atherogenic lipids and quorum sensing lactones, in PON2 deficient endothelial cells. Using expression profiling and network modeling, identified the unfolded protein response (UPR), cell cycle genes, and the mitogen-activated protein kinase-signaling pathway to be heavily involved in the HAEC response to 3OC12-HSL. The network also showed striking similarities to a network created based on HAEC response to Ox-PAPC, a major component of minimally modified low-density lipoprotein. HAECs in which small interfering RNA silenced PON2 showed increased pro-inflammatory response and UPR when treated with 3OC12-HSL or Ox-PAPC. 3OC12-HSL and Ox-PAPC influence similar inflammatory and UPR pathways. The authors concluded that the antiatherogenic effects of PON2 might include destruction of quorum sensing molecules, such as 3OC12-HSL, which contribute to the proatherogenic effects of chronic infection.

Recent work by Schweikert et al. [54] demonstrated that the anti-oxidative and anti-inflammatory functions of PON2 and PON3 are an important part of innate defense system against *P. aeruginosa* infections. Taking the work described above on PON2 and PON3 [50–54] together with the demonstration by Stoltz et al. [9] that *Drosophila* are protected from *P. aeruginosa* lethality by transgenic expression of PON1, it is very clear that PON proteins are natural quorum quenchers. Similar to PON2, PON3 not only hydrolyzes 3OC12-HSL, but also diminishes the oxidative stress and NF-κb activation induced by pyocyanin [54], a virulent factor produced by *P. aeruginosa*. These studies suggest involvement of PON gene family in innate immunity.

5.4 Cancer

Witte et al. [10] hypothesized that since ER stress is also relevant to cancer and associated with anti-cancer treatment resistance, PON2 may play a role in tumorigenesis. Human tumors had upregulated PON2, and PON2 knockdown caused apoptosis of tumor cells [10]. Schweikert et al. [11] demonstrated that PON3 is overexpressed in human tumors and diminishes mitochondrial superoxide formation by sequestering ubisemiquinone in cancer cells, leading to enhanced cell death resistance. The authors suggest that PON3, similar to PON2, may aid in tumor cell development. In a review of functions and mechanisms of PON2 and PON3 proteins, Witte et al. [55] suggest that although PON2 and PON3 proteins are protective from a cardiovascular standpoint, they may not be protective but may actually promote cancer cells by preventing cell death [55].

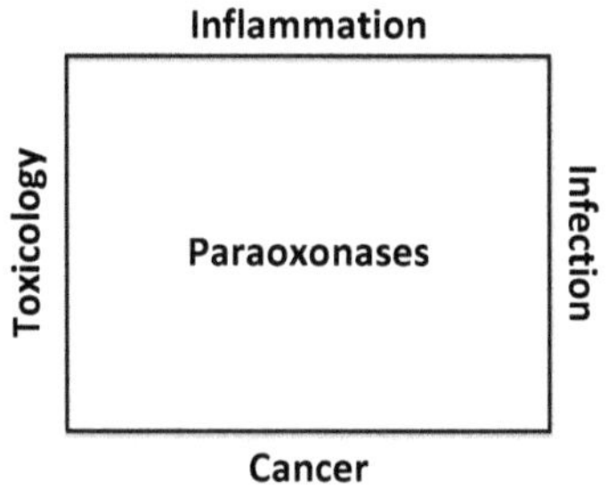

Fig. 5.3 PON genes play significant roles in inflammation, infection, toxicology, and cancer

Currently, *in vivo* proof for a pro-tumorigenic role of PON2 and PON3 proteins is lacking and future studies in animal models will determine whether PON protein family indeed aids in cancer development.

5.5 Conclusions

Expression of *all* three PON genes negatively correlates with a number of inflammatory diseases including atherosclerosis and cancer. Based on published studies, it is evident that all three PON proteins play important roles in inflammation, infection, toxicology, and cancer (Fig. 5.3). However, there is a gap in our knowledge on the mechanisms of action and function of PON proteins. Future studies aimed at understanding the molecular targets of PON proteins will unravel novel markers and therapeutic targets for the treatment of inflammatory diseases associated with ER and mitochondrial dysfunction mediated oxidative stress including atherosclerosis, bacterial infections, and cancer.

References

1. Reddy ST, Devarajan A, Bourquard N, Shih D, Fogelman AM. Is it just paraoxonase 1 or are other members of the paraoxonase gene family implicated in atherosclerosis? Curr Opin Lipidol. 2008;19:405–8.
2. She ZG, Chen HZ, Yan Y, Li H, Liu DP. The human paraoxonase gene cluster as a target in the treatment of atherosclerosis. Antioxid Redox Signal. 2012; 16:597–632.
3. Primo-Parmo SL, Sorenson RC, Teiber J, La Du BN. The human serum paraoxonase/arylesterase gene (PON1) is one member of a multigene family. Genomics. 1996;33:498–507.
4. Draganov DI, La Du BN. Pharmacogenetics of paraoxonases: a brief review. Naunyn Schmiedebergs Arch Pharmacol. 2004;369:78–88.
5. Aldridge WN. Serum esterases I. Two types of esterase (A and B) hydrolysing p-nitrophenyl acetate, propionate and butyrate and a method for their determination. Biochem J. 1953;53:110–7.
6. Aldridge WN. Serum esterases II. An enzyme hydrolyzing diethyl p-nitrophenyl acetate (E600) and its identity with the A-esterase of mammalian sera. Biochem J. 1953;53:117–24.
7. Bhattacharyya T, Nicholls SJ, Topol EJ, Zhang R, Yang X, Schmitt D, et al. Relationship of paraoxonase 1 (PON1) gene polymorphisms and functional activity with systemic oxidative stress and cardiovascular risk. JAMA. 2008;299:1265–76.
8. Stevens RC, Suzuki SM, Cole TB, Park SS, Richter RJ, Furlong CE. Engineered recombinant human paraoxonase 1 purified from Escherichia coli protects against organophosphate poisoning. Proc Natl Acad Sci U S A. 2008;105:12780–4.
9. Stoltz DA, Ozer EA, Taft PJ, Barry M, Liu L, Kiss PJ, et al. Drosophila are protected from Pseudomonas aeruginosa lethality by transgenic expression of paraoxonase-1. J Clin Invest. 2008;118:3123–31.
10. Witte I, Altenhöfer S, Wilgenbus P, Amort J, Clement AM, Pautz A, et al. Beyond reduction of atherosclerosis: PON2 provides apoptosis resistance and stabilizes tumor cells. Cell Death Dis. 2011;13(2):e112.
11. Schweikert EM, Devarajan A, Witte I, Wilgenbus P, Amort J, Förstermann U, et al. PON3 is upregulated in cancer tissues and protects against mitochondrial superoxide-mediated cell death. Cell Death Differ. 2012;19:1549–60.
12. Shih DM, Gu L, Xia YR, Navab M, Li WF, Hama S, et al. Mice lacking serum paraoxonase are susceptible to organophosphate toxicity and atherosclerosis. Nature. 1998;394:284–7.
13. Shih DM, Xia YR, Wang XP, Miller E, Castellani LW, Subbanagounder G, et al. Combined serum paraoxonase knockout/apolipoprotein E knockout mice exhibit increased lipoprotein oxidation and atherosclerosis. J Biol Chem. 2000;275:17527–35.
14. Tward A, Xia YR, Wang XP, Shi YS, Park C, Castellani LW, et al. Decreased atherosclerotic lesion formation in human serum paraoxonase transgenic mice. Circulation. 2002;106:484–90.
15. Devarajan A, Bourquard N, Hama S, Navab M, Grijalva VR, Morvardi S, et al. Paraoxonase 2 deficiency alters mitochondrial function and exacerbates the development of atherosclerosis. Antioxid Redox Signal. 2011;14:341–51.
16. Shih DM, Xia YR, Wang XP, Wang SS, Bourquard N, Fogelman AM, et al. Decreased obesity and atherosclerosis in human paraoxonase 3 transgenic mice. Circ Res. 2007;100:1200–7.
17. She ZG, Zheng W, Wei YS, Chen HZ, Wang AB, Li HL, et al. Human paraoxonase gene cluster transgenic

overexpression represses atherogenesis and promotes atherosclerotic plaque stability in ApoE-null mice. Circ Res. 2009;104:1160–8.
18. Aviram M, Rosenblat M, Bisgaier CL, Newton RS, Primo-Parmo SL, La Du BN. Paraoxonase inhibits high-density lipoprotein oxidation and preserves its functions. A possible peroxidative role for paraoxonase. J Clin Invest. 1998;101:1581–90.
19. Watson AD, Berliner JA, Hama SY, La Du BN, Faull KF, Fogelman AM, et al. Protective effect of high density lipoprotein associated paraoxonase. Inhibition of the biological activity of minimally oxidized low density lipoprotein. J Clin Invest. 1995;96:2882–91.
20. Deakin SP, Bioletto S, Bochaton-Piallat ML, James RW. HDL-associated PON1 can redistribute to cell membranes and influence sensitivity to oxidative stress. Free Radic Biol Med. 2011;50:102–9.
21. Aviram M. Atherosclerosis: cell biology and lipoproteins – paraoxonases protect against atherosclerosis and diabetes development. Curr Opin Lipidol. 2012; 23:169–71.
22. Kim DS, Marsillach J, Furlong CE, Jarvik GP. Pharmacogenetics of paraoxonase activity: elucidating the role of high-density lipoprotein in disease. Pharmacogenomics. 2013;14:1495–515.
23. Huang Y, Wu Z, Riwanto M, Gao S, Levison BS, Gu X, et al. Myeloperoxidase, paraoxonase-1, and HDL form a functional ternary complex. J Clin Invest. 2013;123:3815–28.
24. Mackness M, Mackness B. Targeting paraoxonase-1 in atherosclerosis. Expert Opin Ther Targets. 2013;17:829–37.
25. Tang WH, Hartiala J, Fan Y, Wu Y, Stewart AF, Erdmann J, et al. Clinical and genetic association of serum paraoxonase and arylesterase activities with cardiovascular risk. Arterioscler Thromb Vasc Biol. 2012;32:2803–12.
26. Charles-Schoeman C, Lee YY, Shahbazian A, Gorn AH, Fitzgerald J, Ranganath VK, et al. Association of paraoxonase 1 gene polymorphisms and enzyme activity with carotid plaque in rheumatoid arthritis. Arthritis Rheum. 2013;65:2765–72.
27. Garcia-Heredia A, Marsillach J, Rull A, Triguero I, Fort I, Mackness B, et al. Paraoxonase-1 inhibits oxidized low-density lipoprotein-induced metabolic alterations and apoptosis in endothelial cells: a non directed metabolomic study. Mediat Inflamm. 2013;2013:156053.
28. Garcia-Heredia A, Kensicki E, Mohney RP, Rull A, Triguero I, Marsillach J, et al. Paraoxonase-1 deficiency is associated with severe liver steatosis in mice fed a high-fat high-cholesterol diet: a metabolomic approach. J Proteome Res. 2013;12:1946–55.
29. Jarvik GP, Rozek LS, Brophy VH, Hatsukami TS, Richter RJ, Schellenberg GD, et al. Paraoxonase (PON1) phenotype is a better predictor of vascular disease than is PON1(192) or PON1(55) genotype. Arterioscler Thromb Vasc Biol. 2000;120:2441–7.
30. Mackness B, Durrington P, McElduff P, Yarnell J, Azam N, Watt M, et al. Low paraoxonase activity predicts coronary events in the Caerphilly Prospective Study. Circulation. 2003;107:2775–9.
31. Tang WH, Wu Y, Mann S, Pepoy M, Shrestha K, Borowski AG, et al. Diminished antioxidant activity of high-density lipoprotein-associated proteins in systolic heart failure. Circ Heart Fail. 2011;4:59–64.
32. Aviram M, Vaya J. Paraoxonase 1 activities, regulation, and interactions with atherosclerotic lesion. Curr Opin Lipidol. 2013;24:339–44.
33. Besler C, Heinrich K, Rohrer L, Doerries C, Riwanto M, Shih DM, et al. Mechanisms underlying adverse effects of HDL on eNOS-activating pathways in patients with coronary artery disease. J Clin Invest. 2011;121:2693–708.
34. Hazen SL, Heinecke JW. 3-Chlorotyrosine, a specific marker of myeloperoxidase-catalyzed oxidation, is markedly elevated in low density lipoprotein isolated from human atherosclerotic intima. J Clin Invest. 1997;99:2075–81.
35. Zhang R, Brennan ML, Shen Z, MacPherson JC, Schmitt D, Molenda CE, et al. Myeloperoxidase functions as a major enzymatic catalyst for initiation of lipid peroxidation at sites of inflammation. J Biol Chem. 2002;277:46116–22.
36. Ng CJ, Bourquard N, Grijalva V, Hama S, Shih DM, Navab M, et al. Paraoxonase-2 deficiency aggravates atherosclerosis in mice despite lower apolipoprotein-B-containing lipoproteins: anti-atherogenic role for paraoxonase-2. J Biol Chem. 2006;281:29491–500.
37. Mabile L, Meilhac O, Escargueil-Blanc I, Troly M, Pieraggi MT, Salvayre R, et al. Mitochondrial function is involved in LDL oxidation mediated by human cultured endothelial cells. Arterioscler Thromb Vasc Biol. 1997;17:1575–82.
38. Altenhöfer S, Witte I, Teiber JF, Wilgenbus P, Pautz A, Li H, et al. One enzyme, two functions: PON2 prevents mitochondrial superoxide formation and apoptosis independent from its lactonase activity. J Biol Chem. 2010;6(285):24398–403.
39. Ng CJ, Wadleigh DJ, Gangopadhyay A, Hama S, Grijalva VR, Navab M, et al. Paraoxonase-2 is a ubiquitously expressed protein with antioxidant properties and is capable of preventing cell-mediated oxidative modification of low density lipoprotein. J Biol Chem. 2001;276:44444–9.
40. Ohnishi T, Trumpower BL. Differential effects of antimycin on ubisemiquinone bound in different environments in isolated succinate. Cytochrome c reductase complex. J Biol Chem. 1980;255:3278–84.
41. Kim SB, Berdanier CD. Oligomycin sensitivity of mitochondrial F(1)F(0)-ATPase in diabetes-prone BHE/Cdb rats. Am J Physiol. 1999;277:E702–7.
42. Horke S, Witte I, Wilgenbus P, Krüger M, Strand D, Förstermann U. Paraoxonase-2 reduces oxidative stress in vascular cells and decreases endoplasmic reticulum stress-induced caspase activation. Circulation. 2007;17(115):2055–64.
43. Horke S, Witte I, Wilgenbus P, Altenhöfer S, Krüger M, Li H, et al. Protective effect of paraoxonase-2 against endoplasmic reticulum stress-induced apoptosis is lost upon disturbance of calcium homoeostasis. Biochem J. 2008;416:395–405.

44. Devarajan A, Grijalva VR, Bourquard N, Meriwether 3rd D, Imaizumi S, Shin BC, et al. Macrophage paraoxonase 2 regulates calcium homeostasis and cell survival under endoplasmic reticulum stress conditions and is sufficient to prevent the development of aggravated atherosclerosis in paraoxonase 2 deficiency/apoE(–/–) mice on a western diet. Mol Genet Metab. 2012;107:416–27.
45. Shih DM, Xia YR, Yu JM, Lusis AJ. Temporal and tissue-specific patterns of PON3 expression in mouse: in situ hybridization analysis. Adv Exp Med Biol. 2009;660:73–87.
46. Reddy ST, Wadleigh DJ, Grijalva V, Ng C, Hama S, Gangopadhyay A, et al. Human paraoxonase-3 is an HDL-associated enzyme with biological activity similar to paraoxonase-1 protein but is not regulated by oxidized lipids. Arterioscler Thromb Vasc Biol. 2001;21:542–7.
47. Draganov DI. Lactonases with organophosphatase activity: structural and evolutionary perspectives. Chem Biol Interact. 2010;187:370–2.
48. Draganov DI, Teiber JF, Speelman A, Osawa Y, Sunahara R, La Du BN. Human paraoxonases (PON1, PON2, and PON3) are lactonases with overlapping and distinct substrate specificities. J Lipid Res. 2005;46:1239–47.
49. Shiner EK, Rumbaugh KP, Williams SC. Interkingdom signaling: deciphering the language of acyl homoserine lactones. FEMS Microbiol Rev. 2005;29:935–47.
50. Chun CK, Ozer EA, Welsh MJ, Zabner M, Greenberg EP. Inactivation of a Pseudomonas aeruginosa quorum-sensing signal by human airway epithelia. Proc Natl Acad Sci U S A. 2004;101:3587–90.
51. Stoltz DA, Ozer EA, Ng CJ, Yu JM, Reddy ST, Lusis A, et al. Paraoxonase-2 deficiency enhances Pseudomonas aeruginosa quorum sensing in murine tracheal epithelia. Am J Physiol Lung Cell Mol Physiol. 2007;292:L852–60.
52. Devarajan A, Bourquard N, Grijalva VR, Gao F, Ganapathy E, Verma J, et al. Role of PON2 in innate immune response in an acute infection model. Mol Genet Metab. 2013;110:362–70.
53. Kim JB, Xia YR, Romanoski CE, Lee S, Meng Y, Shi YS, et al. Paraoxonase-2 modulates stress response of endothelial cells to oxidized phospholipids and a bacterial quorum-sensing molecule. Arterioscler Thromb Vasc Biol. 2011;31:2624–33.
54. Schweikert EM, Amort J, Wilgenbus P, Förstermann U, Teiber JF, Horke S. Paraoxonases-2 and -3 are important defense enzymes against Pseudomonas aeruginosa virulence factors due to their anti-oxidative and anti-inflammatory properties. J Lipids. 2012;2012: 352857.
55. Witte I, Foerstermann U, Devarajan A, Reddy ST, Horke S. Protectors or traitors: the roles of PON2 and PON3 in atherosclerosis and cancer. J Lipids. 2012;2012:342806.

6 Autophagy Is an Inflammation-Related Defensive Mechanism Against Disease

Jorge Joven, Maria Guirro, Roger Mariné-Casadó, Esther Rodríguez-Gallego, and Javier A. Menéndez

Abstract

The inflammatory response is an energy-intensive process. Consequently, metabolism is closely associated with immune function. The autophagy machinery plays a role in metabolism by providing energy but may also be used to attack invading pathogens (xenophagy). The autophagy machinery may function to protect against not only the threats of infection but also the threats of the host's own response acting on the central immunological tolerance and the negative regulation of innate and inflammatory signaling. The balance between too little and too much autophagy is critical for the survival of immune cells because autophagy is linked to type 2-cell death programmed necrosis and apoptosis. Changes in inflammatory cells are driven by extracellular signals; however, the mechanisms by which cytokines mediate autophagy regulation and govern immune cell function remain unknown. Certain cytokines increase autophagy, whereas others inhibit autophagy. The relationship between autophagy and inflammation is also important in the pathogenesis of metabolic, non-communicable diseases. Inflammation *per se* is not the cause of obesity-associated diseases, but it is secondary to both the positive energy balance and the specific cellular responses. In metabolic tissues, the suppression of autophagy increases inflammation with the overexpression of cytokines, resulting in an activation of autophagy. The physiological role of these apparently contradictory findings remains uncertain but exemplifies future challenges in the therapeutic modulation of autophagy in the management of disease.

J. Joven (✉) • M. Guirro • R. Mariné-Casadó
E. Rodríguez-Gallego
Campus of International Excellence Southern Catalonia, Unitat de Recerca Biomèdica, Hospital Universitari de Sant Joan, Institut d'Investigació Sanitària Pere Virgili, Universitat Rovira i Virgili, C. Sant Joan s/n, 43201 Reus, Spain
e-mail: jorge.joven@urv.cat; maria.guirro@gmail.com; roger.marine91@gmail.com; esther.rodriguez@grupsagessa.com

J.A. Menéndez
Metabolism & Cancer Group, Translational Research Laboratory, Catalan Institute of Oncology, Girona, Spain

Molecular Oncology Group, Girona Biomedical Research Institute (IDIBGI), Girona, Spain
e-mail: jmenendez@iconcologia.net; jmenendez@idibgi.org

J. Camps (ed.), *Oxidative Stress and Inflammation in Non-communicable Diseases – Molecular Mechanisms and Perspectives in Therapeutics*, Advances in Experimental Medicine and Biology 824, DOI 10.1007/978-3-319-07320-0_6, © Springer International Publishing Switzerland 2014

Keywords
Autophagy • Cancer • Cardiovascular disease • CCL2 • Immunity • Inflammation • Neurodegeneration

6.1 Basic Concepts and Background

Disturbances in the normal function of cells (cellular stress) result in the accumulation of damaged molecules and dysfunctional organelles that may be deleterious. To mitigate this damage, cells clean up the accumulated end products using the following two major systems: the ubiquitin proteasome system and the autophagy (from the Greek for "self-digestion" or "self-eating") lysosome system. The key elements in the recognition and processing of ubiquitin-protein conjugates have been recently summarized [1], but the exact mechanisms by which cells regulate self-cannibalism are still under investigation. Chaperone-mediated autophagy, microautophagy, and macroautophagy (hereafter referred to as autophagy) are currently considered distinctive forms [2] (Fig. 6.1). Each form differs from the others in its physiological functions and the mode and nature of cargo delivery.

The overall process of autophagy can be separated into the following seven discrete steps: induction or selection/packaging of cargo, nucleation, vesicle expansion, completion, fusion, degradation, and export (Fig. 6.2). A discussion of the known (and supposed) molecular mechanisms is beyond the scope of this manuscript, but it appears that autophagy is generally induced to provide an alternate source of certain basic building blocks required for cell survival (the non-selective pathway). The mechanistic target of rapamycin (MTOR) kinase coordinates nutrient availability and cell growth. When the supply is sufficient, MTOR is active and phosphorylates important proteins for cellular growth. When nutrients are in a limited supply, MTOR is inactivated and limits protein translation, focusing on those required for cell survival [3]. This mechanism might be easily modulated by the action of common and inexpensive drugs, such as metformin [4] and chloroquine [5], but might also be regulated by selected stimuli (*i.e.,* the selective pathway initiated from the cargo itself).

6.1.1 Autophagy: A Tightly Regulated Process

Autophagy is a simple process and is used to provide energy under starvation conditions, but it must be tightly regulated because it has the capacity to be harmful (*i.e.,* it can degrade entire organelles). In mammals, as opposed to yeasts, the regulation of this process is exquisitely complex because it is also used, among other functions, for the purposes of development; consequently, the mechanisms of autophagy must be initiated at precise times. The targeted degradation of a cellular component requires signals derived from either the cargo selected for degradation or the particular function of the cell. The regulation of this process is poorly understood, but it is likely that certain selective autophagy pathways are relevant to disease.

For example, autophagy of mitochondria (mitophagy) is involved in the response of the cell to pernicious stimuli and, particularly, in the avoidance of effects from inflammation or oxidative stress (details are summarized in Fig. 6.3). This response is clearly observed in conditional knockout mouse models [6], which accumulate deformed mitochondria in the liver under defective autophagy conditions. It has also been observed that chemical inhibitors of autophagy and mitochondrial depolarization (cyclosporine) jeopardize the defensive role of

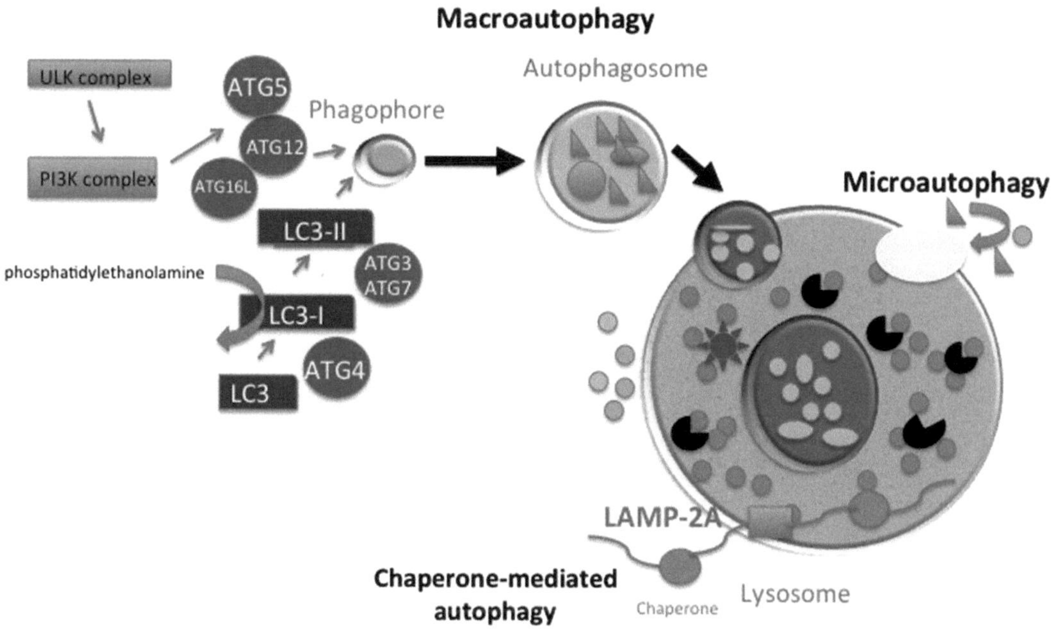

Fig. 6.1 Autophagy was first described in mammalian cells over 50 years ago, but the molecular basis for this process has not been completely elucidated. The following three autophagy forms have been identified: chaperon-mediated autophagy, microautophagy and macroautophagy (autophagy). *LAMP* lysosomal associated membrane protein

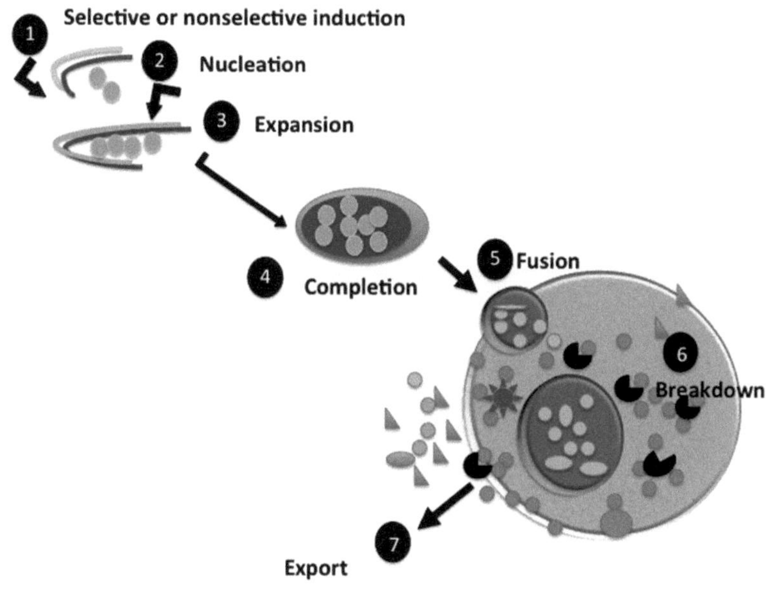

Fig. 6.2 Several steps, including nucleation, expansion, completion, fusion, breakdown, and export, have been identified in autophagy. The signal to initiate autophagy may be selective (changes in nutrient availability) or nonselective (the nature of the cargo itself)

mitophagy in hepatocytes [7]. The autophagy of peroxisomes (pexophagy) has only been explored in yeasts, and it is difficult to differentiate the process from peroxisomal biogenesis; however, a role in mammals is likely to exist. To further complicate the issue, there is cross

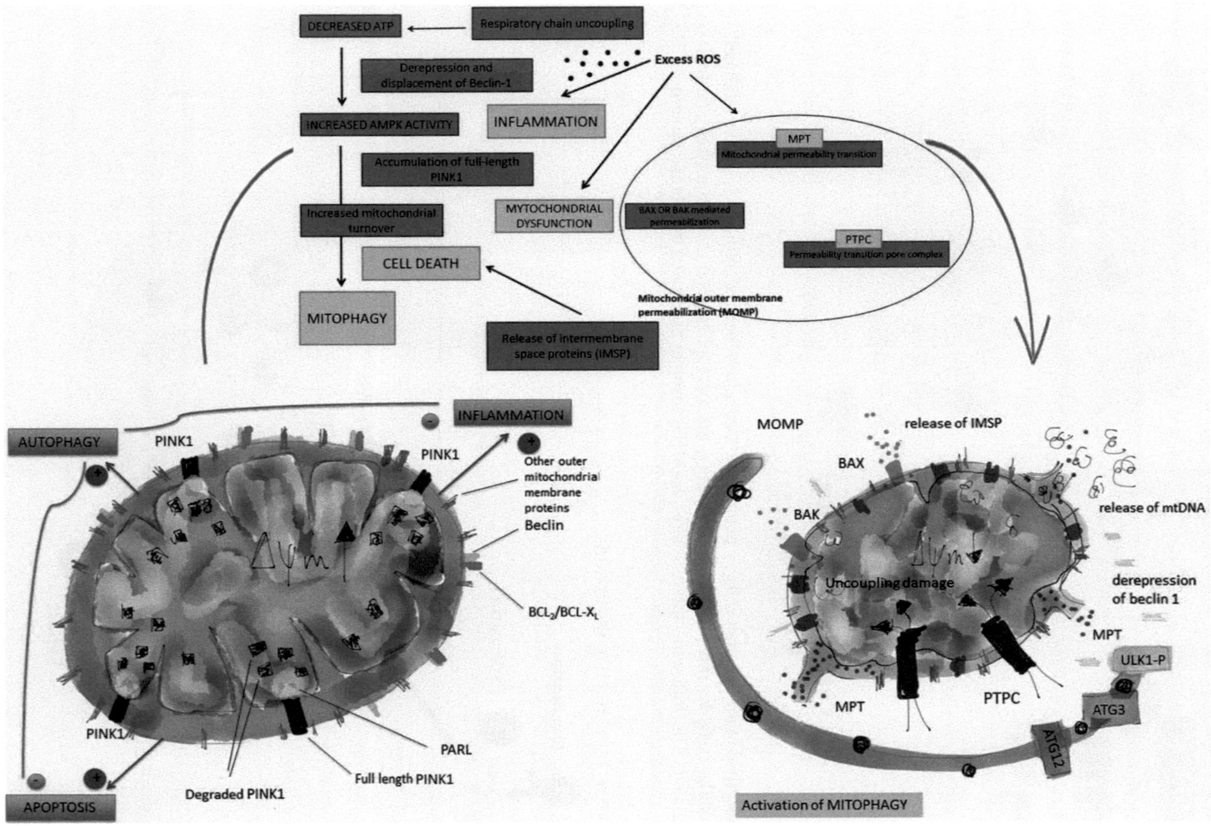

Fig. 6.3 Mitochondria, apoptosis, autophagy and inflammation are closely related. Autophagy limits both inflammation and apoptosis (cell death). Conversely, the loss of autophagy (mitophagy) leads to the accumulation of damaged mitochondria, which promotes inflammation. In healthy mitochondria, BCL2 and BCL-XL bind to Beclin 1, which is inhibited. PTEN-induced putative kinase 1 (PINK1) is a mitochondrial serine/threonine-protein kinase that apparently protects cells from stress-induced mitochondrial dysfunction. PINK1 is imported into healthy mitochondria, according to the mitochondrial transmembrane potential ($\Delta\psi m$), where it is degraded by the protease PARL (presenilins-associated rhomboid-like protein). Mitophagy is activated by a number of factors. For example, Beclin 1 may be displaced by hypoxia. In response to the uncoupling, the mitochondrial permeability transition (MPT) or the mitochondrial damage, there is an accumulation of full-length PINK1 and an excess amount of ROS. Autophagy or increased mitochondrial turnover may be the consequence of AMP accumulation (decreased ATP), leading to the activation by AMPK of proteins that function specifically in mitophagy. Other lethal stimuli lead to BAX- or BAK-mediated mitochondrial outer membrane permeabilization (MOMP) or to the activation of the permeability transition pore complex (PTPC). The net consequence is the release of intermembrane space proteins (IMSP) and other conjugates with pro-apoptotic and/or pro-inflammatory action. Consequently, mitophagy acts as a defensive mechanism to alleviate inflammation, cell death and mitochondrial dysfunction

communication between the two major pathways of protein degradation, *i.e.*, substrates that are normally degraded by the ubiquitin-proteasome pathway can be selectively degraded by autophagy under certain circumstances [8]. The induction of autophagy is also critical in the maintenance of glycogen stores. Glycogen autophagy in the liver, heart and diaphragm is a selective and highly regulated process that is most likely due to the abrupt increases in the energy requirement under critical conditions, such as birth [9].

6.2 The Relevance of Autophagy in Disease

6.2.1 Autophagy Is Important in the Pathogenesis of Cancer, Neurodegeneration and Myopathies

Autophagy has been implicated in the pathogenesis of several diseases, primarily as the consequence of an imbalance in its dual role of adapting to cellular stresses and/or contributing to cell death. Mitochondria are particularly important in the process.

There are at least three different types of programmed cell death as follows: type I (nuclear or apoptotic), type II (autophagic), and type III (cytoplasmic). The autophagic and apoptotic pathways are closely related [10], although autophagy is relatively unknown compared to apoptosis, and it is unclear whether autophagy represents a failed effort to preserve cell viability. This observation is particularly evident in cancer. Several findings have led to the hypotheses that BECN1 (beclin-1) and other genes involved in autophagy may function as tumor suppressors and that defects in autophagy may promote cancer. Similarly, the PI3K (phosphoinositide 3 kinase)/AKT (protein kinase B)/MTOR pathway is frequently activated in response to mutations in genes encoding negative regulators, thereby constraining autophagy in certain tumors [5, 11–13]. Many signals promoting unrestricted cell proliferation also inhibit autophagy, which is normally induced to sustain cells under conditions of nutrient limitation. Interestingly, the uncoupling of the cellular response to the nutrient availability renders cells more susceptible to a metabolic catastrophe; thus, the concept of "oncometabolite" and the requirement for further insights in cancer metabolism are rapidly evolving. This concept may be a first step in understanding why the defensive metabolic and inflammatory responses to tumor necrosis may promote rather than entangle an increase in the overall tumor burden [14]. Cancer in adults, but not in children, arises under chronic inflammation circumstances, in a tumor microenvironment that is characterized by hypoxia, glycolysis, perpetual autophagy and the resultant necrosis under conditions of stress. The regulation of autophagy in tumor cells provides possible therapeutic strategies, although given the complexity of the signaling pathways, these should be tailored to the different classes of tumors. For example, in tumors that do not outgrow their food supply, the inhibition of autophagy may render cells more susceptible to conventional treatments. Conversely, the stimulation of autophagy in tumors with constitutive activation of the PI3K/AKT pathway can render them more susceptible to apoptosis.

Autophagic activity is essential for the normal function of the nervous system. A deficiency of basal autophagy in the mouse brain (with a knockout of autophagic proteins) results in neurodegeneration [15]. Although the role of autophagy may vary in different neurodegenerative diseases, age-related neurodegenerative diseases are generally characterized by the accumulation of protein aggregates in the affected brain. These protein aggregates may be degraded by autophagy [16], suggesting that a regulated and specific increase in autophagy may become a protective response in both Alzheimer's and Parkinson's diseases. Lysosomal storage disorders, which result in alterations of the lysosomal function, should inhibit autophagosome maturation and the subsequent accumulation of autophagic vacuoles in the affected cells. Inefficient autophagolysosomal recycling of mitochondria may generate fragmented mitochondria and an increased sensitivity to apoptosis, thus complicating the course of the disease [17]. Many congenital myopathies are characterized by the presence of myofibrillar disorganization and the accumulation of autophagic vacuoles. In this context, increased autophagy appears to be involved in pathways leading to muscle wasting, an effect that should be considered in the pharmacological modulation of this process [18].

6.2.2 The Role of Autophagy in Infection and Inflammatory Disorders

Recent studies have demonstrated the role of autophagy in processes affecting the immune system, including the coordination of metabolic signals, immune cell differentiation, secretion of cytokines and both innate and adaptive immune defenses against pathogens. Therefore, it is tempting to speculate that detailed insights into the function of autophagy may provide novel therapeutic strategies in the management of inflammatory disorders.

6.2.2.1 The Relationship Between Autophagy and Immunity

Autophagosomes play an important role in the major histocompatibility complex (MHC) class II antigen presentation. The endocytosis of protein antigens from the extracellular space is followed by autophagy and subsequent lysosomal degradation and transfer to the MHC class II loading compartment prior to transport to the cell surface [19]. Other studies have indicated that dendritic cells (DCs) require autophagy to efficiently process and present antigens on the MHC class II molecules. This process is not limited to certain antigens, and autophagy also appears to be important in the MHC class I antigen presentation [20]. Other studies have implicated the signaling lymphocyte activation molecule, a cell adhesion protein, as a bacterial sensor that regulates the degradation of endocytosed gram-negative bacteria in macrophages via NADPH oxidase activity and autophagy [21]. Receptor ligation is also important in the defense against other pathogens and in the function of other cells, such as neutrophils, to induce autophagy, a potential mechanism for clearing pathogens that has also been associated with facilitating the recognition of pathogen-associated molecular patterns. Autophagy is involved in the mechanisms by which neutrophils capture microbes through neutrophil extracellular traps (NETs) that induce a form of cell death, termed NETosis [22]. Several autophagy proteins also play a role in B cell development and survival. Beclin 1–deficient chimeric mice have reduced numbers of lymphoid progenitor cells in the bone marrow, indicating a role for Beclin 1 in early B cell development. These and other available results demonstrate that autophagy plays a role in several aspects of B cell physiology, development, and maintenance, MHC class II presentation, and co-stimulation through surface receptors [23]. Moreover, autophagic pathways intersect with a variety of T cell functions, including the development of CD4+ T cell self-tolerance. Similar to B cells, multiple autophagy proteins are essential for the development, maintenance, and survival of T cells. An investigation of autophagy-deficient peripheral T cells showed that they had excess mitochondria and elevated reactive oxygen species (ROS) production. Autophagy maintains mitochondrial homeostasis, and mitophagy is responsible for removing damaged mitochondria and preventing their accumulation (Fig. 6.3). Proliferating T cells clear the mitochondria as they shift to a more glycolytic metabolism after activation, and blocking autophagy can impair ATP production in T cells [24]. Other results indicate that autophagy plays a role in regulating the endoplasmic reticulum (ER) and calcium homeostasis in T cells and that impaired autophagy can lead to ER stress and impaired calcium homeostasis. Alternatively, autophagy can be induced by ER stress–signaling pathways independent of the unfolded protein response that activates NF-κB (nuclear factor kappa-light-chain-enhancer of activated B cells). Therefore, these roles in maintaining the ER and mitochondrial homeostasis and in controlling the ROS levels could explain the importance of autophagy proteins in the survival of immune cells [25, 26].

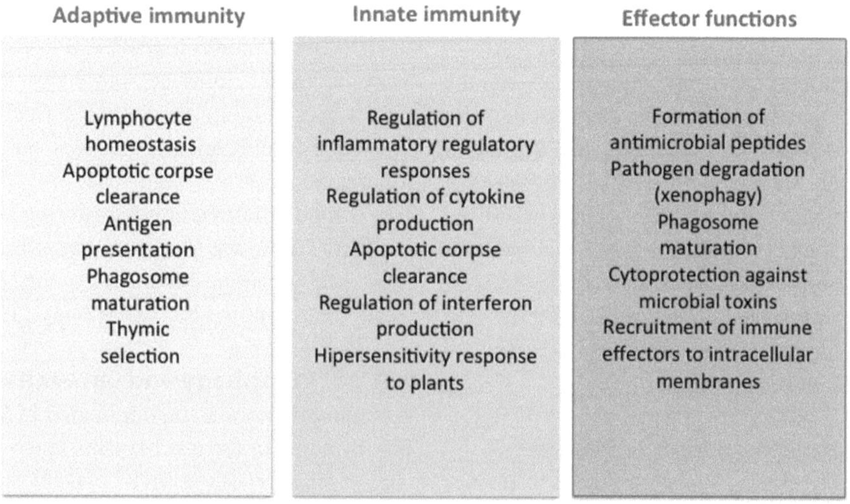

Fig. 6.4 The relationship between autophagy and immunity is likely to involve a combination of xenophagy, the activation of innate and adaptive immune responses, and alterations in pathogen-induced cell death

6.2.2.2 The Role of Autophagy Against Pathogens (Xenophagy)

The mechanisms by which autophagy mediates resistance to infection are not fully understood but are likely to involve a combination of the effects summarized in Fig. 6.4. The process of removing large amounts of protein aggregates may be selective and may be used to clear intracellular bacteria, parasites, and viruses. Autophagy appears to be the only pathway to address this function, and if it fails, the only other option for infected cells is cell death and clearance by phagocytes.

The first line of defense during pathogen infection is the activation of innate immune responses. The autophagy machinery is also used to attack invading pathogens (first described in 2004, [27]). This fact and other findings could explain the evolution in certain pathogens of mechanisms that aim to evade, inhibit, and usurp the host cell autophagy to promote survival and replication.

The Antibacterial Role of Autophagy

Bacteria may enter mammalian cells by mechanisms that are different for gram-negative and gram-positive bacteria [28]. Once inside the host cells, the predominant strategy against bacterial pathogens appears to involve compartmentalization within membrane-bound vacuoles, including autophagosomes [29]. The cellular autophagic response differs according to the nature of the intracellular bacterial pathogen and is involved in sequestering and eliminating intracellular bacteria such as Group A *Streptococcus*, *Salmonella typhimurium*, *Shigella flexneri*, and *Listeria monocytogenes*. The exact mechanisms are unknown, but the role of autophagy in macrophages compared with that in epithelial cells differs. In epithelial cells, autophagy limits bacterial growth, and in macrophages, mitophagy leads to cell death [30]. Alveolar macrophages ingest the *M. tuberculosis* bacilli and enclose them in phagosomes, which are arrested in maturation. The bacilli survive and grow within the phagosomes until the macrophages die. Alternatively, the activation of infected macrophages by an interferon induces autophagy, and phagosomes fuse with autophagosomes with the subsequent degradation of the bacilli [31]. In contrast, infection with *Legionella pneumophila* or *Porphyromonas gingivalis* activates autophagy, a mechanism that provides a replicative, noxious niche. At least *in vitro*, the suppression of autophagy results in the trafficking of internalized bacteria to phagolysosomes and degradation of the organism [32].

The Autophagic Responses to Fungal Infections

Although knowledge in this area is limited, the chemical or genetic disruption of autophagy in murine macrophages resulted in decreased *Cryptococcus neoformans* uptake, replication, and escape from host cells [33]. The goal of the ongoing studies is to understand the mechanisms by which autophagy proteins determine the ability of *C. neoformans* to establish a replicative niche and how these functions are coordinated with other host factors.

The Autophagic Responses to Parasites

Knowledge regarding the autophagic responses to parasites is also limited. However, following infection in macrophages, *Toxoplasma gondii*, an obligate intracellular protozoan parasite, replicates in specialized vacuoles that are protected from fusion with lysosomes. Experiments in animal models have indicated a role for autophagy involving cell-cell receptor-ligand interactions in activated macrophages that eliminate the parasite [34].

The Autophagic Responses to Viruses

In this particular type of infection, the virus and the host cell type determine the functions of the autophagy-mediated responses. For example, autophagy is antiviral in response to herpes simplex virus type 1, but viruses have evolved mechanisms to perturb autophagy, most likely using Beclin 1 as a preferential target. Antagonistic interactions between viral proteins and Beclin 1 can subvert autophagosome formation and ultimately promote virulence. Certain viruses, including hepatitis B, hepatitis C, and the Dengue viruses, promote autophagy as a means to encourage viral replication. The exact mechanisms are poorly understood but have been extensively studied concerning infection with human immunodeficiency virus (HIV-1). For example, HIV-1 blocks autophagy in dendritic cells by activating the MTOR pathway, most likely via the interaction of envelope glycoproteins with the CD4 receptor, which leads to an increased cellular viral content, an increased transfer of HIV-1 to CD4+ T cells, an increased virus survival and reduced antigen presentation [35, 36]. Other studies have indicated a major role for autophagy in HIV-1-mediated cell death in uninfected CD4+ T cells. HIV-1 inhibits autophagy in infected CD4+ T cells by downregulating Beclin 1 at the transcriptional level and induces autophagy in uninfected CD4+T cells [37, 38]. Therefore, HIV-1 manipulates autophagy during infection to evade the immune response and illustrates the risks and benefits induced by the exogenous manipulation of the host autophagy [39, 40].

6.2.2.3 Autophagy and Immunity

Autophagy may also be implicated in autoimmunity by some of the mechanisms listed in Fig. 6.4 (*i.e.,* promotion of the MHC class II presentation of antigens, control of T lymphocyte homeostasis, induction by Th1 cytokines and possibly by facilitation of specific serum autoantibodies) [41]. The role of autophagy in the innate immune system has been extensively studied but remains poorly understood. Pathogen recognition and intracellular killing can be controlled by autophagy, but this process might also act beyond pathogen control and be related to the general function of the clearance of dead cells. In the absence of autophagy, apoptotic cells fail to signal for efficient removal [42], and LC3 II (Microtubule-associated protein 1 light chain 3 alpha)-associated phagocytosis is a requirement for dead cell clearance [43]. This defect may be linked to susceptibility to autoimmunity in mice [42, 43], and it is clinically evident that a number of autoimmune diseases are precipitated or aggravated after infections.

Induced autophagy may exacerbate the process of presentation on MHC-II of peptides from intracellular sources [44]. Dendritic cell-mediated antigen presentation in the context of MHC-II is most likely an area in which autophagy might influence immune diseases because this is the major activator of the adaptive immune cells. Dendritic cells are also important in the pathogenesis of Crohn's disease because impaired autophagy facilitates the replication of adherent-invasive *Escherichia coli*, an important factor in the pathogenesis of the disease [45].

Diseases resulting from increased activation of the immune system comprise the following two different categories: autoinflammatory diseases and autoimmune diseases in which the adaptive immune cells target the self-antigens. Inflammasome activation modulates cytokine secretion, and autophagy plays a negative role with respect to inflammasome activation; autophagy deficiency leads to the increased production of IL-1β and IL-18. The IL-1β receptor blockade has beneficial effects in rheumatoid arthritis and has been suggested as a therapy for autoinflammatory diseases [46]. Autophagy is also involved in a number of secretory processes in immune cells (and non-immune cells). Proteins with a signal peptide are secreted through a canonical pathway involving the endoplasmic reticulum. A secretion of proteins without a signal peptide appears to be mediated by autophagy [47]. Interestingly, autophagy is involved in interferon (IFN) secretion. Elevated levels of IFN-α are the hallmark of systemic lupus erythematosus, and ongoing clinical trials are testing the effectiveness of monoclonal antibodies against IFN-α [48]. The paucity of data in humans using agents that modulate autophagy is intriguing. However, the antimalarial drug chloroquine has been used for decades as an anti-inflammatory agent for the treatment of systemic lupus erythematosus and rheumatoid arthritis. A mechanistic explanation has been recently reported indicating that chloroquine promotes the transrepression of proinflammatory cytokines by the glucocorticoid receptor. The fact that glucocorticoid signaling is regulated by lysosomes provides a mechanistic basis for treating autoimmune diseases with a combination of glucocorticoids and lysosomal inhibitors [49].

6.3 Nutritional Stress, Autophagy and Inflammation

6.3.1 Influence of Changes in the Metabolic State on the Immune System

Nutritional factors (*i.e.*, obesity, metabolic syndrome, nutritional restriction, among others) have an important regulatory role in modulating the function of the immune system. Possible mechanisms include changes in secretion of immune mediators by metabolic tissues, alterations in the composition of resident populations of T cells or macrophages, direct effects of metabolites on immune cells and intrinsic defects in signaling pathways. Changes in the autophagic activity of different cell populations are likely to influence the quality of the immune responses that occur in response to nutritional stress or nutritional intervention, particularly in the context of a high fat diet [50–54]. Metabolic complications are associated with a state of subclinical chronic inflammation that has been identified as a pathogenic risk. Because inflammation is not the cause of obesity, we speculate that it may represent a superimposed mechanism that may be defensive and is related to autophagy, oxidation, dietary factors and the consequent metabolic response in specific types of cells.

6.3.2 Chemokine (C-C motif) Ligand 2, Autophagy and Inflammation

Obese adipose tissue, characterized by adipocyte hyperplasia and hypertrophy, is a mediator of low-grade inflammation of the white adipose tissue that is due to a chronic activation of the innate immune system leading to an increase in cytokines, which mediate the recruitment of macrophages and T cells. Certain cytokines, primarily chemokine (C-C motif) ligand 2 (CCL2), also support vasculogenesis.

Autophagy modifies pancreatic beta cell functions regulating glucose homeostasis [55]. Because ER stress is involved in insulin resistance, autophagy might also be involved in insulin resistance by modulating the ER stress response [56]. Inflammation apparently represents an indicator of the failure of adipose tissue in its major function of energy storage. We have recently developed an animal model overexpressing CCL2 that is designed to characterize the relationship between autophagy and cytokine gene expression. The preliminary data suggest that metabolism (*i.e.*, dietary variations) plays a contributory factor; animals overexpressing CCL2 displayed an

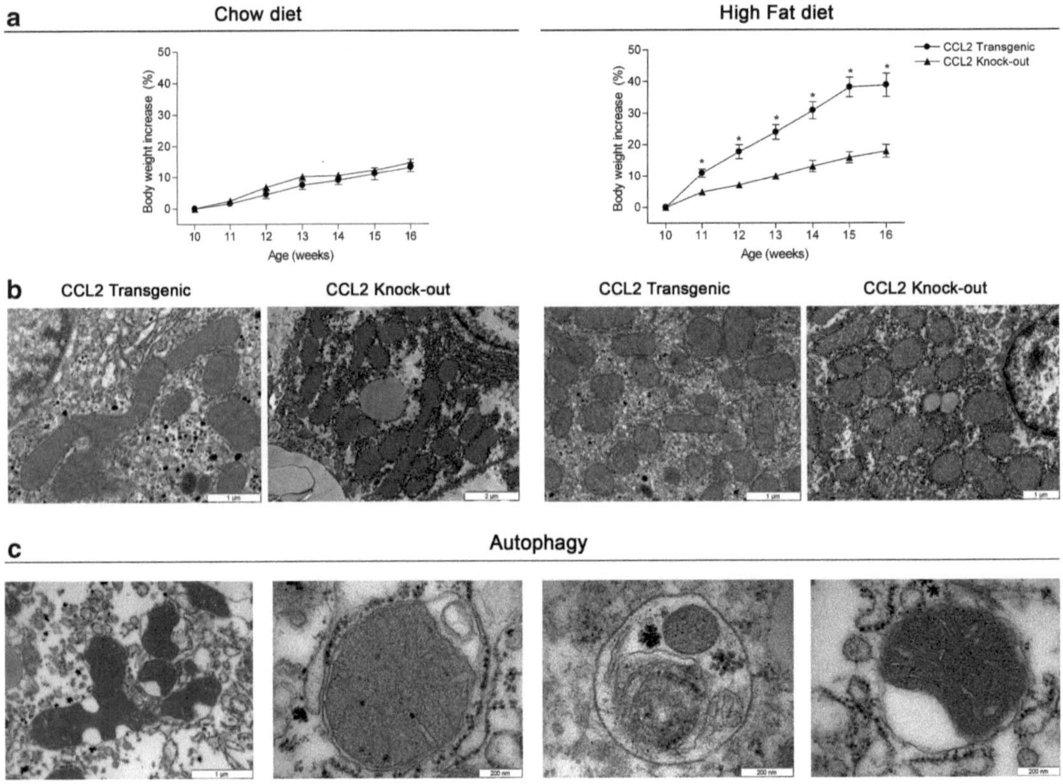

Fig. 6.5 The ingestion of highly caloric, fat-rich, food elicited significant differences in weight increase among mice overexpressing CCL2 when compared to CCL2 knockout animals (**a**). Additionally, dietary conditions were also responsible for significant changes in the fusion-fission balance in mitochondrial biogenesis (**b**) and the amount of autophagosomes in the liver (**c**)

increased number of autophagosomes and a differential rate of mitophagy as compared with CCL2 deficient animals (Fig. 6.5) [57]. However, the role of CCL2 may vary in response to different environments [58]. Ongoing studies in our model may add information about the regulation of autophagy by cytokines, which is based on relatively limited data obtained in mammalian cells. IFN-γ, a Th1 cytokine, induces autophagy [59], but Th2 cytokines (IL-4 and IL-13) act as suppressors of autophagy and most likely activate the MTOR cascade [60]. More importantly, in vascular smooth muscle cells (VSMC) from atherosclerotic plaques, the expression of autophagy genes in VSMC was controlled by insulin growth factor like-1 (IGF-1) and TNF-α; TNF-α may induce autophagy [61]. This result suggests that IGF-1 can induce cell survival by inhibiting autophagy in plaque-associated VSMC [62, 63]. This effect could be important to provide novel strategies to increase stability of atherosclerotic plaques.

This finding is also relevant in obesity, a condition associated with a frightening disease burden. Although the search for a simple and acceptable nutritionally based therapeutic approach continues, other strategies are clearly needed. The effect on liver morphology and function of obesity and the consequent development of fatty liver disease has received substantial attention in our laboratory [64–66]. It has been repeatedly invoked that autophagy may play a dual role in cell fate, depending on the stimuli and cell type. Therefore, it is essential to consider the effect of autophagy in every metabolic tissue involved in obesity-associated disturbances. Most likely, the effect has been predominantly investigated in adipose tissue and the liver, especially in nonalcoholic fatty liver disease (NAFLD).

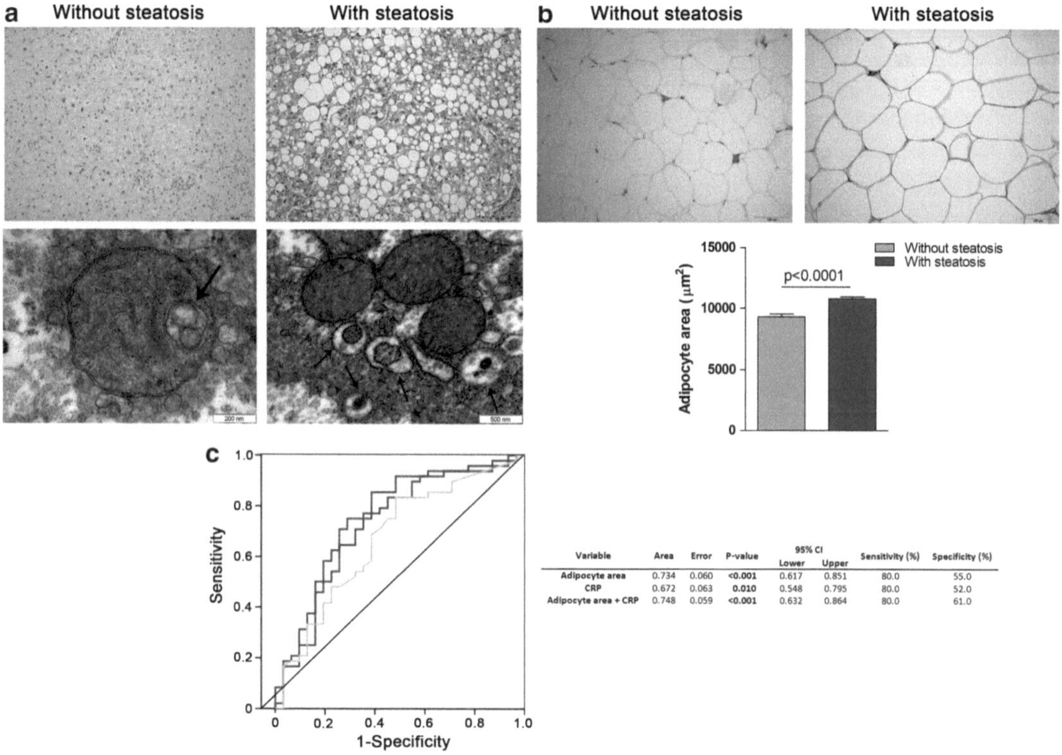

Variable	Area	Error	P-value	95% CI Lower	95% CI Upper	Sensitivity (%)	Specificity (%)
Adipocyte area	0.734	0.060	**<0.001**	0.617	0.851	80.0	55.0
CRP	0.672	0.063	**0.010**	0.548	0.795	80.0	52.0
Adipocyte area + CRP	0.748	0.059	**<0.001**	0.632	0.864	80.0	61.0

Fig. 6.6 The degree of steatosis is considerably heterogeneous among obese patients and this may be associated with the presence of autophagy (mostly mitophagy) (**a**). In addition, the adipocytes in subcutaneous adipose tissue are significantly higher in patients with steatosis (**b**). It remains to be ascertained whether this measurement could be qualified as a clinical biomarker but it is apparently associated with circulating C-reactive protein concentration (**c**)

Intriguingly, the degree of steatosis is heterogeneous among obese patients and a substantial number of patients do not develop this complication. Obese patients with steatosis display a higher number of autophagosomes in the liver and toroidal mitochondria are constantly observed in obese patients without steatosis (Fig. 6.6). Apparently, in the liver, these effects are not related to inflammation but autophagy affects the inflammatory status of the adipocyte. The reported data are conflicting and even counterintuitive. Patients with steatosis displayed higher adipocytes than those without and may constitute a predictive factor, which is associated with circulating C-reactive protein (Fig. 6.6).

The absence of autophagy in adipocytes, as identified in adipocyte-specific ATG7 knockout mice, or the mitigation observed in CCL2 deficient mice, protect against high-fat diet-induced obesity and insulin resistance in mice [67–69]. However, other data report that autophagy may function to limit excessive inflammation in adipose tissue during obesity [70]. In addition, obese patients do not necessarily display insulin resistance. The inhibition of autophagic action by insulin could partly explain an expected increase of autophagy in obese patients with insulin resistance, and consequently steatosis should be more likely associated with metabolic disturbances. In these patients, however, changes in autophagy could also play a role in both inflammatory gene expression and secretion of cytokines. This role may not be exactly the case for other types of cells (*i.e.*, macrophages) that are present in expanded adipose tissue. Therefore, it remains plausible that the inhibition of autophagy in adipose tissue might worsen insulin sensitivity via its effects on inflammatory cytokine production, and that inflammation is associated with autophagy as part of the induced effects of obesity. Other findings [71, 72] suggest that obesity-associated inflammation in adipose tissue up-regulates autophagy to mitigate the production of proinflammatory cytokines (*i.e.,* autophagy could be a consequence rather than a cause of obesity-induced adipose tissue inflammation).

6.3.3 Lipophagy and the Lipolytic Role of Macrophages in Adipose Tissue

The function of autophagy in regulating intracellular lipid droplets (LDs) is also relevant for obesity, especially in tissues that are not designed for energy storage, such as the liver or muscle (lipophagy). Whether impaired lipophagy is a sufficient condition to cause NAFLD is debatable [73]. Among other data, the existence of lipophagy has been suggested, via different mechanisms, to degrade LDs through autophagy, indicating that lysosomes do not fuse directly with LDs but with LD-containing autophagosomes [74]. In the context of NAFLD, we have previously shown the synergic roles of both dietary energy ingestion and the tissue CCL2 response [75–77]. LDs were initially viewed as storage lipid depots, but it is now known that these organelles may participate in different cell functions [78]. Moreover, when induced by oxidized lipids, LDs are active in the arachidonic acid conversion into leukotrienes through the 5-LO pathway, a key component in the pathogenesis of atherosclerosis [79–81]. Prostaglandins and leukotrienes are important inflammation mediators; consequently, LDs may act in the coordination of immune responses. Different subpopulations of LDs that may differ in their lipid and protein concentrations have been observed in different type cells within the same cell and/or in response to different stimuli, suggesting maturation and integration in multiple cellular processes [82]. Curiously, emerging evidence suggests a linkage between LDs and the regulation of immune responses in the context of host-pathogen interactions, which have been previously reviewed [83]. The regulatory network of autophagy in the development of NAFLD may also involve signaling through AMP-activated protein kinase (AMPK), which is a central regulator of cellular metabolism [84]. AMPK regulates metabolism through its effects on glucose homeostasis, lipid metabolism, protein synthesis, and oxidative metabolism. However, ongoing studies in a model of hyperlipidemia suggest that the modulation of AMPK by metformin may have different effects according to the baseline metabolic context. AMPK may up-regulate glycolysis and decrease the synthesis of glycogen [85]. The control of lipid metabolism by AMPK is most likely the result of both a decrease in lipogenesis and the stimulation of mitochondrial fatty acid oxidation. The possible mechanisms remain unclear, but the former is most likely due to the inactivation of 3-hydroxy-3-methylglutaryl-coenzyme A reductase and acetyl CoA carboxylase 1. The ability of AMPK to promote fatty acid oxidation may be due to the decreased synthesis of malonyl CoA or a decrease in transcription of lipogenic genes via the inhibition of the sterol regulatory element-binding protein-1c. AMPK most likely regulates PGC-1α, which subsequently increases mitochondrial biogenesis [86]. All of these findings are particularly important because metformin has been considered an agent against fatty liver disease. Moreover, the invoked mechanisms of action are continuously revisited [87–89]. A certain alteration in the mitochondrial function is likely and deserves future research. Metformin may also attenuate inflammatory responses by suppressing the production of TNF-α, suppressing the expressions of scavenger receptors in macrophages [90], and exacerbating the allergic eosinophilic inflammation in high fat-diet-induced obesity in mice [91].

To further complicate the issue, macrophages in adipose tissue (ATMs) play a role in lipolysis that is independent of inflammatory action. In an attempt to identify the cellular functions of ATMs that are regulated by adiposity, a program of obesity-associated lysosome biogenesis has been recently identified, which suggests that ATM lysosomes are important in lipid metabolism [92]. A class of antimalarial drugs, including chloroquine, inhibits the acidification of lysosomes (*i.e.,* decreased autophagy), consequently leading to the accumulation of LDs in

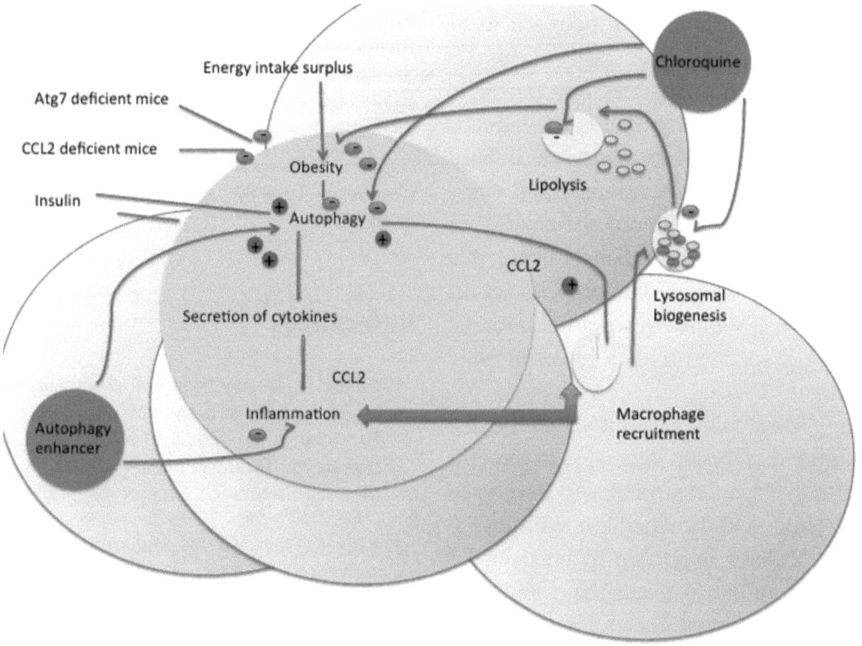

Fig. 6.7 Despite their apparent complexity and the possible variations among metabolic tissues, inflammation, lipolysis and autophagy are closely related in metabolic diseases. Treatment with available drugs may require further elucidation of the uptake of lipids by adipose tissue and cells in the environment, particularly macrophages, the signaling effects of lipid sensing, the fate of lysosomal products of lipid hydrolysis, and the anabolic factors released by autophagy

macrophages. Macrophages are cells that possess distinct noninflammatory trophic functions that are regulated by the cellular and/or metabolic context. These new findings indicate that ATMs, which are the dominant immune cell in AT, respond to increases in adiposity and are closely coupled to lipid accumulation (*i.e.*, the net effect is to buffer increased concentrations of lipids). Moreover, the trafficking of lipids differs between lean and obese individuals. In large adipocytes, the released lipid is taken up by macrophages and targeted to lysosomes for metabolism. This process may be central to the development of metabolic diseases [93]; however, attempts to target autophagy and lysosome function require further extensive research (Fig. 6.7).

6.4 Conclusion and Perspectives

The crosstalk between the signaling pathways and autophagy, particularly the link between MTOR downregulation and autophagy induction, requires further exploration considering the possible regulatory effects of inflammation. Caloric restriction apparently delays the onset of chronic diseases, which are more common later in life. Caloric restriction also favors autophagy, which provides energy to cells from self-degraded components and removes harmful compounds that play a role in oxidative stress and immune function. In older organisms, autophagic activity decreases, which may possibly favor the genesis of chronic diseases and partially explain the relationship between these diseases and an excessive energy intake. Although caloric restriction does not appear to be feasible in the human population, the induced decrease in chronic systemic low-grade inflammation may be focal in the development of many typical Western diseases centered on obesity. For example, epidemiological studies indicate that the consumption of polyphenol-rich foods, which may have a caloric restriction mimetic effect, may be associated with a lower chronic disease risk.

Autophagy research may lead to either the discovery of future therapeutic targets or to the efficient use of either marketed or newly developed drugs under safe and beneficial conditions. It should be particularly important to uncover the multiple interactions between autophagy and cytokines, suggesting important roles for autophagy in the control of inflammation and fine-tuning of the immune response. In this setting, atherosclerosis appears to be a relatively novel target. Certain experiments in mice demonstrate defective autophagy in macrophages of atherosclerotic plaques. Complete deficiency of macrophage autophagy resulted in a pro-inflammatory state and enhanced plaque progression via an inflammasome-dependent mechanism mediated by cholesterol crystals [94]. The inflammasomes, recently discovered multi-protein complexes, seem to be crucial to understand how the immune system sense perturbations of cellular homeostasis, and at which stage of metabolic disease the subsequent interactions impact disease progression.

Knowledge on the mechanisms and pathways involved in autophagy might reveal a considerable potential for pharmacological modulation of the process in the management of inflammation and related conditions. Noninflammatory functions of immune cells should also be considered in the context of potential improvements in the therapeutic approach for the devastating metabolic diseases.

Acknowledgements We would like to acknowledge the contribution of the staff members that assisted in the clinical management, laboratory measurements, statistical assessment, and data collection during the last years. The Unitat de Recerca Biomèdica is currently being supported by the program of consolidated groups from the Universitat Rovira iVirgili and grants from the Fondo de Investigación Sanitaria (FIS PI08/1032, PI08/1381 and PI11/00130). E. Rodríguez-Gallego is the recipient of a fellowship from the Generalitat de Catalunya (2012FI B 00389).

Conflict of Interest The authors declare no competing interests.

References

1. Finley D. Recognition and processing of ubiquitin-protein conjugates by the proteasome. Annu Rev Biochem. 2009;78:477–513.
2. Kundu M, Thompson CB. Autophagy: basic principles and relevance to disease. Annu Rev Pathol. 2008;3:427–55.
3. Kim J, Guan KL. Amino acid signaling in TOR activation. Annu Rev Biochem. 2011;80:1001–32.
4. Menendez JA, Cufí S, Oliveras-Ferraros C, Vellon L, Joven J, Vazquez-Martin A. Gerosuppressant metformin: less is more. Aging (Albany NY). 2011; 3:348–62.
5. Cufí S, Vazquez-Martin A, Oliveras-Ferraros C, Corominas-Faja B, Cuyàs E, López-Bonet E, et al. The anti-malarial chloroquine overcomes primary resistance and restores sensitivity to trastuzumab in HER2-positive breast cancer. Sci Rep. 2013;3:2469. doi:10.1038/srep02469.
6. Komatsu M, Waguri S, Ueno T, Iwata J, Murata S, Tanida I, et al. Impairment of starvation-induced and constitutive autophagy in Atg7-deficient mice. J Cell Biol. 2005;169:425–34.
7. Rodriguez-Enriquez S, Kim I, Currin RT, Lemasters JJ. Tracker dyes to probe mitochondrial autophagy (mitophagy) in rat hepatocytes. Autophagy. 2006; 2:39–46.
8. Li D. Selective degradation of the IkappaB kinase (IKK) by autophagy. Cell Res. 2006;16:855–6.
9. Mizushima N, Yamamoto A, Matsui M, Yoshimori T, Ohsumi Y. In vivo analysis of autophagy in response to nutrient starvation using transgenic mice expressing a fluorescent autophagosome marker. Mol Biol Cell. 2004;15:1101–11.
10. Yousefi S, Perozzo R, Schmid I, Ziemiecki A, Schaffner T, Scapozza L, et al. Calpain-mediated cleavage of Atg5 switches autophagy to apoptosis. Nat Cell Biol. 2006;8:1124–32.
11. Liang XH, Jackson S, Seaman M, Brown K, Kempkes B, Hibshoosh H, Levine B. Induction of autophagy and inhibition of tumorigenesis by beclin 1. Nature. 1999;402:672–6.
12. Hay N. The Akt-mTOR tango and its relevance to cancer. Cancer Cell. 2005;8:179–83.
13. Joven J, Rull A, Rodriguez-Gallego E, Camps J, Riera-Borrull M, Hernández-Aguilera A, et al. Multifunctional targets of dietary polyphenols

in disease: a case for the chemokine network and energy metabolism. Food Chem Toxicol. 2013; 51:267–79.
14. Jin S, DiPaola RS, Mathew R, White E. Metabolic catastrophe as a means to cancer cell death. J Cell Sci. 2007;120:379–83.
15. Hara T, Nakamura K, Matsui M, Yamamoto A, Nakahara Y, Suzuki-Migishima R, et al. Suppression of basal autophagy in neural cells causes neurodegenerative disease in mice. Nature. 2006;441:885–9.
16. Taylor JP, Hardy J, Fischbeck KH. Toxic proteins in neurodegenerative disease. Science. 2002;296: 1991–5.
17. Jennings Jr JJ, Zhu JH, Rbaibi Y, Luo X, Chu CT, et al. Mitochondrial aberrations in mucolipidosis type IV. J Biol Chem. 2006;281:39041–50.
18. Tanaka Y, Guhde G, Suter A, Eskelinen EL, Hartmann D, et al. Accumulation of autophagic vacuoles and cardiomyopathy in LAMP-2-deficient mice. Nature. 2000;406:902–6.
19. Schmid D, Pypaert M, Munz C. Antigen-loading compartments for major histocompatibility complex class II molecules continuously receive input from autophagosomes. Immunity. 2007;26:79–92.
20. English L, Chemali M, Duron J, Rondeau C, Laplante A, et al. Autophagy enhances the presentation of endogenous viral antigens on MHC class I molecules during HSV-1 infection. Nat Immunol. 2009; 10:480–7.
21. Berger SB, Romero X, Ma C, Wang G, Faubion WA, et al. SLAM is a microbial sensor that regulates bacterial phagosome functions in macrophages. Nat Immunol. 2010;11:920–7.
22. Brinkmann V, Reichard U, Goosmann C, Fauler B, Uhlemann Y, et al. Neutrophil extracellular traps kill bacteria. Science. 2004;303:1532–5.
23. Watanabe K, Tsubata T. Autophagy connects antigen receptor signaling to costimulatory signaling in B lymphocytes. Autophagy. 2009;5:108–10.
24. Hubbard VM, Valdor R, Patel B, Singh R, Cuervo AM, Macian F. Macroautophagy regulates energy metabolism during effector T cell activation. J Immunol. 2010;185:7349–57.
25. Hildeman DA, Mitchell T, Teague TK, Henson P, Day BJ, et al. Reactive oxygen species regulate activation-induced T cell apoptosis. Immunity. 1999;10:735–44.
26. Cardenas C, Miller RA, Smith I, Bui T, Molgo J, et al. Essential regulation of cell bioenergetics by constitutive InsP3 receptor Ca2+ transfer to mitochondria. Cell. 2010;142:270–83.
27. Nakagawa I, Amano A, Mizushima N, Yamamoto A, Yamaguchi H, Kamimoto T, et al. Autophagy defends cells against invading group A Streptococcus. Science. 2004;306:1037–40.
28. Inohara N, Chamaillard M, McDonald C, Nunez G. NOD-LRR proteins: role in host-microbial interactions and inflammatory disease. Annu Rev Biochem. 2005;74:355–83.
29. Delbridge LM, O'Riordan MX. Innate recognition of intracellular bacteria. Curr Opin Immunol. 2007; 19:10–6.
30. Rich KA, Burkett C, Webster P. Cytoplasmic bacteria can be targets for autophagy. Cell Microbiol. 2003; 5:455–68.
31. Deretic V, Singh S, Master S, Harris J, Roberts E, et al. Mycobacterium tuberculosis inhibition of phagolysosome biogenesis and autophagy as a host defence mechanism. Cell Microbiol. 2006;8:719–27.
32. Amer AO, Swanson MS. Autophagy is an immediate macrophage response to Legionella pneumophila. Cell Microbiol. 2005;7:765–78.
33. Qin QM, Luo J, Lin X, Pei J, Li L, et al. Functional analysis of host factors that mediate the intracellular lifestyle of Cryptococcus neoformans. PLoS Pathog. 2011;7:e1002078.
34. Portillo JA, Okenka G, Reed E, Subauste A, Van Grol J, et al. The CD40-autophagy pathwayis needed for host protection despite IFN-γ-dependent immunity and CD40 induces autophagy via control of P21 levels. PLoS One. 2010;5:e14472.
35. Kyei GB, Dinkins C, Davis AS, Roberts E, Singh SB, et al. Autophagy pathway intersects with HIV-1 biosynthesis and regulates viral yields in macrophages. J Cell Biol. 2009;186:255–68.
36. Joven J, Menéndez JA, Fernandez-Sender L, Espinel E, Rull A, Beltrán-Debón R, et al. Metformin: a cheap and well-tolerated drug that provides benefits for viral infections. HIV Med. 2013;14:233–40.
37. Denizot M, Varbanov M, Espert L, Robert-Hebmann V, Sagnier S, et al. HIV-1 gp41 fusogenic function triggers autophagy in uninfected cells. Autophagy. 2008;4:998–1008.
38. Zhou D, Spector SA. Human immunodeficiency virus type-1 infection inhibits autophagy. AIDS. 2008; 22:695–9.
39. Paton NI, Goodall RL, Dunn DT, Franzen S, Collaco-Moraes Y, Gazzard BG, et al. Effects of hydroxychloroquine on immune activation and disease progression among HIV-infected patients not receiving antiretroviral therapy: a randomized controlled trial. JAMA. 2012;308:353–61.
40. Alonso-Villaverde C, Menéndez JA, Joven J. Metabolic stress in infected cells may represent a therapeutic target for human immunodeficiency virus infection. Med Hypotheses. 2013;81:125–30.
41. Deretic V. Autophagy: an emerging immunological paradigm. J Immunol. 2012;189:15–20.
42. Qu X, Zou Z, Sun Q, Luby-Phelps K, Cheng P, et al. Autophagy gene-dependent clearance of apoptotic cells during embryonic development. Cell. 2007; 128:931–46.
43. Martinez J, Almendinger J, Oberst A, Ness R, Dillon CP, Fitzgerald P, et al. Microtubule-associated protein1 light chain 3 alpha (LC3)-associated phagocytosis is required for the efficient clearance of dead cells. Proc Natl Acad Sci U S A. 2011; 108:17396–401.

44. Dengjel J, Schoor O, Fischer R, Reich M, Kraus M, et al. Autophagy promotes MHC class II presentation of peptides from intracellular source proteins. Proc Natl Acad Sci U S A. 2005;102:7922–7.
45. Lapaquette P, Glasser AL, Huett A, Xavier RJ, Darfeuille-Michaud A. Crohn's disease-associated adherent-invasive E.coli are selectively favoured by impaired autophagy to replicate intracellularly. Cell Microbiol. 2010;12:99–113.
46. Goldbach-Mansky R. Blocking interleukin-1 in rheumatic diseases. Ann N Y Acad Sci. 2009; 1182:111–23.
47. Giuliani F, Grieve A, Rabouille C. Unconventional secretion: a stress on GRASP. Curr Opin Cell Biol. 2011;23:498–504.
48. Lichtman EI, Helfgott SM, Kriegel MA. Emerging therapies for systemic lupus erythematosus–focus on targeting interferon-alpha. Clin Immunol. 2012; 143:210–21.
49. He Y, Xu Y, Zhang C, Gao X, Dykema KJ, Martin KR, et al. Identification of a lysosomal pathway that modulates glucocorticoid signaling and the inflammatory response. Sci Signal. 2011;4:ra44.
50. Dorshkind K, Swain S. Age-associated declines in immune system development and function: causes, consequences, and reversal. Curr Opin Immunol. 2009;21:404–7.
51. Olefsky JM, Glass CK. Macrophages, inflammation, and insulin resistance. Annu Rev Physiol. 2010; 72:219–46.
52. Wen H, Gris D, Lei Y, Jha S, Zhang L, Huang MT, et al. Fatty acid-induced NLRP3-ASC inflammasome activation interferes with insulin signaling. Nat Immunol. 2011;12:408–15.
53. O'Neill LA, Hardie DG. Metabolism of inflammation limited by AMPK and pseudo-starvation. Nature. 2013;493:346–55.
54. Rull A, Camps J, Alonso-Villaverde C, Joven J. Insulin resistance, inflammation, and obesity: role of monocyte chemoattractant protein-1 (or CCL2) in the regulation of metabolism. Mediators Inflamm. 2010;2010:pii: 326580. doi: 10.1155/2010/326580.
55. Ebato C, Uchida T, Arakawa M, Komatsu M, Ueno T, Komiya K, et al. Autophagy is important in islet homeostasis and compensatory increase of beta cell mass in response to high-fat diet. Cell Metab. 2008;8:325–32.
56. Quan W, Lim YM, Lee MS. Role of autophagy in diabetes and endoplasmic reticulum stress of pancreatic beta cells. Exp Mol Med. 2012;44:81–8.
57. Rodríguez-Gallego E, Riera-Borrull M, Hernández-Aguilera A, Mariné-Casadó R, Rull A, Beltrán-Debón R, et al. Ubiquitous transgenic overexpression of C-C Chemokine Ligand 2: a model to assess the combined effect of high energy intake and continuous low-grade inflammation. Mediat Inflamm. 2013;2013:953841. doi: 10.1155/2013/953841.
58. Tous M, Ferré N, Rull A, Marsillach J, Coll B, Alonso-Villaverde C, Camps J, Joven J. Dietary cholesterol and differential monocyte chemoattractant protein-1 gene expression in aorta and liver of apo E-deficient mice. Biochem Biophys Res Commun. 2006;340:1078–84.
59. Gutierrez MG, Master SS, Singh SB, Taylor GA, Colombo MI, Deretic V. Autophagy is a defense mechanism inhibiting BCG and Mycobacterium tuberculosis survival in infected macrophages. Cell. 2004;119:753–66.
60. Wright K, Ward SG, Kolios G, Westwick J. Activation of phosphatidylinositol 3-kinase by interleukin-13. An inhibitory signal for inducible nitric-oxide synthase expression in epithelial cell line HT-29. J Biol Chem. 1997;272:12626–33.
61. Jia G, Cheng G, Gangahar DM, Agrawal DK. Insulin-like growth factor-1 and TNF-alpha regulate autophagy through c-jun N-terminal kinase and Akt pathways in human atherosclerotic vascular smooth cells. Immunol Cell Biol. 2006;84:448–54.
62. Hall JL, Gibbons GH, Chatham JC. IGF-I promotes a shift in metabolic flux in vascular smooth muscle cells. Am J Physiol Endocrinol Metab. 2002; 283:E465–71.
63. Coll B, Alonso-Villaverde C, Joven J. Monocyte chemoattractant protein-1 and atherosclerosis: is there room for an additional biomarker? Clin Chim Acta. 2007;383:21–9.
64. Must A, Spadano J, Coakley EH, Field AE, Colditz G, Dietz WH. The disease burden associated with overweight and obesity. JAMA. 1999;282:1523–9.
65. Joven J, Micol V, Segura-Carretero A, Alonso-Villaverde C, Menendez JA. Polyphenols and the modulation of gene expression pathways: can we eat our way out of the danger of chronic disease? Crit Rev Food Sci Nutr. 2014;54(8):985–1001. doi: 10.1080/10408398.2011.621772.
66. Segura-Carretero A, Puertas-Mejía MA, Cortacero-Ramírez S, Beltrán R, Alonso-Villaverde C, Joven J, Dinelli G, Fernández-Gutiérrez A. Selective extraction, separation, and identification of anthocyanins from Hibiscus sabdariffa L. using solid phase extraction-capillary electrophoresis-mass spectrometry (time-of-flight/ion trap). Electrophoresis. 2008;29:2852–61.
67. Goldman S, Zhang Y, Jin S. Autophagy and adipogenesis: implications in obesity and type II diabetes. Autophagy. 2010;6:179–81.
68. Singh R, Xiang Y, Wang Y, Baikati K, Cuervo AM, Luu YK, et al. Autophagy regulatesadipose mass and differentiation in mice. J Clin Invest. 2009;119:3329–39.
69. Zhang Y, Goldman S, Baerga R, Zhao Y, Komatsu M, Jin S. Adipose-specific deletion of autophagy-related gene 7 (atg7) in mice reveals a role in adipogenesis. Proc Natl Acad Sci U S A. 2009;106:19860–5.
70. Jansen HJ, van Essen P, Koenen T, Joosten LA, Netea MG, Tack CJ, Stienstra R. Autophagy activity is up-regulated in adipose tissue of obese individuals and modulates proinflammatory cytokine expression. Endocrinology. 2012;153:5866–74.
71. Yoshizaki T, Kusunoki C, Kondo M, Yasuda M, Kume S, Morino K, et al. Autophagy regulates inflammation in adipocytes. Biochem Biophys Res Commun. 2012;417:352–7.

72. Marsillach J, Camps J, Ferré N, Beltran R, Rull A, Mackness B, Mackness M, Joven J. Paraoxonase-1 is related to inflammation, fibrosis and PPAR delta in experimental liver disease. BMC Gastroenterol. 2009;9:3. doi:10.1186/1471-230X-9-3.
73. Liu K, Czaja MJ. Regulation of lipid stores and metabolism by lipophagy. Cell Death Differ. 2013; 20:3–11.
74. Singh R, Kaushik S, Wang Y, Xiang Y, Novak I, Komatsu M, et al. Autophagy regulates lipid metabolism. Nature. 2009;458:1131–5.
75. Vinaixa M, Rodríguez MA, Rull A, Beltrán R, Bladé C, Brezmes J, et al. Metabolomic assessment of the effect of dietary cholesterol in the progressive development of fatty liver disease. J Proteome Res. 2010;9:2527–38.
76. Rull A, Rodríguez F, Aragonès G, Marsillach J, Beltrán R, Alonso-Villaverde C, Camps J, Joven J. Hepatic monocyte chemoattractant protein-1 is upregulated by dietary cholesterol and contributes to liver steatosis. Cytokine. 2009;48:273–9.
77. Joven J, Espinel E, Rull A, Aragonès G, Rodríguez-Gallego E, Camps J, et al. Plant-derived polyphenols regulate expression of miRNA paralogs miR-103/107 and miR-122 and prevent diet-induced fatty liver disease in hyperlipidemic mice. Biochim Biophys Acta. 1820;2012:894–9.
78. Bozza PT, Magalhães KG, Weller PF. Leukocyte lipid bodies – biogenesis and functions in inflammation. Biochim Biophys Acta. 2009;1791:540–51.
79. Silva AR, Pacheco P, Vieira-de-Abreu A, Maya-Monteiro CM, D'Alegria B, Magalhães KG, et al. Lipid bodies in oxidized LDL-induced foam cells are leukotriene-synthesizing organelles: a MCP-1/CCL2 regulated phenomenon. Biochim Biophys Acta. 2009;1791:1066–75.
80. Spanbroek R, Grabner R, Lotzer K, Hildner M, Urbach A, Ruhling K, et al. Expanding expression of the 5-lipoxygenase pathway within the arterial wall during human atherogenesis. Proc Natl Acad Sci U S A. 2003;100:1238–43.
81. Ferré N, Martínez-Clemente M, López-Parra M, González-Périz A, Horrillo R, Planagumà A, et al. Increased susceptibility to exacerbated liver injury in hypercholesterolemic ApoE-deficient mice: potential involvement of oxysterols. Am J Physiol Gastrointest Liver Physiol. 2009;296:G553–62.
82. Hodges BD, Wu CC. Proteomic insights into an expanded cellular role for cytoplasmic lipid droplets. J Lipid Res. 2010;51:262–73.
83. Saka HA, Valdivia R. Emerging roles for lipid droplets in immunity and host-pathogen interactions. Annu Rev Cell Dev Biol. 2012;28:411–37.
84. Hardie DG. AMP-activated protein kinase: an energy sensor that regulates all aspects of cell function. Genes Dev. 2011;25:1895–908.
85. Jørgensen SB, Nielsen JN, Birk JB, Olsen GS, Viollet B, Andreelli F, et al. The alpha2-5′AMP-activated protein kinase is a site 2 glycogen synthase kinase in skeletal muscle and is responsive to glucose loading. Diabetes. 2004;53:3074–81.
86. Narkar VA, Downes M, Yu RT, Embler E, Wang YX, Banayo E, et al. AMPK and PPARdelta agonists are exercise mimetics. Cell. 2008;134:405–15.
87. Zhou G, Myers R, Li Y, Chen Y, Shen X, Fenyk-Melody J, et al. Role of AMP-activated protein kinase in mechanism of metformin action. J Clin Invest. 2001;108:1167–74.
88. Zang M, Zuccollo A, Hou X, Nagata D, Walsh K, Herscovitz H, et al. AMP-activated protein kinase is required for the lipid-lowering effect of metformin in insulin-resistant human HepG2 cells. J Biol Chem. 2004;279:47898–905.
89. Menendez JA, Joven J. One-carbon metabolism: an aging-cancer crossroad for the gerosuppressant metformin. Aging (Albany NY). 2012;4:894–8.
90. Hyun B, Shin S, Lee A, Lee S, Song Y, Ha NJ, Cho KH, Kim K. Metformin down-regulates TNF-α secretion via suppression of scavenger receptors in macrophages. Immune Netw. 2013;13:123–32.
91. Calixto MC, Lintomen L, André DM, Leiria LO, Ferreira D, Lellis-Santos C, et al. Metformin attenuates the exacerbation of the allergic eosinophilic inflammation in high fat-diet-induced obesity in mice. PLoS One. 2013;8:e76786.
92. Xu X, Grijalva A, Skowronski A, van Eijk M, Serlie MJ, Ferrante Jr AW. Obesity activates a program of lysosomal-dependent lipid metabolism in adipose tissue macrophages independently of classic activation. Cell Metab. 2013;18:816–30.
93. Camps J, Rodríguez-Gallego E, García-Heredia A, Triguero I, Riera-Borrull M, Hernández-Aguilera A, Luciano-Mateo F, Fernández-Arroyo S, Joven J. Paraoxonases and chemokine (C–C Motif) ligand-2 in noncommunicable diseases. Adv Clin Chem. 2014; 63:247–308.
94. Razani B, Feng C, Coleman T, Emanuel R, Wen H, Hwang S, et al. Autophagy links inflammasomes to atherosclerotic progression. Cell Metab. 2012; 15:534–44.

Delta-5 and Delta-6 Desaturases: Crucial Enzymes in Polyunsaturated Fatty Acid-Related Pathways with Pleiotropic Influences in Health and Disease

7

Federica Tosi, Filippo Sartori, Patrizia Guarini, Oliviero Olivieri, and Nicola Martinelli

Abstract

Polyunsaturated fatty acids (PUFA) play pleiotropic and crucial roles in biological systems. Both blood and tissue levels of PUFA are influenced not only by diet, but to a large extent also by genetic heritability. Delta-5 (D5D) and delta-6 desaturases (D6D), encoded respectively by FADS1 and FADS2 genes, are the rate-limiting enzymes for PUFA conversion and are recognized as main determinants of PUFA levels. Alterations of D5D/D6D activity have been associated with several diseases, from metabolic derangements to neuropsychiatric illnesses, from type 2 diabetes to cardiovascular disease, from inflammation to tumorigenesis. Similar results have been found by investigations on FADS1/FADS2 genotypes. Recent genome-wide association studies showed that FADS1/FADS2 genetic locus, beyond being the main determinant of PUFA, was strongly associated with plasma lipids and glucose metabolism. Other analyses suggested potential link between FADS1/FADS2 polymorphisms and cognitive development, immunological illnesses, and cardiovascular disease. Lessons from both animal models and rare disorders in humans further emphasized the key role of desaturases in health and disease. Remarkably, some of the above mentioned associations appear to be influenced by the environmental context/PUFA dietary intake, in particular the relative prevalence of ω-3 and ω-6 PUFA. In this narrative review we provide a summary of the evidences linking FADS1/FADS2 gene variants and D5D/D6D activities with various traits of human physiopathology. Moreover,

---

Federica Tosi, Filippo Sartori, and Nicola Martinelli contributed equally to this work.

F. Tosi • F. Sartori • P. Guarini • O. Olivieri
N. Martinelli (✉)
Department of Medicine, University of Verona,
Piazzale Luduvico Antonio Scuro, 10,
37134 Verona, Italy
e-mail: fedetosi84@gmail.com; filippo.sartori@gmail.com; patrizia.guarini@univr.it; oliviero.olivieri@univr.it; nicola.martinelli@univr.it

J. Camps (ed.), *Oxidative Stress and Inflammation in Non-communicable Diseases – Molecular Mechanisms and Perspectives in Therapeutics*, Advances in Experimental Medicine and Biology 824,
DOI 10.1007/978-3-319-07320-0_7, © Springer International Publishing Switzerland 2014

we focus also on the potentially useful therapeutic application of D5D/D6D activity modulation, as suggested by anti-inflammatory and tumor-suppressing effects of D6D inhibition in mice models.

Keywords
Cancer • Cardiovascular disease • Delta-5 desaturase • Delta-6 desaturase • FADS • FADS2 • Inflammation • Polyunsaturated fatty acids

7.1 Introduction: From the Pleiotropic Effects of PUFA to the Increasing Interest on Desaturases

Polyunsaturated fatty acids (PUFA) have heterogeneous and crucial functions in the human body (Fig. 7.1). First of all, they are fundamental structural components of cell membranes, in particular at level of central nervous system, and their incorporation into the phospholipid layer may alter membrane's fluidity and selective permeability (e.g. an increase of PUFA incorporation enhances membrane's fluidity, while saturated fatty acids make the membrane more rigid). By this way they can influence the function and activity of membrane-associated receptors and enzymes [1–4]. As paradigmatic example, the increase of membrane's fluidity has been associated with improved insulin sensitivity due to both augmented number and higher affinity of insulin receptors on cellular surface [5]. PUFA are also well recognized to be a major fuel source for energy metabolism and are beta-oxidized within mitochondria by most of cells in the body [6].

PUFA can influence cellular function by regulating several metabolic pathways, acting as both direct and second messengers. Long-chain PUFA may act directly as ligands for transcription factors like sterol regulatory element binding protein 1 (SREBP-1), nuclear factor κB (NF-κB), hepatocyte nuclear factor 4α (HNF-4α), and peroxisome proliferator-activated receptors (PPARs), which are involved in lipogenesis, steroid hormones synthesis, and FA oxidation [7]. For instance, ω-3 PUFA are known to influence several nuclear receptors modulating lipid metabolism (e.g. PPARs, HNF-4α, liver X receptor (LXR), and farnesol X receptor (FXR)) with hypotriglyceridemic effects [8]. As second messengers, PUFA regulate the synthesis of several inflammatory mediators, like eicosanoids that represent the key link between PUFA and inflammation. Eicosanoids are lipid mediators of inflammation and include a variety of compounds (prostaglandins, thromboxanes, leukotrienes, lipoxins, isoprostanes, hydroxyl and epoxy fatty

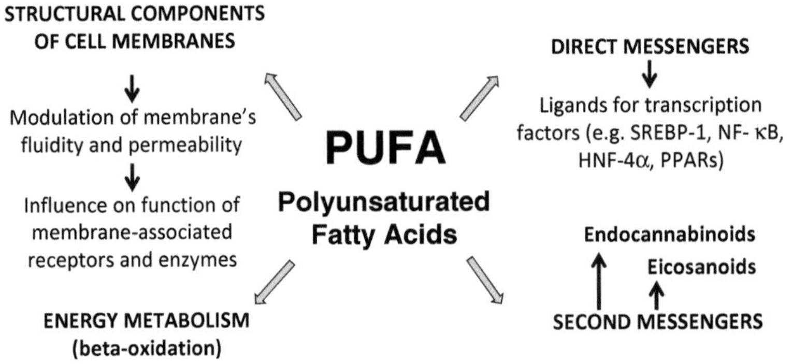

Fig. 7.1 The heterogeneous functions and pleiotropic effects of polyunsaturated fatty acids (PUFA)

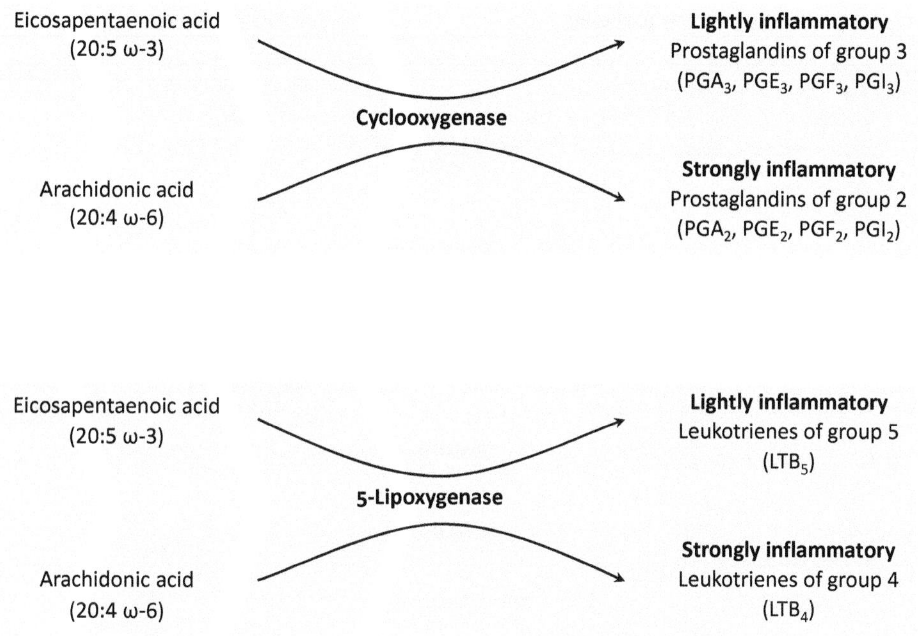

Fig. 7.2 Eicosanoids derived from ω-3 or ω-6 polyunsaturated fatty acids present different proinflammatory activities

acid). In particular arachidonic acid (AA, 20:4 ω-6) is the most important substrate for the synthesis of the strongest pro-inflammatory eicosanoids [9]. AA-derived metabolites have been demonstrated to play crucial roles in chronic inflammation, as well as in cardiovascular disease (CVD) and cancer [10–13]. Noteworthy, AA is also the substrate for the synthesis of anandamide and other endocannabinoids, which are neuromodulatory mediators with a very broad range of biological effects involved in several pathophysiological conditions, from neurodegenerative to inflammatory disorders, metabolic derangements, CVD, cancer, and cachexia [14–16]. A recent study showed that the activation of Nlrp3 inflammasome by endocannabinoids promotes beta cell failure in type 2 diabetes [17].

From a chemical point view, fatty acids are carboxylic acids with a long aliphatic chains, which may be either saturated or unsaturated. Long-chain fatty acids having 16–20 carbon units are the most abundant cellular fatty acids, while very long-chain fatty acids having more than 20 carbon units are much less abundant. PUFA are chemically characterized by the presence of two or more double bonds in their hydrocarbon chain. They are classified according to (i) the number of carbon atoms, (ii) the number of double bonds and (iii) the position of the double bond nearest to the terminal methyl group. Four families of PUFA have been identified on the basis of these structural characteristic, that are ω-3, ω-6, ω-7, and ω-9 PUFA. However, ω-3 and ω-6 PUFA are universally recognized as the most important series [3, 4]. From this point of view it should be emphasized that both ω-3 and ω-6 long-chain PUFA are particularly important in humans in order to maintain the function of brain and central nervous system [18, 19]. On the other hand, ω-3 and ω-6 PUFA appear to differ profoundly for some of their biological and clinical consequences. It is worth noting that ω-6 PUFA-derived eicosanoids (e.g. prostaglandin E_2, thromboxane A_2, leukotriene B_4) have strong pro-inflammatory effects, while those derived from ω-3 PUFA (e.g. prostaglandin E_3, thromboxane A_3, leukotriene B_5) may have anti-inflammatory action (Fig. 7.2) [20, 21]. The different properties of ω-3 and ω-6 PUFA are particularly evident in cardiovascular field, where ω-3 PUFA – in

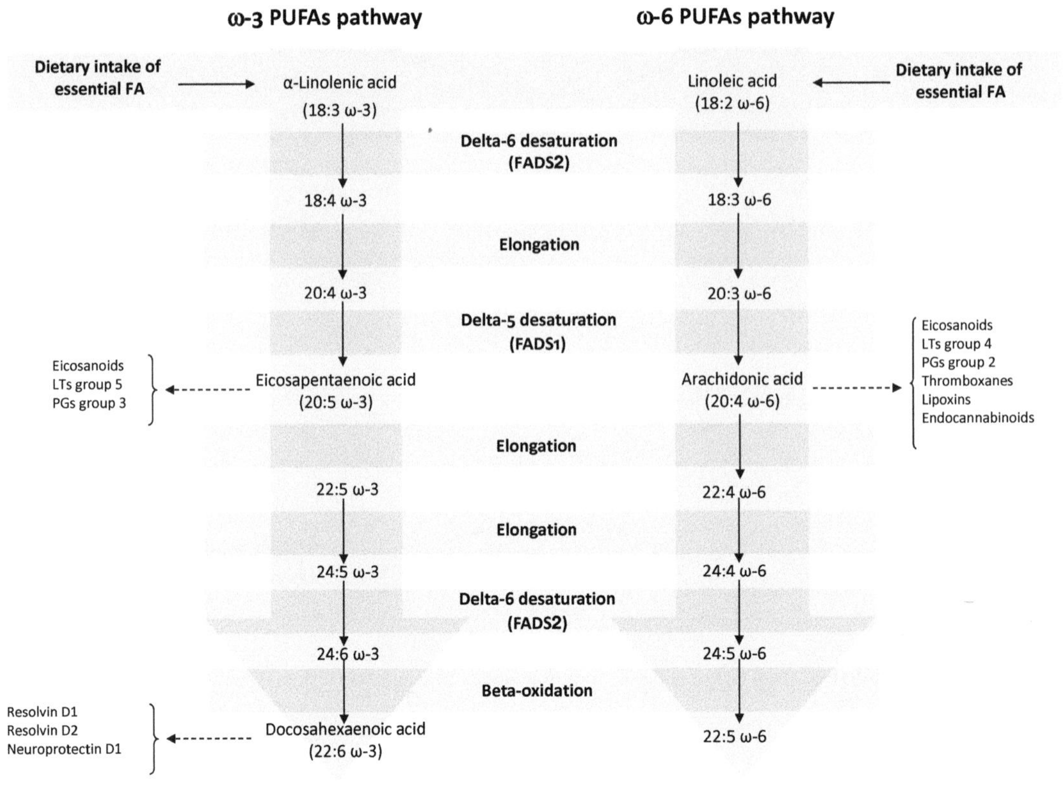

Fig. 7.3 The ω-3 and ω-6 fatty acid metabolism pathways

particular the long chain eicosapentenoic (EPA, 20:5 ω-3) and docosahexaenoic acid (DHA, 22:6 ω-3) – are usually considered as good friends for cardiovascular health. The favorable effects of ω-3 PUFA are complex and pleiotropic, including preventing cardiac arrhythmias, lowering plasma triglycerides, decreasing blood pressure, and reducing platelet aggregability [22, 23]. The pioneering GISSI-Prevention trial showed that 1 g daily consumption of ω-3 PUFA reduced cardiovascular mortality after myocardial infarction, mainly by preventing sudden death [24]. Already in 2002, the American Heart Association (AHA) recommended ω-3 PUFA for patients with ischemic heart disease and subjects with hypertriglyceridemia [25]. In 2004, the United States Food and Drug Administration (FDA) also approved ω-3 PUFA for hypertriglyceridemia therapy [23, 26]. In contrast, ω-6 PUFA with their proinflammatory effects appear as a potentially dangerous bad company for heart and vessel. Consistently, high levels of AA in adipose tissue have been associated with an increased risk of myocardial infarction [27, 28].

The PUFA profile in human body reflects both dietary intake and endogenous metabolism. The precursors of both ω-3 and ω-6 PUFA (α-linolenic (ALA, 18:3 ω-3) and linoleic acid (LA, 18:2 ω-6), respectively) are essential fatty acids and therefore cannot be synthesized in mammals. Dietary sources for ALA are canola, flaxseed and rapeseed oils, walnuts and leafy green vegetables, while LA is contained in eggs, poultry, cereals, margarine, sunflower and corn oils [29]. In the most of Western diets, ALA and in particular LA contribute more than 95 % of dietary PUFA intake, that in turn represents up to 20 % of dietary fat [30]. Long chain-PUFA are then synthesized endogenously from ALA and LA by reactions of both insertion of additional double bonds, catalyzed by desaturases, and elongation of the acyl chain, catalyzed by elongases (Fig. 7.3). The conversion of LA and ALA into longer PUFA involves only a relatively small proportion of the essential fatty acids introduced with the diet, since the majority undergoes beta-oxidation reactions to accomplish the needs of energy metabolism [31]. Noteworthy, both ω-3 and ω-6 PUFA compete for the same set of enzymes in this pathway, although a preferential affinity for ω-3 rather than ω-6 PUFA has been demonstrated [3].

Elongation involves the addition of two carbon units to aliphatic chain and this process takes place mainly in the endoplasmic reticulum catalyzed by membrane-bound enzymes [32]. Seven isozymes of elongation of very long chain fatty acid proteins (ELOVL 1–7) have been identified so far, each characterized by specific substrate specificity and function [32, 33]. ELOVL1-3-6-7 show a selective preference for saturated and monounsaturated fatty acids, whereas ELOVL2-4-5 have that for PUFA. ELOVL1-5-6 genes are ubiquitously expressed, while ELOVL2-3-4-7 genes are characterized by specific tissue expression (although the physiological role of such differentiated distribution remains still unknown) [34].

Desaturases are microsomal enzymes. They are thought to be a component of a three-enzyme system which includes also cytochrome b5 and NADH-cytochrome b5 reductase [35, 36]. The delta-5 (D5D) and delta-6 desaturases (D6D) are key enzymes in both ω-3 and ω-6 PUFA metabolism, allowing the formation of long chain metabolites from dietary ALA and LA. Noteworthy, the D6D-mediated conversion of either ALA to stearidonic acid (SA, 18:4 ω-3) or LA to γ-linolenic acid (GLA, 18:3 ω-6) is the rate limiting step of both ω-3 and ω-6 PUFA metabolic pathways.

D5D and D6D are encoded by FADS1 and FADS2 genes, respectively. The FADS gene cluster is located on chromosome 11 (11q12-13.1) and includes a third gene, i.e. FADS3, that shares 52–62 % sequence identity with FADS1/FADS2 genes and encodes for a yet unidentified protein [37]. Both blood and tissue levels of PUFA are influenced to a large extent by genetic heritability. Up to 28 % of variation of blood levels AA is due to genetic variation, while such value is about 10 % for AA precursors [38]. Preliminary studies in the German and Italian populations showed very strong associations between FADS1-FADS2 polymorphisms and PUFA levels in both serum phospholipids and red blood cell membranes [38, 39]. Minor alleles of single nucleotide polymorphisms (SNPs) in FADS1/FADS2 genes were usually associated with higher LA and lower AA levels, suggesting a corresponding impairment of desaturases activity. Subsequent analyses replicated such associations in several different populations of European, Asian, and African descent [40–45]. Consistently with these results, genome-wide association studies (GWAS) confirmed FADS locus as the strongest genetic predictor of plasma phospholipid PUFA [46]. If the relationship between FADS genetic variants and PUFA levels is unquestionable, it should be noted that the mechanisms by which FADS polymorphisms may influence desaturase activity and PUFA concentration remain largely unknown. Only few functional studies have been performed on this topic so far. A recent study showed an influence of polymorphism rs968567 on FADS2 promoter activity by luciferase reporter gene assays [47]. On the other hand, in another study the polymorphism rs3834458 did not appear to directly affect FADS2 promoter activity [48].

PUFA status has been related with several outcomes in human health, from neuronal development to psychiatric illnesses [49, 50], from inflammatory to immunologic response [51], from metabolic disorders to CVD [13, 23]. Both FADS polymorphisms and desaturase activities have been accordingly associated with the same diseases during the last decades [13, 52]. Hence, the interest on FADS genes and desaturases exponentially increased, as testified by the progressively higher number of publications on these issues (Fig. 7.4). In this narrative review we try to summarize the main evidences linking FADS genes, D5D, and D6D with human diseases, addressing also the potential therapeutic applications.

7.2 Delta-5 and Delta-6 Desaturase Activities in Human Diseases

Desaturase activity is assessed in vitro or in animal studies by measuring the rate of the conversion of radiolabeled fatty acids to their respective

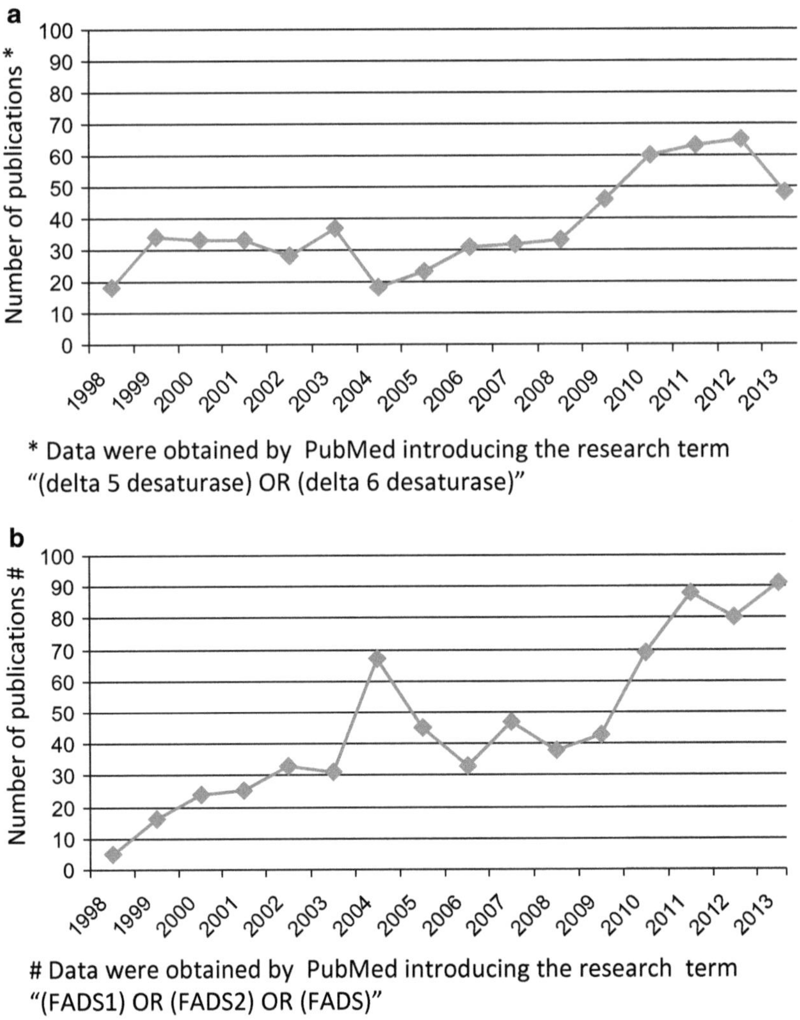

Fig. 7.4 Publications on delta-5/delta-6 desaturase (**a**) and FADS1/FADS2 (**b**) over 15 years

products [53], but ethical and practical reasons preclude this possibility in humans. On the other hand, FADS gene expression could be determined in liver cells, but this approach would require obtaining liver biopsies. Therefore, the direct measurement of desaturase activity is not feasible for the use in large scale epidemiological studies [54]. Nonetheless, the use of product to precursor ratio as a surrogate measure to estimate desaturase activity is well established. Indirect information can be acquired by the analysis of lipid composition of either human plasma or cell membranes (usually from the easily collectable red blood cells) [13, 53]. The product to precursor ratios are commonly performed on ω-6 fatty acids (owing to the ω-3 final products, like EPA and DHA, may be more strongly influenced by dietary intake) and can estimate the global desaturase activity (with AA/LA ratio), as well as the specific D5D (with AA/DGLA ratio) or D6D activity (with GLA/LA). The product to precursor ratio approach has been used in many clinical studies linking desaturase activities and diseases in humans (Fig. 7.5) [13].

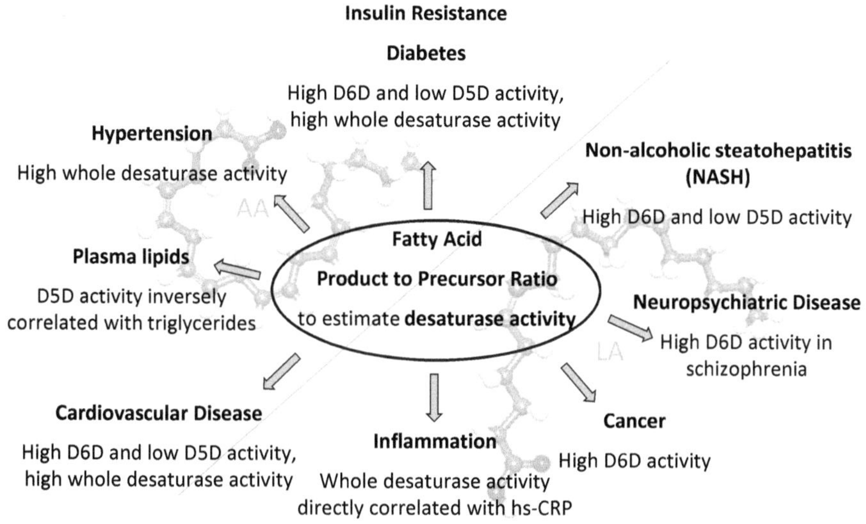

Fig. 7.5 Associations of delta-5 (D5D)/delta-6 desaturase (D6D) activity with clinical phenotypes and human diseases. D5D and D6D activities were generally estimated by fatty acid product to precursor ratio. "Whole desaturase activity" refers to arachidonic to linoleic acid ratio, thus including the metabolic passages of both D5D and D6D

Several studies have associated desaturase activity with dysmetabolic phenotypes and cardiovascular risk factors. An increased whole desaturase activity was found in patients with essential hypertension [55], as well as in subjects with insulin resistance, obesity, and metabolic syndrome [56, 57].

The relationship with type 2 diabetes is particularly impressive. Several studies have investigated the potential link between desaturase activity and alteration in glucose metabolism (for a review, see reference n. [54]). Most of these studies showed independent and significant correlations of estimated desaturase activities with diabetes risk, inverse for D5D and direct for D6D.

High D6D activity and low D5D activity were shown to be associated with insulin resistance [58] and, most importantly, to predict the development of diabetes [59–62]. Moreover, an increased D6D and a decreased D5D activity have been found in Japanese children with abdominal obesity [63] and both could predict the long-term development of the metabolic syndrome, which is a cluster of metabolic abnormalities with a physiopathological core represented by insulin resistance [64]. All these results prompt to hypothesize that desaturase activity plays a role in the pathogenesis of diabetes and may influence the individual susceptibility to diabetes.

Non-alcoholic steatohepatitis (NASH) is a disease characterized by inflammation and fat accumulation in the liver. It is usually associated with metabolic derangements, like obesity, plasma lipids abnormalities, insulin resistance, and type 2 diabetes. NASH is increasingly recognized as a common liver condition, may progress to cirrhosis, and potentially would become the leading cause of liver transplantation worldwide [65]. Increased D6D and decreased D5D estimated activities have been found in patients with non-alcoholic fatty liver disease as compared with normal subjects [66]. Consistently, mice with high-fat diet-induced obesity and NASH showed an increased expression of D5D and D6D at both mRNA and protein level in the cells of the liver. Noteworthy, a combined D5D/D6D inhibitor, CP-24879, significantly reduced intracellular lipid accumulation and inflammatory injury in hepatocytes [67]. Moreover, CP-24879 exhibited superior antisteatotic and anti-inflammatory actions in hepatocytes from fat-1 mice, that express an ω-3 desaturase allowing the endogenous conversion of ω-6 into ω-3 fatty acids and restoring hepatic ω-3 content [68].

A relationship between desaturases activities and CVD has been also proposed. In a community-based prospective population of 50-year old men D6D activity showed a direct association with cardiovascular mortality (HR 1.12 with 95% CI 1.00–1.24), while D5D had an inverse correlation (HR 0.88 with 95% CI 0.80–0.98) [69]. Accordingly, high D6D activity was associated with most of cardiovascular and metabolic risk factors, while triglycerides and fasting insulin were beneficially related to D5D activity [70]. In a previous work of our group within the angiographically-controlled Verona Heart Study, using the AA/LA ratio as a surrogate measure of "overall" desaturase activity, we found that this ratio was higher in patients with coronary artery disease (CAD). The proportion of CAD increased progressively from the lowest to the highest AA/LA tertile and after adjustment by multiple logistic regression AA/LA remained associated with CAD independently from all the traditional cardiovascular risk factors (the highest *versus* the lowest tertile OR 2.55 with 95 % CI 1.61–4.05). Moreover, the serum concentration of C-reactive protein measured by high sensitivity methods (hs-CRP) also increased progressively across AA/LA tertile [71]. Thus, it appeared reasonable to hypothesize that inappropriately high desaturase activity may indicate a peculiar susceptibility to the inflammatory stimuli involving the arterial wall during the atherosclerotic process. Actually, an increased desaturase activity – especially in an ω-6 rich environment, like that generated by the current Western diet – may lead to a greater AA bioavailability, thus favoring the synthesis of AA-derived mediators and vascular inflammatory damage involved in the pathogenesis of atherosclerosis [13, 71]. There is a great interest in the role of AA-derived eicosanoids in atherosclerosis [9]. The leukotriene pathway has been associated with CVD in both mice and humans [9, 72, 73]. Consistently with such point of view, accumulation of AA in adipose tissue has been associated with a greater risk of myocardial infarction (MI), suggesting a proatherosclerotic/prothrombotic role of AA excess [27, 28], although the results were sometimes object of controversy [74]. Remarkably, a recent study in Chinese Han population confirmed that AA/LA ratio level was higher in CAD patients [75].

Our result suggesting that high "overall" desaturase activity may be harmful for cardiovascular health [71] is consistent with those about D6D activity [69, 70]. Taking into account that D6D is the rate-limiting step of the whole PUFA pathway, this concordance appears as biologically plausible. On the other hand, D5D is the key enzyme in synthesizing long-chain PUFA and increased D5D activity has been associated with high plasma levels of EPA and DHA [76]. Accordingly, D5D activity has been inversely associated with triglycerides, fasting insulin [70], cardiovascular mortality [69], and incident risk of ischemic heart disease [76].

Inflammation and AA-derived mediators play crucial roles also in cancer biology. Increased AA metabolism and the related eicosanoid formation are characteristic in various types of cancer cells [77, 78]. AA-derived pro-inflammatory and pro-angiogenic eicosanoids, which are produced by tumor cells and their surrounding stromal cells, are key mediators in their crosstalk and can accelerate tumor growth and metastasis development through several mechanisms [10]. Selective inhibition of D6D with SC-26196 (a highly selective inhibitor of D6D) has been shown to inhibit tumorigenesis in two mice models of intestinal cancer. Noteworthy, this effect on tumorigenesis was abrogated by concomitant treatment with dietary AA, suggesting that such influence was due to the interference with AA-related pathways [79]. More recently, the group of Kang showed that D6D activity was upregulated during melanoma and lung tumor growth in mice and AA/LA ratio was positively correlated with tumor size. Most importantly, the suppression of D6D activity, either by knocking down D6D expression with RNAi or by inhibiting D6D enzyme activity with SC-26196, reduced tumor growth. Accordingly, the content of AA and AA-derived eicosanoids was significantly decreased in tumor tissues. Both D6D-RNAi and SC-26196 did not show any significant tumor growth suppression in vitro, suggesting that blocking D6D is not directly toxic for cancer cells. The authors rather proposed that the anti-tumor effect of D6D inhibition could be due to its impact on the tumor microenvironment, modulating inflammation and angiogenesis [80].

The composition of phospholipid fatty acids of separated mononuclear blood cells was altered and an increased D6D activity has been found in patients with acute lymphoblastic leukemia, but not in acute myeloid leukemia [81]. Moreover, in vitro studies showed that the inhibition of D6D could reduce the growth of different leukemia cell types [82]. A study evaluating PUFA composition of subcellular fractions from healthy and cancerous kidney tissues, revealed lower LA content and D5D activity, and higher AA, ALA, EPA contents and D6D activity in renal cell carcinoma than in healthy renal tissue [83]. An increased D6D activity was found in human breast cancer tissue [84]. On the other hand, a previous case–control study evaluating fatty acid composition of erythrocyte membrane could not clearly support that desaturase activity influence the development of breast cancer [58].

Long-chain PUFA are fundamental constituents of human central nervous system. The availability of both ω-3 and ω-6 PUFA influences brain development and growth, particularly during the perinatal period and in early life. Alterations in PUFA composition have been associated with different neurological and neuropsychiatric illnesses [85, 86]. Therefore, also desaturase activities have been hypothesized to influence neurological and psychiatric functions in health and disease. Post-mortem studies on prefrontal cortex of patients with schizophrenia showed a greater D6D activity index. Moreover, FADS2 mRNA expression was higher in patients with schizophrenia than in controls and such difference was independent of antipsychotic medications [87, 88]. FADS2 mRNA expression was significantly elevated also in the prefrontal cortex of patients with bipolar disorder [89]. On the other hand, D6D and D5D activities were not different between patients with recurrent major depressive disorders and age- and sex-matched healthy controls [90].

Finally, a reduction of D5D activity has been estimated from analysis of whole blood fatty acids in patients with cystic fibrosis, especially in the case of severe disease [91]. In contrast, further studies showed that cystic fibrosis cells in culture had an increased expression of D5D and D6D enzymes [92, 93].

7.3 FADS1/FADS2 Gene Variants in Human Diseases: From the Earlier Reports to Genome-Wide Association Studies (GWAS)

FADS1 and FADS2 genes encode D5D and D6D, respectively. They are located on chromosome 11 (11q12-13.1), in reverse orientation, and separated by <11 kb region. Cloning of both D5D and D6D was performed in 1999 [35, 36]. A third gene, named FADS3, is included in this cluster, but no functional role has been yet attributed to the FADS3 putative transcriptional product [94]. A recent study showed that FADS3 does exist under multiple protein isoforms depending on the mammalian tissues [95]. Since all the three FADS genes share a common location and had a similar structure, it has been hypothesized that they arose evolutionary from gene duplication, acquiring substrate specificity during the evolution [37, 52].

Earlier studies on FADS1-FADS2 polymorphisms disclosed a very strong associations with PUFA levels in both serum phospholipids and red blood cell membranes [38, 39]. As previously mentioned, minor alleles of single nucleotide polymorphisms (SNPs) in FADS1/FADS2 genes were usually associated with higher precursor and lower product levels, suggesting an impairment of desaturases activity. Remarkably, GWAS confirmed FADS locus as the strongest genetic predictor of plasma phospholipid PUFA [46, 96].

Taking into account the pleiotropic effects of PUFA, it appears biologically plausible that FADS gene variants associated with different desaturase activity can have an impact on several pathological conditions. Therefore, although the mechanisms by which FADS polymorphisms influence desaturase activity and PUFA concentration remain largely unknown, studies analyzing

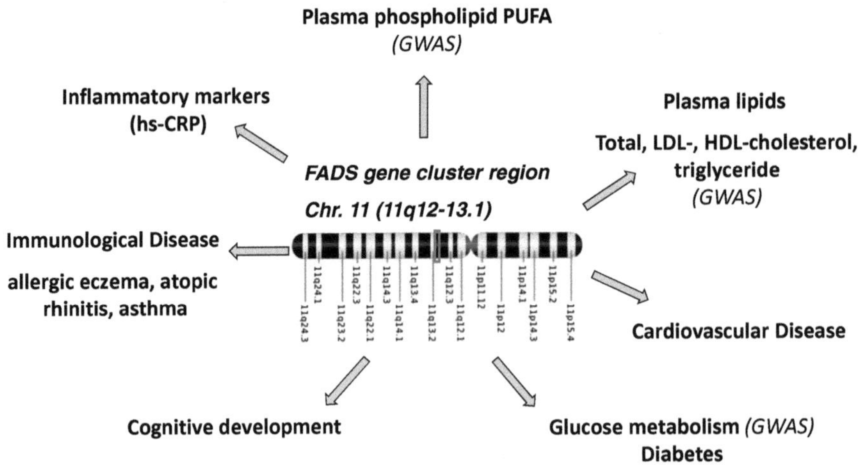

Fig. 7.6 Associations of FADS genotypes with clinical phenotypes and human diseases. Those relations which have been confirmed by genome-wide studies are accordingly indicated (GWAS)

the possible relationship between FADS genotype and human diseases flourished (Fig. 7.6).

The pioneering study of Schaeffer et al. suggested that FADS genotypes associated with low desaturase activity (and thus low AA bioavailability) may have a lower risk of immunological disease, like allergic eczema and atopic rhinitis [38]. Noteworthy, about 20 years ago the results of genome-wide search for linkage to traits associated with allergic asthma identified one linkage on chromosome 11q13, i.e. where FADS gene cluster is located [97]. More recent studies suggested that the association between dietary intake of PUFA and allergic diseases, as well as the association between breastfeeding and asthma, might be modulated by FADS gene variants in children [98, 99]. Furthermore, a recent study found that FADS polymorphisms may influence hs-CRP plasma concentration in young adults [100].

According to the concept of the potentially harmful effects of an excess of AA-derived pro-inflammatory mediators, we showed within the Verona Heart Study population that FADS polygenic models (including rs174545, rs174570, rs17583, and rs1000778 SNPs) and FADS haplotypes associated with high AA/LA ratio had increased hs-CRP plasma concentration and were more represented in CAD patients than in CAD-free subjects [71]. Noteworthy, we emphasized the deleterious effects of FADS genotypes associated with high desaturase activity could be particularly evident in an ω-6 rich diet, like it was that of Verona Heart Study. Previously, another study investigated the possible relationship between FADS genes and CVD. In the Costa Rican population a common deletion variant in the FADS2 promoter (rs3834458) was associated with a decrease in serum AA and EPA, as well as triglyceride, but not with MI. Interestingly, the authors of this work suggested that the results may have been masked by the high availability of the ω-3 ALA in the diet of Costa Rican population [101]. It should be noted that GWAS did not found so far any significant association between FADS gene locus and CAD/MI [102], speculatively because of the diet variability in the different populations included.

In contrast, GWAS approach identified FADS locus as an important genetic determinant of plasma concentration of all plasma lipids, from triglyceride to HDL- and LDL-cholesterol [103–105], even in populations with different ethnic background [106]. FADS1 variant associated with high transcription level were linked with both high production of long chain ω-3 PUFA and favorable effects on plasma lipids, like lowering triglyceride and increasing HDL-cholesterol plasma concentration [103]. Noteworthy, a very recent study examining 188,577 individuals using genome-wide and custom genotyping arrays identified FADS locus as one of the four (the others were CETP, TRIB1, and APOA1) which were associated with all lipid traits, i.e. total, HDL-, and LDL-cholesterol, and triglyceride [107]. Overall, these data suggest a direct genetic influence on lipid parameters, but not so strong to affect secondarily also the clinical (cardiovascular) outcomes.

A relationship between FADS gene locus and glucose metabolism has been also disclosed. In a GWAS involving more than 120,000 subjects FADS1 rs174550 polymorphism was associated with fasting glucose and had a weak relation with diabetes risk [108]. In another study FADS1 rs174550 was associated with abnormalities in early insulin secretion [109]. In the EPIC-Potsdam Study FADS1 rs174546, which is in complete linkage with the above mentioned rs174550, was associated with both D5D and D6D activities (with minor allele carriers having both lower D5D and D6D activities) [40]. As previously cited in the paragraph on D5D and D6D activities in human diseases, most of studies investigating the association between desaturase activity and diabetes disclosed opposite trends of correlation for either D5D or D6D activity. More precisely, there was an inverse correlation of D5D and a direct correlation of D6D with diabetes risk [59–62]. Analysis of FADS polymorphisms by adjusting for the estimated D5D and D6D activities confirmed contrasting influences on diabetes risk [62]. Consequently, it was suggested that a reciprocal counterbalance (e.g. FADS genotypes associated with both high D5D and high D6D activity, or vice versa) might result in an overall weak association of FADS genotypes with diabetes [54, 62]. This concept may provide a possible explanation for some controversial results about the relationship between FADS gene locus and glucose metabolism traits [109–114].

The key role of desaturases in human metabolism has been further emphasized by a comprehensive analysis of phenotypes using a GWAS with non-targeted metabolomics. In such analysis FADS gene locus presented one of the strongest association with metabolic traits ($P=8.5\times10^{-116}$ for FADS1 rs174547 association with 1-arachidonoyl-glycerophosphoethanolamine/1-linoleoyl-glycerophosphoethanolamine ratio) [115].

Long-chain PUFA, like AA and DHA, are crucial to enhance cognitive development. Breastfeeding exposes babies to increased concentration of AA and DHA and has been found to improve significantly the cognitive development with higher later intelligence quotient (IQ) [116]. Remarkably, the association between breastfeeding and IQ was moderated by FADS2 rs174575 polymorphism in two birth cohorts. More precisely, there was no effect of breastfeeding on IQ in GG homozygotes, while breastfed children carrying the C allele had a striking IQ advantage over those who were no breastfed [116]. In contrast with these data, in a subsequent study rs174575 GG homozygotes exhibited the greatest difference between feeding methods with no breastfed GG children performing worse than other no breastfed children [117]. It is worthy to note that FADS1 and FADS2 genetic variants have been shown to influence the PUFA composition of breast milk in pregnancy and lactation [118]. Subsequent studies indicated that FADS1/FADS2 gene locus controls brain expression of FADS1 and that its genetic variance in combination with breastfeeding and/or food intake might alter PUFA composition in the brain, thereby potentially influencing cognition [119]. Another recent study confirmed the favorable role of maternal AA and DHA on fetal neural development and suggested that the endogenous synthesis of long-chain PUFA regulated by FADS genes, in particular FADS2, may also be important [120]. Therefore, although some results are controversial and no definitive proof has been provided, FADS1/FADS2 and D5D/D6D may have a role in the complex and still shadowy field of cognitive development.

7.4 Lessons from Animal Models

Mice models have provided precious information about the biological role of desaturases and their various effects in different pathological conditions

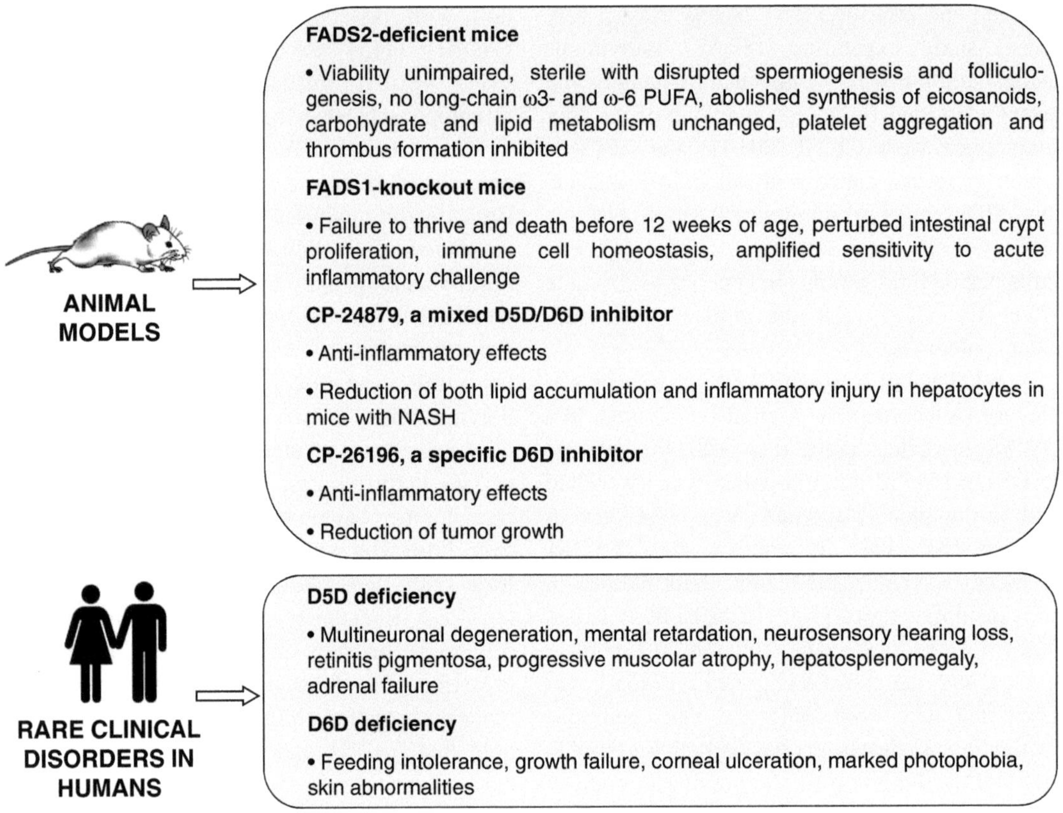

Fig. 7.7 Lessons on desaturases from animal models and rare clinical disorders in humans

(Fig. 7.7, top panel). Stoffel and colleagues investigated a model of genetically defined FADS2-deficient (fads2$^{-/-}$) mice [121]. FADS2 deficiency was confirmed to prevent the processing of LA and ALA to long-chain ω3- and ω6-PUFA. Consistently, eicosanoids synthesis was abolished and macrophages failed to synthesize leukotrienes in immune response. Surprisingly, the viability of fads2$^{-/-}$ mice was unimpaired. Key parameters of carbohydrate and lipid metabolism remained unchanged in spite of the lack of long-chain PUFA. Male and female mice were sterile with disrupted spermiogenesis and folliculogenesis, respectively. Platelet aggregation and thrombus formation were inhibited. Thus, bleeding time was increased, while fads2$^{-/-}$ mice were protected from vessel thrombosis inducted by vascular injury [121].

Fan and colleagues investigated a model of FADS1 knockout mice [122]. As expected, the levels of DGLA and AA were reciprocally altered in all the tissues. The lack of AA-derived eicosanoids was associated with perturbed intestinal crypt proliferation, immune cell homeostasis, and an amplified sensitivity to acute inflammatory challenge. Mice failed to thrive and died before 12 weeks of age. Noteworthy, dietary supplementation with AA extended the longevity of FADS1 knockout mice to levels comparable to wild-type mice [122].

Pharmacological interventions on mice furnished intriguing data. Studies in mice treated with CP-24879, a mixed D5D/D6D inhibitor, suggested that desaturase inhibition may have anti-inflammatory properties by decreasing the levels of AA and, consequently, of AA-derived eicosanoids [123]. The specific D6D inhibitor SC-26196 had marked anti-inflammatory properties and decreased edema to the same extent as indomethacin did in mice [123]. All these results supported the hypothesis of a significant role of desaturases in inflammation and suggested desaturases as a target for the development of new anti-inflammatory drugs. Moreover, the mixed inhibitor CP-24879 was shown to reduce lipid accumulation and inflammatory injury in hepatocytes of mice with NASH induced by high-fat diet [67].

Finally, the selective inhibition of D6D with SC-26196 has been shown to block tumorigenesis and/or reduce tumor growth in different mice models of neoplasia, such intestinal cancer, lung cancer, and melanoma (see also the previous paragraph "Delta-5 and delta-6 desaturase activities in human diseases") [79, 80]. Thus, D6D inhibition could be a potential target as well for the therapy and the prevention of cancer.

7.5 Lessons from Rare Clinical Disorders in Humans

The thorough observation of rare, genetic and metabolic disorders has often represented a useful source of clinical and scientific information in the history of medicine. Very low levels of D6D activity have been described in patients with the Sjögren-Larsson syndrome, a genetic disease which is caused by mutations in ALDH3A2 gene codifying for fatty aldehyde dehydrogenase and is characterized by ichthyosis, spastic diplegia, and mental retardation [124, 125]. About 10 years ago, a nucleotide insertion in the transcriptional regulatory region of the human FADS2 gene (i.e. an insertion of thymidine between positions −942 and −941 upstream of the translation start site) was described [126]. This mutation caused a sixfold decrease in promoter activity and thus D6D deficiency with severe impairment of LA to AA conversion. The subject carrying this mutation was clinically characterized by feeding intolerance, growth failure, corneal ulcerations, marked photophobia, and skin abnormalities. Noteworthy, all these clinical signs and symptoms significantly improved after diet supplementation with a mixture of LA and DHA [126]. D5D deficiency has been suggested in two brothers with multineuronal degeneration, mental retardation, neurosensory hearing loss, retinitis pigmentosa, progressive muscular atrophy, hepatosplenomegaly, and adrenal failure (Fig. 7.7, bottom panel) [127].

7.6 A "Desaturase Hypothesis" for Atherosclerosis

The influence of desaturases on biological systems is extremely complex and pleiotropic, so that every oversimplification about clinical consequences of such activities (e.g. high desaturase activity is dangerous, while low desaturase activity is advantageous, or vice versa) looks as inadequate. Most importantly, the biological effects of desaturase activity could be modulated by the different balance between ω-3 and ω-6 PUFA. The case of CVD may represent a paradigmatic example from this point of view. A high desaturase activity in a ω-3 PUFA-rich environment would increase the cardioprotective EPA and DHA. On the other hand, a high desaturase activity in a ω-6 PUFA-rich environment would increase the AA bioavailability, potentially favoring the synthesis of proinflammatory eicosanoids and the consequent inflammatory vascular damage. In the Verona Heart Study population, within a clinical context characterized by a larger prevalence of ω-6 PUFA, a higher desaturase activity estimated by means of AA/LA ratio was associated with both hs-CRP concentration and increased risk of CAD. Moreover, FADS haplotypes associated with higher AA/LA ratio had also higher hs-CRP concentration and were more represented within the CAD population [71].

Desaturases appear therefore as Janus-faced enzymes with both favorable and unfavorable effects according the relative balance/imbalance of PUFA [13]. In the current Western diet the ω-6/ω-3 PUFA ratio is estimated to be about 15/1, instead of 1/1 as in prehistoric or tribal human populations. Noteworthy, a balance between ω-6 and ω-3 PUFA is thought to have existed for millions of years during the evolution of mankind [128] and only recently the imbalance to ω-6 PUFA may have unraveled the proinflammatory and potentially harmful consequences of high

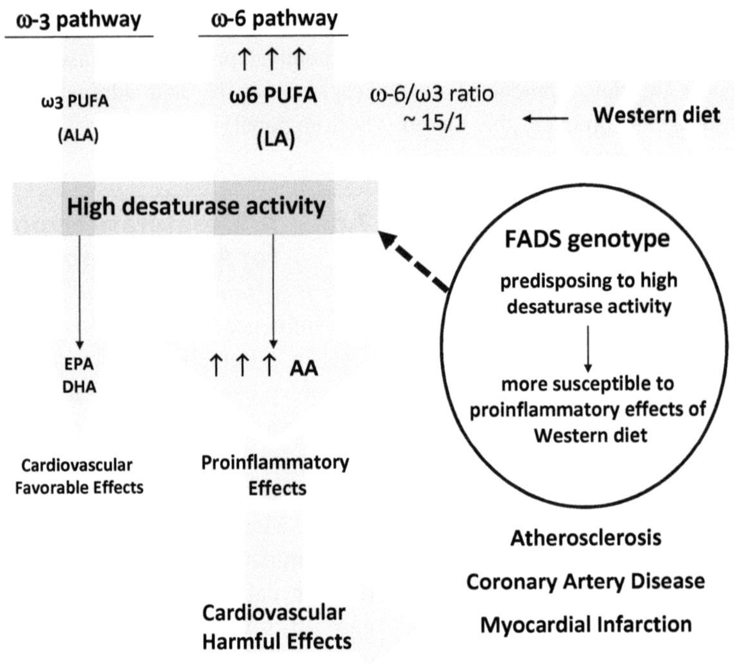

Fig. 7.8 A "desaturase hypothesis" for atherosclerosis. Desaturases could be considered as Janus-faced enzymes with both favorable and harmful effects for cardiovascular health. In subjects following a Western diet rich in ω-6 PUFA, a high desaturase activity would predispose to more pronounced vascular inflammatory damage. Accordingly, subjects with FADS genotype predisposing to high desaturase activity may be more susceptible to the proinflammatory effects of diets rich in ω-6 PUFA and deficient in ω-3 PUFA. *AA* arachidonic acid, *ALA* α-linolenic, *DHA* docosahexaenoic acid, *EPA* eicosapentaenoic acid, *LA* linoleic acid

desaturase activity [13]. Finally, the variants in FADS genes suggest that individuals with different genotype may require different amount of dietary PUFA to achieve comparable biological effects [129]. Accordingly with this hypothesis, subjects with FADS genotype predisposing to high desaturase activity may be more susceptible to the proinflammatory effects of diets rich in ω-6 PUFA and deficient in ω-3 PUFA (Fig. 7.8), but at the same time may have particular benefits from approaches reducing the ω-6/ω-3 PUFA ratio imbalance, like ω-3 PUFA supplementation [13].

7.7 An Evolutionary Hypothesis for Desaturases

The synthesis of long-chain PUFA is particularly important in humans. Actually, humans are extremely dependent on long-chain PUFA because of their large brain, which requires high amounts of long-chain PUFA, like DHA and AA, in order to maintain its function [18, 19]. A recent and fascinating study of Gyllensten's group performed genome-wide genotyping and targeted resequencing of FADS region in different human populations, as well as analyzed genomic data from archaic hominids and other primates [130]. Present-day humans were shown to have mainly two common haplotypes associated with different desaturase activity: the haplotype D (frequency 62.1 %) associated with a higher ability to generate long-chain PUFA and the haplotype A (frequency 33.0 %). Noteworthy, the results suggested that the haplotype D appeared after the split from Neanderthals – around 500,000 years ago – but prior to the exodus of modern humans from Africa – 50,000–100,000 years ago. Moreover, this haplotype showed evidence of positive selection, possibly by providing advantage

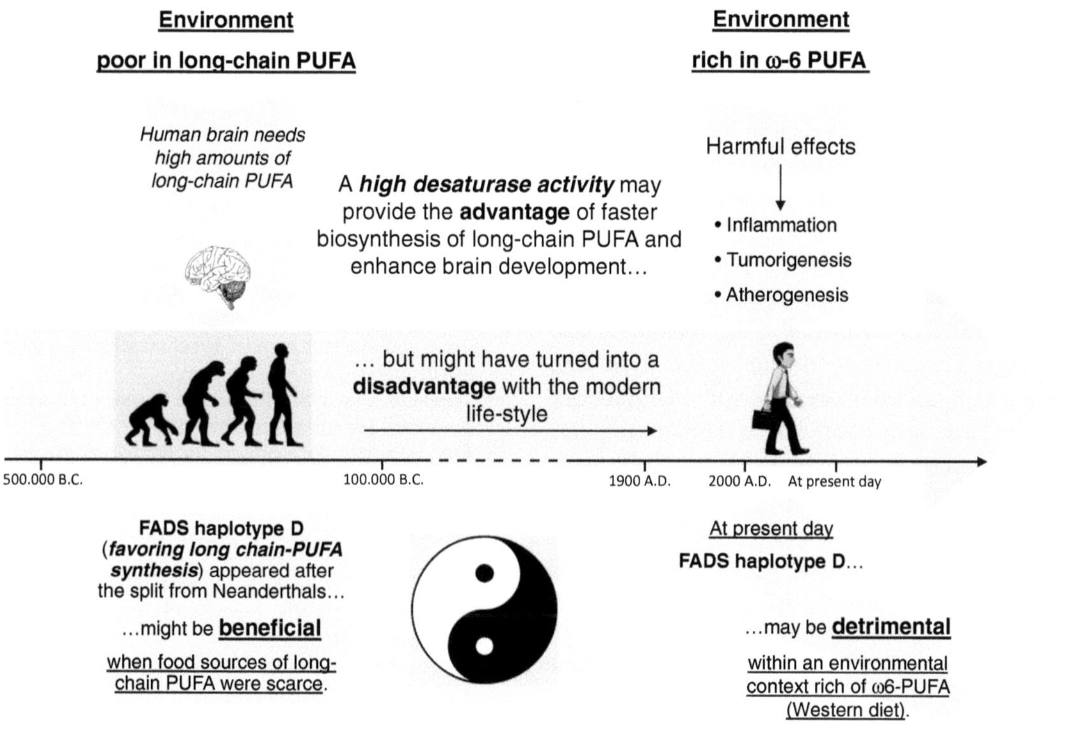

Fig. 7.9 An evolutionary hypothesis for desaturases. Humans are extremely dependent on long-chain PUFA because of their large brain, which requires high amounts of long-chain PUFA. FADS haplotype associated with a higher ability to generate long-chain PUFA (haplotype D) might have been beneficial when food sources of long-chain PUFA were scarce. On the other hand, within an environmental context characterized by high ω-6 PUFA intake the haplotype D may favor the synthesis of eicosanoids leading to inflammatory damage and its harmful clinical consequences. Therefore, the advantage of having a faster biosynthesis of long-chain PUFA might have turned into a disadvantage with the modern Western diet, like in a sort of evolutionistic Yin-Yang

in environments with limited availability of dietary long-chain PUFA. The authors hypothesized that the advantage of having a faster biosynthesis of long-chain PUFA might have turned into a disadvantage with the modern Western diet, like in a sort of evolutionistic Yin-Yang. Actually, haplotype D might have been beneficial when food sources of long-chain PUFA were scarce. On the other hand, within an environmental context characterized by high ω-6 PUFA intake the haplotype D might favor the synthesis of AA-derived eicosanoids leading to inflammatory damage and its harmful clinical consequences, including CVD (Fig. 7.9) [130]. This idea seems to fit perfectly with the above mentioned "desaturase hypothesis" for atherosclerosis [13] according to an evolutionistic and holistic point of view.

In summary, the more efficient ability to generate long-chain PUFA could have been selected as a favorable trait during the evolution of mankind. Only during the last century (the genetic selection pressure needs much more time than one century to be effective), the dramatic and rapid changes in environmental context and dietary intake have unraveled the potentially harmful effects of high desaturase activity. More precisely, the skill to generate long-chain PUFA (and the related FADS genotypes) may change into detrimental if ω-6 PUFA are largely prevalent, favoring eicosanoids-mediated inflammatory damage [11, 13].

7.8 Concluding Remarks

PUFA metabolism is crucial in humans. D5D and D6D are well recognized as main predictors of PUFA variability, thereby influencing several biological mechanisms [13, 30, 52]. Both alterations of D5D/D6D activities and FADS1/FADS2 polymorphisms have been associated with a lot of different diseases. However, the complex and heterogeneous effects of desaturase activity, as well as the multifaceted interactions with PUFA dietary intake, have made it difficult so far to moving this group of enzymes from the bench to the bedside.

Nonetheless, the modulation of D5D/D6D activity appears as a potentially promising target in different pathological conditions, as testified by both anti-inflammatory and tumor-suppressing effects of D6D inhibition in mice models [79, 80, 123]. On the other hand, the deleterious consequences which have been observed in case of severe lack of desaturase activity in both animal models [121, 122] and humans [124–127], advise against an indiscriminate inhibition of desaturases. Rather they address the need for search of tissue-specific modulators of D5D/D6D activity. To this aim, we require a better comprehension of the processes, like the epigenetic ones, differently regulating FADS1/FADS2 expression. Moreover, the specificity of either D5D or D6D activity should be taken into account (e.g. D6D and D5D activities have been differently associated with different diseases), as well as the environmental context in which the enzymes are working (e.g. ω-6-rich diets may enhance the pro-inflammatory effect of high desaturase activity). Further studies addressing all these issues are strongly warranted and may contribute to the development of both tailored dietary strategies for reducing the risk of illness and specific modulations of D5D/D6D activity for treating diseases associated with alterations in PUFA homeostasis.

References

1. Jump D. The biochemistry of n-3 polyunsaturated fatty acids. J Biol Chem. 2002;277:8755–8.
2. Leonard AE, Pereira SL, Sprecher H, Huang YS. Elongation of long-chain fatty acids. Prog Lipid Res. 2004;43:36–54.
3. Das UN. Essential fatty acids: biochemistry, physiology and pathology. Biotechnol J. 2006;1:420–39.
4. Das UN. Essential fatty acids – a review. Curr Pharm Biotechnol. 2006;7:467–82.
5. Risérus U. Fatty acids and insulin sensitivity. Curr Opin Clin Nutr Metab Care. 2008;11:100–5.
6. Zhang L, Keung W, Samokhvalov V, Wang W, Lopaschuk GD. Role of fatty acid uptake and fatty acid beta-oxidation in mediating insulin resistance in heart and skeletal muscle. Biochim Biophys Acta. 2010;1801:1–22.
7. Jump DB, Clarke SD. Regulation of gene expression by dietary fat. Annu Rev Nutr. 1999;19:63–90.
8. Davidson MH. Mechanisms for the hypotriglyceridemic effect of marine omega-3 fatty acids. Am J Cardiol. 2006;98:27i–33.
9. De Caterina R, Zampolli A. From asthma to atherosclerosis-5-lipoxygenase, leukotrienes, and inflammation. N Engl J Med. 2004;350:4–7.
10. Wang D, Dubois RN. Eicosanoids and cancer. Nat Rev Cancer. 2010;10:181–93.
11. Weylandt KH, Kang JX. Rethinking lipid mediators. Lancet. 2005;366:618–20.
12. Kang JX, Weylandt KH. Modulation of inflammatory cytokines by omega-3 fatty acids. Subcell Biochem. 2008;49:133–43.
13. Martinelli N, Consoli L, Olivieri O. A 'desaturase hypothesis' for atherosclerosis: Janus-faced enzymes in omega-6 and omega-3 polyunsaturated fatty acid metabolism. J Nutrigenet Nutrigenomics. 2009;2:129–39.
14. Pacher P, Kunos G. Modulating the endocannabinoid system in human health and disease-successes and failures. FEBS J. 2013;280:1918–43.
15. Silvestri C, Di Marzo V. The endocannabinoid system in energy homeostasis and the etiopathology of metabolic disorders. Cell Metab. 2013;17:475–90.
16. Pisanti S, Picardi P, D'Alessandro A, Laezza C, Bifulco M. The endocannabinoid signaling system in cancer. Trends Pharmacol Sci. 2013;34:273–82.
17. Jourdan T, Godlewski G, Cinar R, Bertola A, Szanda G, Liu J, et al. Activation of the Nlrp3 inflammasome in infiltrating macrophages by endocannabinoids mediates beta cell loss in type 2 diabetes. Nat Med. 2013;19:1132–40.
18. Darios F, Davletov B. Omega-3 and omega-6 fatty acids stimulate cell membrane expansion by acting on syntaxin 3. Nature. 2006;440:813–7.
19. Marszalek JR, Lodish HF. Docosahexaenoic acid, fatty acid-interacting proteins, and neuronal function: breastmilk and fish are good for you. Annu Rev Cell Dev Biol. 2005;21:633–57.
20. Thies F, Miles EA, Nebe-von-Caron G, Powell JR, Hurst TL, Newsholme EA, et al. Influence of dietary supplementation with long-chain n-3 or n-6 polyunsaturated fatty acids on blood inflammatory cell populations and functions and on plasma soluble

adhesion molecules in healthy adults. Lipids. 2001;36:1183–93.
21. Harris WS, Assaad B, Poston WC. Tissue omega-6/omega-3 fatty acid ratio and risk for coronary artery disease. Am J Cardiol. 2006;21(98):19i–26.
22. Breslow JL. n-3 fatty acids and cardiovascular disease. Am J Clin Nutr. 2006;83:1477S–82.
23. De Caterina R. n-3 fatty acids in cardiovascular disease. N Engl J Med. 2011;23(364):2439–50.
24. GISSI-Prevenzione Investigators (Gruppo Italiano per lo Studio della Sopravvivenza nell'Infarto miocardico). Dietary supplementation with n-3 polyunsaturated fatty acids and vitamin E after myocardial infarction: results of the GISSI-Prevenzione trial. Lancet. 1999;354:447–55.
25. Kris-Etherton PM, Harris WS, Appel LJ. Fish consumption, fish oil, omega-3 fatty acids, and cardiovascular disease. Circulation. 2002;106:2747–57.
26. McKenney JM, Sica D. Role of prescription omega-3 fatty acids in the treatment of hypertriglyceridemia. Pharmacotherapy. 2007;27:715–28.
27. Kark JD, Kaufmann NA, Binka F, Goldberger N, Berry EM. Adipose tissue n-6 fatty acids and acute myocardial infarction in a population consuming a diet high in polyunsaturated fatty acids. Am J Clin Nutr. 2003;77:796–802.
28. Baylin A, Campos H. Arachidonic acid in adipose tissue is associated with nonfatal acute myocardial infarction in the central valley of Costa Rica. J Nutr. 2004;134:3095–9.
29. Meyer BJ, Mann NJ, Lewis JL. Dietary intakes and food sources of omega-6 and omega-3 polyunsaturated fatty acids. Lipids. 2003;38:391–8.
30. Glaser C, Heinrich J, Koletzko B. Role of FADS1 and FADS2 polymorphisms in polyunsaturated fatty acid metabolism. Metabolism. 2010;59:993–9.
31. Cunnane SC, Anderson MJ. The majority of dietary linoleate in growing rats is beta-oxidized or stored in visceral fat. J Nutr. 1997;127:146–52.
32. Jakobsson A, Westerberg R, Jacobsson A. Fatty acid elongases in mammals: their regulation and roles in metabolism. Progr Lipids Res. 2009;45:237–49.
33. Naganuma T, Sato Y, Sassa T, Ohno Y, Kihara A. Biochemical characterization of the very long-chain fatty acid elongase ELOVL7. FEBS Lett. 2011;585:3337–41.
34. Guillou H, Zadravec D, Martin PG, Jacobsson A. The key roles of elongases and desaturases in mammalian metabolism: insight from transgenic mice. Progr Lipids Res. 2010;49:186–99.
35. Cho HP, Nakamura M, Clarke SD. Cloning, expression, and fatty acid regulation of the human Δ-5 desaturase. J Biol Chem. 1999;274:37335–9.
36. Cho HP, Nakamura M, Clarke SD. Cloning, expression, and fatty acid regulation of the human Δ-6 desaturase. J Biol Chem. 1999;274:471–7.
37. Nakamura MT, Nara TY. Structure, function, and dietary regulation of delta6, delta5, and delta9 desaturases. Annu Rev Nutr. 2004;24:345–76.
38. Schaeffer L, Gohlke H, Müller M, Heid IM, Palmer LJ, Kompauer I, Demmelmair H, Illig T, Koletzko B, Heinrich J. Common genetic variants of the FADS1 FADS2 gene cluster and their reconstructed haplotypes are associated with the fatty acid composition in phospholipids. Hum Mol Genet. 2006;15:1745–56.
39. Malerba G, Schaeffer L, Xumerle L, Klopp N, Trabetti E, Biscuola M, et al. SNPs of the FADS gene cluster are associated with polyunsaturated fatty acids in a cohort of patients with cardiovascular disease. Lipids. 2008;43:289–99.
40. Zietemann V, Kröger J, Enzenbach C. Genetic variation of the FADS1 FADS2 gene cluster and n-6 PUFA composition in erythrocyte membranes in the European Prospective Investigation into Cancer and Nutrition-Potsdam study. Br J Nutr. 2010;104:1748–59.
41. Merino DM, Johnston H, Clarke S. Polymorphisms in FADS1 and FADS2 alter desaturase activity in young Caucasian and Asian adults. Mol Genet Metab. 2011;103:171–8.
42. Kwak JH, Paik JK, Kim OY. FADS gene polymorphisms in Koreans: association with ω6 polyunsaturated fatty acids in serum phospholipids, lipid peroxides, and coronary artery disease. Atherosclerosis. 2011;214:94–100.
43. Mathias RA, Vergara C, Gao L. FADS genetic variants and omega-6 polyunsaturated fatty acid metabolism in a homogeneous island population. J Lipid Res. 2010;51:2766–74.
44. Mathias RA, Sergeant S, Ruczinski I. The impact of FADS genetic variants on ω6 polyunsaturated fatty acid metabolism in African Americans. BMC Genet. 2011;20:12–50.
45. Sergeant S, Hugenschmidt CE, Rudock ME, Ziegler JT, Ivester P, Ainsworth HC, et al. Differences in arachidonic acid levels and fatty acid desaturase (FADS) gene variants in African Americans and European Americans with diabetes/metabolic syndrome. Br J Nutr. 2012;107:547–55.
46. Lemaitre RN, Tanaka T, Tang W, Manichaikul A, Foy M, Kabagambe EK, et al. Genetic loci associated with plasma phospholipid n-3 fatty acids: a meta-analysis of genome-wide association studies from the CHARGE Consortium. PLoS Genet. 2011;7:e1002193.
47. Lattka E, Eggers S, Moeller G, Heim K, Weber M, Mehta D, et al. A common FADS2 promoter polymorphism increases promoter activity and facilitates binding of transcription factor ELK1. J Lipid Res. 2010;5:182–91.
48. Gregory MK, Lester SE, Cook-Johnson RJ. Fatty acid desaturase 2 promoter mutation is not responsible for Δ6-desaturase deficiency. Eur J Hum Genet. 2011;19:1202–4.
49. Koletzko B, Cetin I, Brenna JT. Dietary fat intakes for pregnant and lactating women. Br J Nutr. 2007;98:873–7.
50. Muskiet FA, Kemperman RF. Folate and long-chain polyunsaturated fatty acids in psychiatric disease. J Nutr Biochem. 2006;17:717–27.

51. Trak-Fellermeier MA, Brasche S, Winkler G, Koletzko B, Heinrich J. Food and fatty acid intake and atopic disease in adults. Eur Respir J. 2004;23:575–82.
52. Merino DM, Ma DW, Mutch DM. Genetic variation in lipid desaturases and its impact on the development of human disease. Lipids Health Dis. 2010;9:63.
53. Narce M, Asdrubal P, Delachambre MC. Age-related changes in linoleic acid bioconversion by isolated hepatocytes from spontaneously hypertensive and normotensive rats. Mol Cell Biochem. 1999;141:9–13.
54. Kroger J, Schulze MB. Recent insights into the relation of D5 desaturase and D6 desaturase activity to the development of type 2 diabetes. Curr Opin Lipidol. 2012;23:4–10.
55. Russo C, Olivieri O, Girelli D. Increased membrane ratios of metabolite to precursor fatty acid in essential hypertension. Hypertension. 1997;29:1058–63.
56. Vessby B. Dietary fat, fatty acid composition in plasma and the metabolic syndrome. Curr Opin Lipidol. 2003;14:15–9.
57. Warensjö E, Ohrvall M, Vessby B. Fatty acid composition and estimated desaturase activities are associated with obesity and lifestyle variables in men and women. Nutr Metab Cardiovasc Dis. 2006;6:128–36.
58. Wirfält E, Vessby B, Mattisson I, Gullberg B, Olsson H, Berglund G. No relations between breast cancer risk and fatty acids of erythrocyte membranes in postmenopausal women of the Malmö Diet Cancer cohort (Sweden). Eur J Clin Nutr. 2004;58:761–70.
59. Hodge AM, English DR, O'Dea K, Sinclair AJ, Makrides M, Gibson RA, et al. Plasma phospholipid and dietary fatty acids as predictors of type 2 diabetes: interpreting the role of linoleic acid. Am J Clin Nut. 2007;86:189–97.
60. Krachler B, Norberg M, Eriksson JW, Hallmans G, Johansson I, et al. Fatty acid profile of the erythrocyte membrane preceding development of type 2 diabetes mellitus. Nutr Metab Cardiovasc Dis. 2008;18:503–10.
61. Patel PS, Sharp SJ, Jansen E, Luben RN, Khaw KT, Wareham NJ, et al. Fatty acids measured in plasma and erythrocyte-membrane phospholipids and derived by food-frequency questionnaire and the risk of new-onset type 2 diabetes: a pilot study in the European Prospective Investigation into Cancer and Nutrition (EPIC)-Norfolk cohort. Am J Clin Nutr. 2010;92:1214–22.
62. Kröger J, Zietemann V, Enzenbach C. Erythrocyte membrane phospholipid fatty acids, desaturase activity, and dietary fatty acids in relation to risk of type 2 diabetes in the European Prospective Investigation into Cancer and Nutrition (EPIC)-Potsdam study. Am J Clin Nutr. 2011;93:127–42.
63. Saito E, Okada T, Abe Y, Odaka M, Kuromori Y, Iwata F, et al. Abdominal adiposity is associated with fatty acid desaturase activity in boys: implications for C-reactive protein and insulin resistance. Prostaglandins Leukot Essent Fatty Acids. 2013;88:307–11.
64. Warensjö E, Risérus U, Vessby B. Fatty acid composition of serum lipids predicts the development of the metabolic syndrome in men. Diabetologia. 2005;48:1999–2005.
65. Armutcu F, Akyol S, Ucar F. Markers in nonalcoholic steatohepatitis. Adv Clin Chem. 2013;61: 67–125.
66. Park H, Hasegawa G, Shima T, Fukui M, Nakamura N, Yamaguchi K, et al. The fatty acid composition of plasma cholesteryl esters and estimated desaturase activities in patients with nonalcoholic fatty liver disease and the effect of long-term ezetimibe therapy on these levels. Clin Chim Acta. 2010;411:1735–40.
67. López-Vicario C, González-Périz A, Rius B, Morán-Salvador E, García-Alonso V, Lozano JJ, et al. Molecular interplay between Δ5/Δ6 desaturases and long-chain fatty acids in the pathogenesis of nonalcoholic steatohepatitis. Gut. 2014;63:344–55.
68. Kang JX, Wang J, Wu L, Kang ZB. Transgenic mice: fat-1 mice convert n-6 to n-3 fatty acids. Nature. 2004;427:504.
69. Warensjö E, Sundström J, Vessby B, Cederholm T, Risérus U. Markers of dietary fat quality and fatty acid desaturation as predictors of total and cardiovascular mortality: a population-based prospective study. Am J Clin Nutr. 2008;88:203–9.
70. Steffen LM, Vessby B, Jacobs Jr DR, Cederholm T, Risérus U. Serum phospholipid and cholesteryl ester fatty acids and estimated desaturase activities are related to overweight and cardiovascular risk factors in adolescents. Int J Obes (Lond). 2008;32:1297–304.
71. Martinelli N, Girelli D, Malerba G, Guarini P, Illig T, Trabetti E, et al. FADS genotypes and desaturase activity estimated by the ratio of arachidonic acid to linoleic acid are associated with inflammation and coronary artery disease. Am J Clin Nutr. 2008;88:941–9.
72. Dwyer JH, Allayee H, Dwyer KM, Fan J, Wu H, Mar R, et al. Arachidonate 5-lipoxygenase promoter genotype, dietary arachidonic acid, and atherosclerosis. N Engl J Med. 2004;350:29–37.
73. Helgadottir A, Manolescu A, Thorleifsson G, Gretarsdottir S, Jonsdottir H, Thorsteinsdottir U, et al. The gene encoding 5-lipoxygenase activating protein confers risk of myocardial infarction and stroke. Nat Genet. 2004;36:233–9.
74. Harris WS, Mozaffarian D, Rimm E, Kris-Etherton P, Rudel LL, Appel LJ, et al. Omega-6 fatty acids and risk for cardiovascular disease: a science advisory from the American Heart Association Nutrition Subcommittee of the Council on Nutrition, Physical Activity, and Metabolism; Council on Cardiovascular Nursing; and Council on Epidemiology and Prevention. Circulation. 2009;19:902–7.
75. Li SW, Lin K, Ma P, Zhang ZL, Zhou YD, Lu SY, et al. FADS gene polymorphisms confer the risk of coronary artery disease in a Chinese Han population through the altered desaturase activities: based on high-resolution melting analysis. PLoS One. 2013;8:e55869.

76. Lu Y, Vaarhorst A, Merry AH, Dollé ME, Hovenier R, Imholz S, et al. Markers of endogenous desaturase activity and risk of coronary heart disease in the CAREMA cohort study. PLoS One. 2012;7:e41681.
77. Eberhart CE, Coffey RJ, Radhika A, Giardiello FM, Ferrenbach S, DuBois RN. Up-regulation of cyclooxygenase 2 gene expression in human colorectal adenomas and adenocarcinomas. Gastroenterology. 1994;107:1183–8.
78. Mazhar D, Ang R, Waxman J. COX inhibitors and breast cancer. Br J Cancer. 2006;94:346–50.
79. Hansen Petrik MB, McEntee MF, Johnson BT, Obukowicz MG, Masferrer J, Zweifel B, et al. Selective inhibition of delta-6 desaturase impedes intestinal tumorigenesis. Cancer Lett. 2002;175:157–63.
80. He C, Qu X, Wan J, Rong R, Huang L, Cai C, et al. Inhibiting delta-6 desaturase activity suppresses tumor growth in mice. PLoS One. 2012;7:e47567.
81. Agatha G, Häfer R, Zintl F. Fatty acid composition of lymphocyte membrane phospholipids in children with acute leukemia. Cancer Lett. 2001;173:139–44.
82. Agatha G, Voigt A, Kauf E, Zintl F. Conjugated linoleic acid modulation of cell membrane in leukemia cells. Cancer Lett. 2004;209:87–103.
83. Hoffmann K, Blaudszun J, Brunken C, Höpker WW, Tauber R, Steinhart H. Distribution of polyunsaturated fatty acids including conjugated linoleic acids in total and subcellular fractions from healthy and cancerous parts of human kidneys. Lipids. 2005;40:309–15.
84. Pender Cudlip MC, Krag KJ, Martini D, Yu J, Guidi A, Skinner SS, et al. Delta-6-desaturase activity and arachidonic acid synthesis are increased in human breast cancer tissue. Cancer Sci. 2013;104:760–4.
85. Lukiw WJ, Cui JG, Marcheselli VL, Bodker M, Botkjaer A, Gotlinger K, et al. A role for docosahexaenoic acid-derived neuroprotectin D1 in neural cell survival and Alzheimer disease. J Clin Invest. 2005;115:2774–83.
86. Calon F, Lim GP, Yang F, Morihara T, Teter B, Ubeda O, et al. Docosahexaenoic acid protects from dendritic pathology in an Alzheimer's disease mouse model. Neuron. 2004;43:633–45.
87. Liu Y, Jandacek R, Rider T, Tso P, McNamara RK. Elevated delta-6 desaturase (FADS2) expression in the postmortem prefrontal cortex of schizophrenic patients: relationship with fatty acid composition. Schizophr Res. 2009;109:113–20.
88. McNamara RK, Jandacek R, Rider T, Tso P, Dwivedi Y, Pandey GN. Adult medication-free schizophrenic patients exhibit long-chain omega-3 fatty acid deficiency: implications for cardiovascular disease risk. Cardiovasc Psychiatry Neurol. 2013;2013:796462.
89. Liu Y, McNamara RK. Elevated delta-6 desaturase (FADS2) gene expression in the prefrontal cortex of patients with bipolar disorder. J Psychiatr Res. 2011;45:269–72.
90. Mocking RJ, Assies J, Bot M, Jansen EH, Schene AH, Pouwer F. Biological effects of add-on eicosapentaenoic acid supplementation in diabetes mellitus and co-morbid depression: a randomized controlled trial. PLoS One. 2012;7:e49431.
91. Risé P, Volpi S, Colombo C, Padoan RF, D'Orazio C, Ghezzi S, et al. Whole blood fatty acid analysis with micromethod in cystic fibrosis and pulmonary disease. J Cyst Fibros. 2010;9:228–33.
92. Thomsen KF, Laposata M, Njoroge SW, Umunakwe OC, Katrangi W, Seegmiller AC. Increased elongase 6 and Δ9-desaturase activity are associated with n-7 and n-9 fatty acid changes in cystic fibrosis. Lipids. 2011;46:669–77.
93. Njoroge SW, Seegmiller AC, Katrangi W, Laposata M. Increased Δ5- and Δ6-desaturase, cyclooxygenase-2, and lipoxygenase-5 expression and activity are associated with fatty acid and eicosanoid changes in cystic fibrosis. Biochim Biophys Acta. 2011;1811:431–40.
94. Marquarrdt A, Stohr H, White K. cDNA cloning, genomic structure, and chromosomal localization of the three members of the human desaturase family. Genomics. 2000;66:175–83.
95. Pédrono F, Blanchard H, Kloareg M, D'andréa S, Daval S, Rioux V, et al. The fatty acid desaturase 3 gene encodes for different FADS3 protein isoforms in mammalian tissues. J Lipid Res. 2010;51:472–9.
96. Tanaka T, Shen J, Abecasis GR, Kisialiou A, Ordovas JM, Guralnik JM, et al. Genome-wide association study of plasma polyunsaturated fatty acids in the InCHIANTI study. PLoS Genet. 2009;5:e1000338.
97. Daniels SE, Bhattacharrya S, James A, Leaves NI, Young A, Hill MR, et al. A genome-wide search for quantitative trait loci underlying asthma. Nature. 1996;383:247–50.
98. Standl M, Sausenthaler S, Lattka E, Koletzko S, Bauer CP, Wichmann HE, et al. FADS gene variants modulate the effect of dietary fatty acid intake on allergic diseases in children. Clin Exp Allergy. 2011;41:1757–66.
99. Standl M, Sausenthaler S, Lattka E, Koletzko S, Bauer CP, Wichmann HE, et al. FADS gene cluster modulates the effect of breastfeeding on asthma. Results from the GINIplus and LISAplus studies. Allergy. 2012;67:83–90.
100. Roke K, Ralston JC, Abdelmagid S, Nielsen DE, Badawi A, El-Sohemy A, et al. Variation in the FADS1/2 gene cluster alters plasma n-6 PUFA and is weakly associated with hsCRP levels in healthy young adults. Prostaglandins Leukot Essent Fatty Acids. 2013;89:257–63.
101. Baylin A, Ruiz-Narvaez E, Kraft P, Campos H. Alpha-Linolenic acid, delta6-desaturase gene polymorphism, and the risk of nonfatal myocardial infarction. Am J Clin Nutr. 2007;85:554–60.
102. CARDIoGRAMplusC4D Consortium, Deloukas P, Kanoni S, Willenborg C, Farrall M, Assimes TL, et al. Large-scale association analysis identifies new risk loci for coronary artery disease. Nat Genet. 2013;45:25–33.
103. Kathiresan S, Willer CJ, Peloso GM, Demissie S, Musunuru K, Schadt EE, et al. Common variants at

30 loci contribute to polygenic dyslipidemia. Nat Genet. 2009;41:56–65.
104. Teslovich TM, Musunuru K, Smith AV, Edmondson AC, Stylianou IM, Koseki M, et al. Biological, clinical and population relevance of 95 loci for blood lipids. Nature. 2010;466:707–13.
105. Dumitrescu L, Carty CL, Taylor K, Schumacher FR, Hindorff LA, Ambite JL, et al. Genetic determinants of lipid traits in diverse populations from the population architecture using genomics and epidemiology (PAGE) study. PLoS Genet. 2011;7:e1002138.
106. Nakayama K, Bayasgalan T, Tazoe F, Yanagisawa Y, Gotoh T, Yamanaka K, et al. A single nucleotide polymorphism in the FADS1/FADS2 gene is associated with plasma lipid profiles in two genetically similar Asian ethnic groups with distinctive differences in lifestyle. Hum Genet. 2010;127:685–90.
107. Global Lipids Genetics Consortium, Willer CJ, Schmidt EM, Sengupta S, Peloso GM, Gustafsson S, et al. Discovery and refinement of loci associated with lipid levels. Nat Genet. 2013;45:1274–83.
108. Dupuis J, Langenberg C, Prokopenko I, Saxena R, Soranzo N, Jackson AU, et al. New genetic loci implicated in fasting glucose homeostasis and their impact on type 2 diabetes risk. Nat Genet. 2010;42:105–16.
109. Ingelsson E, Langenberg C, Hivert MF, Prokopenko I, Lyssenko V, Dupuis J, et al. Detailed physiologic characterization reveals diverse mechanisms for novel genetic loci regulating glucose and insulin metabolism in humans. Diabetes. 2010;59:1266–75.
110. Boesgaard TW, Grarup N, Jørgensen T, Borch-Johnsen K, Meta-Analysis of Glucose and Insulin-Related Trait Consortium (MAGIC), Hansen T, et al. Variants at DGKB/TMEM195, ADRA2A, GLIS3 and C2CD4B loci are associated with reduced glucose-stimulated beta cell function in middle-aged Danish people. Diabetologia. 2010;53:1647–55.
111. Hu C, Hoene M, Zhao X, Häring HU, Schleicher E, Lehmann R, et al. Lipidomics analysis reveals efficient storage of hepatic triacylglycerides enriched in unsaturated fatty acids after one bout of exercise in mice. PLoS One. 2010;5:e13318.
112. Kim OY, Lim HH, Yang LI, Chae JS, Lee JH. Fatty acid desaturase (FADS) gene polymorphisms and insulin resistance in association with serum phospholipid polyunsaturated fatty acid composition in healthy Korean men: cross-sectional study. Nutr Metab (Lond). 2011;8:24.
113. Park MH, Kim N, Lee JY, Park HY. Genetic loci associated with lipid concentrations and cardiovascular risk factors in the Korean population. J Med Genet. 2011;48:10–5.
114. Liu C, Li H, Qi L, Loos RJ, Qi Q, Lu L, Gan W, Lin X. Variants in GLIS3 and CRY2 are associated with type 2 diabetes and impaired fasting glucose in Chinese Hans. PLoS One. 2011;6:e21464.
115. Suhre K, Shin SY, Petersen AK, Mohney RP, Meredith D, Wägele B, et al. Human metabolic individuality in biomedical and pharmaceutical research. Nature. 2011;477:54–60.
116. Caspi A, Williams B, Kim-Cohen J, Craig IW, Milne BJ, Poulton R, et al. Moderation of breastfeeding effects on the IQ by genetic variation in fatty acid metabolism. Proc Natl Acad Sci U S A. 2007;104:18860–5.
117. Steer CD, Davey Smith G, Emmett PM, Hibbeln JR, Golding J. FADS2 polymorphisms modify the effect of breastfeeding on child IQ. PLoS One. 2010;5:e11570.
118. Xie L, Innis SM. Genetic variants of the FADS1 FADS2 gene cluster are associated with altered (n-6) and (n-3) essential fatty acids in plasma and erythrocyte phospholipids in women during pregnancy and in breast milk during lactation. J Nutr. 2008;138:2222–8.
119. Rizzi TS, van der Sluis S, Derom C, Thiery E, van Kesteren RE, Jacobs N, et al. FADS2 genetic variance in combination with fatty acid intake might alter composition of the fatty acids in brain. PLoS One. 2013;8:e68000.
120. Steer CD, Lattka E, Koletzko B, Golding J, Hibbeln JR. Maternal fatty acids in pregnancy, FADS polymorphisms, and child intelligence quotient at 8 y of age. Am J Clin Nutr. 2013;98:1575–82.
121. Stoffel W, Holz B, Jenke B, Binczek E, Günter RH, Kiss C. Delta6-desaturase (FADS2) deficiency unveils the role of omega3- and omega6-polyunsaturated fatty acids. EMBO J. 2008; 27:2281–92.
122. Fan YY, Monk JM, Hou TY, Callway E, Vincent L, Weeks B, et al. Characterization of an arachidonic acid-deficient (Fads1 knockout) mouse model. J Lipid Res. 2012;53:1287–95.
123. Obukowicz MG, Raz A, Pyla PD, Rico JG, Wendling JM, Needleman P. Identification and characterization of a novel delta6/delta5 fatty acid desaturase inhibitor as a potential anti-inflammatory agent. Biochem Pharmacol. 1998;55:1045–58.
124. Holmgren G, Jagell SF, Johnson SB, Holman RT. Suspected faulty essential fatty acid metabolism in Sjögren-Larsson syndrome. Pediatr Res. 1982;16:45–9.
125. Taube B, Billeaud C, Labrèze C, Entressangles B, Fontan D, Taïeb A. Sjögren-Larsson syndrome: early diagnosis, dietary management and biochemical studies in two cases. Dermatology. 1999;198:340–5.
126. Nwankwo JO, Spector AA, Domann FE. A nucleotide insertion in the transcriptional regulatory region of FADS2 gives rise to human fatty acid delta-6-desaturase deficiency. J Lipid Res. 2003;44:2311–9.
127. Yao JK, Cannon KP, Holman RT, Dyck PJ. Effects of polyunsaturated fatty acid diets on plasma lipids of patients with adrenomultineuronal degeneration, hepatosplenomegaly and fatty acid derangement. J Neurol Sci. 1983;62:67–75.

128. Simopoulos AP. Evolutionary aspects of diet, the omega-6/omega-3 ratio and genetic variation: nutritional implications for chronic diseases. Biomed Pharmacother. 2006;60:502–7.
129. Simopoulos AP. Commentary. Genetic variants and omega-6, omega-3 fatty acids: their role in the determination of nutritional requirements and chronic disease risk. J Nutrigenet Nutrigenomics. 2009;2:117–8.
130. Ameur A, Enroth S, Johansson A, Zaboli G, Igl W, et al. Genetic adaptation of fatty-acid metabolism: a human-specific haplotype increasing the biosynthesis of long-chain omega-3 and omega-6 fatty acids. Am J Hum Genet. 2012;90:809–20.

Systemic Inflammation, Intestine, and Paraoxonase-1

8

Ladan Vakili, Kaveh Daniel Navab, Maryam Shabihkhani, Nasim Pourtabatabaei, Samra Vazirian, Zarina Barseghian, Seyedehsara Seyedali, and Greg Hough

Abstract

Serum paraoxonase 1 (PON1) has been shown to act as an important guardian against cellular damage from oxidized lipids in low-density lipoprotein (LDL), plasma membrane, against toxic agents such as pesticide residues including organophosphates and against bacterial endotoxin. PON1 associated with circulating high-density lipoprotein (HDL) has the ability to prevent the generation of pro inflammatory oxidized phospholipids by reactive oxygen species. The activities of the HDL-associated PON1 and several other anti-inflammatory factors in HDL are in turn negatively regulated by these oxidized lipids. In rabbits, mice, and humans there appears to be an increase in the formation of these oxidized lipids during the acute phase response. This results in the association of acute phase proteins with HDL and inhibition of the HDL-associated PON1 that renders HDL pro inflammatory.

In populations, low serum HDL-cholesterol is a risk factor for atherosclerosis and efforts are directed toward therapies to improve the quality and the relative concentrations of LDL and HDL. Apolipoprotein A-I

L. Vakili (✉)
Atheroclerosis Research Unit, Division of Cardiology, David Geffen School of Medicine, University of California Los Angeles, Los Angeles 90095, CA, USA
e-mail: lvakili@mednet.ucla.edu

K.D. Navab • N. Pourtabatabaei • S. Vazirian
Z. Barseghian • S. Seyedali • G. Hough
Department of Medicine, David Geffen School of Medicine, University of California Los Angeles, Los Angeles, CA, USA
e-mail: Kavehnavab@mednet.ucla.edu; nasimpt@gmail.com; vazirian@ucla.edu; zara1645@yahoo.com; ssaraa20@yahoo.com; ghough@ucla.edu

M. Shabihkhani (✉)
Department of Pathology and Laboratory Medicine, David Geffen School of Medicine, University of California Los Angeles, Los Angeles, CA, USA
e-mail: mshabihkhani@mednet.ucla.edu

J. Camps (ed.), *Oxidative Stress and Inflammation in Non-communicable Diseases – Molecular Mechanisms and Perspectives in Therapeutics*, Advances in Experimental Medicine and Biology 824, DOI 10.1007/978-3-319-07320-0_8, © Springer International Publishing Switzerland 2014

(apoA-I) has been shown to reduce atherosclerotic lesions in laboratory animals. ApoA-I, however, is a large protein that is costly and needs to be administered parenterally. Our group has developed apoA-I mimetic peptides that are much smaller than apoA-I (18 amino acids long vs 243 in ApoA-I itself). These HDL mimetic peptides are much more effective in removing the oxidized phospholipids and other oxidized lipids. They improve LDL and HDL composition and function and reduce lesion formation in animal models of atherogenesis. Following is a brief description of some of the HDL mimetic peptides that can improve HDL and the effect of the peptide on PON1 activity.

Keywords
Apolipoprotein A-I • D-4F • Dysfunctional HDL • HDL • Inflammation • Intestine • L-4F • Proinflammatory HDL

8.1 In Search of Agents to Improve HDL Function

Experiments in animal models of atherosclerosis [1, 2] and preliminary human studies [3] have made apolipoprotein A-I (apoA-I), the main protein in high-density lipoprotein (HDL), an attractive therapeutic target. The preliminary human studies [3] suggested that therapeutic benefit might be achieved by administering weekly intravenous doses over a period of 5–6 weeks. Subsequent larger clinical trials [4], however, suggested that longer periods of intravenous administration will be required for significant improvements to be achieved. This made the idea an unlikely therapy for the large number of patients with atherosclerosis.

8.2 The Search for an Ideal Agent: HDL Mimetic Peptide

Anantharamaia and colleagues originally designed an 18 amino acid peptide that mimicked the class A amphipathic helixes contained in apoA-I while it did not have any sequence homology with apoA-I but [5–7]. Because it contained 18 amino acids and formed a class A amphipathic helix the peptide was called 18A. The lipid-binding properties and stability of the 18A peptide were improved by blocking the carboxy terminus using an amide group and blocking the amino terminus through the addition of an acetyl group. Because of the presence of the two phenylalanine residues on the hydrophobic face this peptide was named 2F. The 2F peptide failed to alter lesions in a mouse model of atherosclerosis [8] while mimicking many of the lipid-binding properties of apoA-I. We tested a series of peptides for their ability to inhibit low-density lipoprotein (LDL)-induced monocyte chemotactic activity, which is primarily due to the production of monocyte chemoattractant-1 (MCP-1). This was achieved using a human artery wall cell culture assay. Subsequently, the peptide 4F was shown to be superior [8]. The peptide 4F has the same structure as 2F except for two additional phenylalanine residues on the hydrophobic face of the peptide (replacing two leucine residues).

8.3 Animal Models of Atherosclerosis and the Effects of HDL Mimetic Peptides

ApoAI mimetic peptides reduced the development of atherosclerotic lesions in young mice [9, 10]. Lesion regression in old apolipoprotein E null mice was produced when 4F was given together with a statin [11]. Synergy between D-4F (made from all D amino acids) and pravastatin was then tested in these studies. The oral doses for each that were ineffective when given as single agent was

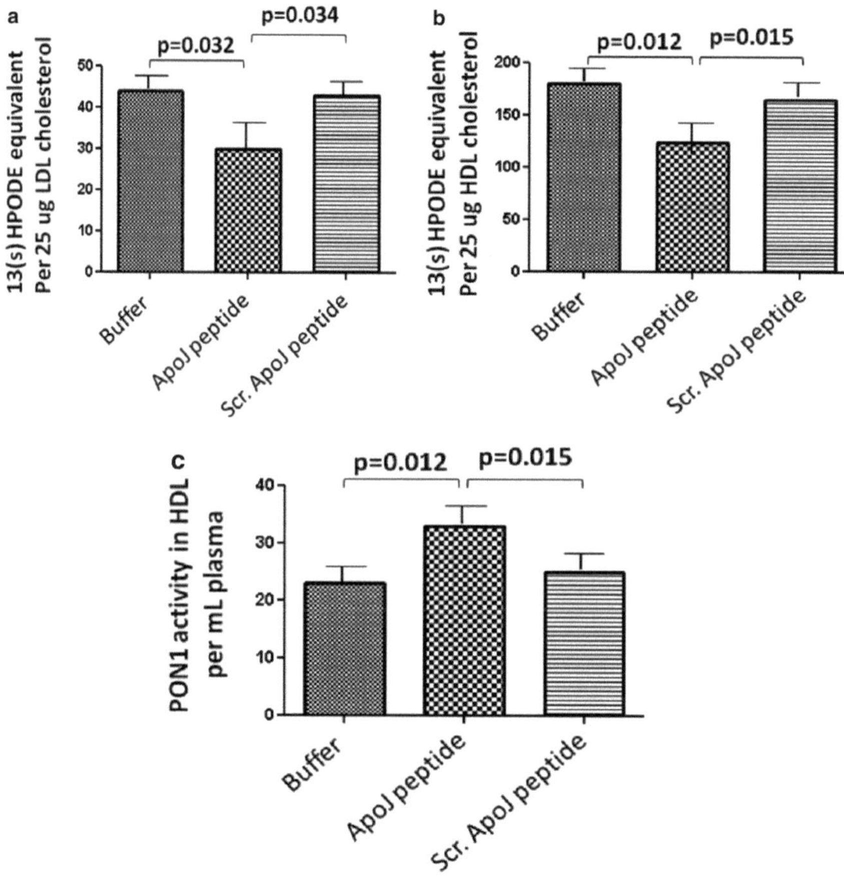

Fig. 8.1 D-[113–122]apoJ in apoE-null mouse plasma in vitro. D-[113–122]apoJ or scrambled D-[113–122]apoJ (Scr.[113–122]apoJ peptide) were added to apoE-null mouse plasma at a concentration of 250 ng/mL. The solutions were gently mixed, layered with argon gas, and the tubes were sealed and incubated for 1 h at 37 °C with gentle mixing. The samples were then fractionated by FPLC and lipid hydroperoxide content of LDL (**a**), lipid hydroperoxide content of HDL (**b**), and PON1 activity (**c**) were determined. The values shown are the Mean ± SD. The data shown are representative of two separate experiments

first determined. Significant increase in HDL cholesterol levels, apoA-I levels, and paraoxonase-1 (PON1) activity was achieved with oral administration of the combination. Rendering HDL anti-inflammatory, significantly preventing lesion formation in young apolipoprotein E null mice, and causing regression of established lesions in old apolipoprotein E null mice were among additional effects observed. Mice that received the combination for 6 months had lesion areas that were much smaller when compared to those before the start of treatment [11]. In mice maintained on chow alone en face lesion area was three times higher than that after 6 months of treatment with the combination. In addition in the remaining lesions there was a significant 22 % reduction in macrophage content, indicating an overall 79 %. Reduction in macrophages. The combination of statin and D-4F increased intestinal apoA-I synthesis by a significant 60 %. In studies in Cynomolgous monkeys also, oral administration of D-4F plus pravastatin rendered HDL anti-inflammatory [11]. These results suggested that the combination of an HDL-based therapy and a statin might be a useful anti-atherosclerosis treatment strategy. The fact that the benefit of apolipoprotein mimetic peptides in atherosclerosis was not limited to 4F was unraveled subsequently. We found that peptides including D-[113–122] apoJ, an apolipoprotein J mimetic peptide [12], and peptides too small to form a helical structure [13] were efficacious as well (Fig. 8.1). The efficacy of these peptides was demonstrated in a rabbit model of atherosclerosis as well [14].

8.4 The Mechanism of Action of HDL Mimetic Peptides

The mechanism of action of the HDL mimetic peptides seems to be based to their ability to remove oxidized lipids from lipoproteins [15–17], promote reverse cholesterol transport from macrophages, and render HDL anti-inflammatory [15, 16]. Atherosclerosis is a long-term inflammatory process that is mediated in part by the oxidation of phospholipids, which induce vascular cells to express cytokines, adhesion molecules, and procoagulant molecules [18, 19]. The mechanism of action of the HDL mimetic peptides appears to be related to their ability to bind and remove these pro inflammatory oxidized lipids [15, 16]. In addition, the HDL mimetic peptides are efficacious in inflammatory processes that have an infectious etiology and in models of vascular diseases that are not classified as atherosclerosis. This suggests that oxidized lipids may be important mediators of a variety of inflammatory conditions other than atherosclerosis as they have been shown to effective in preventing diabetic manifestations, graft rejection, cognitive deficiency in mouse model of hyperlipemia, in collagen induced arthritis model and in animal model of Alzheimer's Disease.

8.5 HDL Mimetic Peptides, Oxidized Lipids and PON1

On an atherogenic diet, PON1 activity decreased by 52 % in atherosclerosis susceptible mice, C57BL/6 J (BL/6), but not in fatty streak resistant mice, C3H/HeJ (C3H) [20]. In apolipoprotein E knockout mice on the chow diet also plasma PON1 activity was significantly decreased as compared to controls. Furthermore, in LDL receptor knockout mice when they were fed a 0.15 %-cholesterol-enriched diet a significant decrease in PON1 activity was observed. Injection of mildly oxidized LDL but not native LDL resulted in a 59 % decrease in PON1 activity in BL/6 mice but not in C3H mice on a chow diet. A threefold reduction in mRNA levels for PON1 was observed following treatment of HepG2 cells in culture with mildly oxidized LDL (but not native LDL). We additionally reported that, the total cholesterol/HDL cholesterol ratio was 3.1 ± 0.9 as compared to 2.9 ± 0.4 in the controls in normolipidemic patients with angiographically documented coronary artery disease who did not have diabetes and were not on lipid-lowering medication [20]. This difference was however not statistically significant. In a subset of these normolipidemic patients, despite similar normal HDL levels, the PON1 activity was low (48 ± 6.6 versus 98 ± 17 U/ml for controls; $p = 0.009$). As predicted the HDL from these patients failed to protect against LDL oxidation in co-cultures of human artery wall cells. Forte and colleagues showed [21] that compared with controls on the chow diet, ApoE$^{-/-}$ mice had elevated lysophosphatidylcholine and bioactive oxidized phospholipids (1-palmitoyl- 2-oxovaleryl-*sn*-*glycero*-3-phosphocholine and 1-palmitoyl-2-glutaryl-*sn*-glycero- 3-phosphocholine). In these mice on a chow diet elevated oxidized phospholipids may, in part, contribute to spontaneous lesions. PON1 activity in these mice decreased by 38 %. It has been suggested that removal of oxidized fatty acids from HDL might cause the return of PON1 activity. We have observed that when ApoE-deficient mouse plasma is incubated with PON1 and FPLC fractionation of lipoproteins is performed, PON1 is found in the post HDL region in addition to in HDL-containing fractions. We have shown that early during the incubation, PON is present in HDL-like particles that contain cholesterol, apoA-I, and other HDL constituents and co-elute with albumin. It is possible that 4F accelerates apoA-I cycling, pre-beta formation and remodeling of HDL. We have also observed that when human plasma is incubated with 4F in vitro, PON1 activity is increased. As for the underlying mechanism, one possibility could be the removal of oxidized lipids from HDL and reactivation of PON1 by 4F. Another might be the changes in lipid–protein interaction such as that in phospholipid–apoA-I–PON1 interactions. The peptide 4F therefore might have beneficial effects supporting HDL function in individuals under conditions that result in low PON1 by removing oxidized lipids and reactivating antioxidant enzymes including PON1.

8.6 The Role of Intestine in Systemic Inflammation and Effect on PON1

In recent years our group has focused on the role of small intestine in systemic inflammation, HDL protective capacity and cardiovascular function. This came about when we observed that oral administration of the peptide 4F reduced systemic inflammation while the plasma levels of the peptide were very low [22]. To test the hypothesis that intestine is a major site of action for D-4F, we fed LDLR$^{-/-}$ mice a Western diet (WD) and administered the peptide orally or subcutaneously (SQ). Whereas peptide levels in small intestine only varied by 1.66 ± 0.33-fold, plasma and liver D-4F levels were 298-fold and 96-fold higher, respectively, after SQ administration. Levels of metabolites of arachidonic and linoleic acids known to bind with high affinity to D-4F were significantly reduced liver, hepatic bile and in intestine, to a similar degree whether administered orally or SQ. Levels of 20-HETE however, which is known to bind the peptide with low affinity, were unchanged. D-4F treatment increased HDL-cholesterol levels, PON 1 activity and reduced plasma serum amyloid A (SAA) and triglyceride level similarly after oral or SQ administration. SAA levels correlated significantly with plasma levels of metabolites of arachidonic and linoleic acids. Feeding 15-HETE in chow (without WD) significantly decreased HDL-cholesterol and PON1 activity and increased plasma SAA and triglyceride levels, all of which were significantly ameliorated by SQ D-4F. Administering 4F along with 15-HETE prevented the reduction in PON 1 activity (Fig. 8.2). We therefore concluded that D-4F administration reduced levels of free metabolites of arachidonic and linoleic acids in the small intestine and this was associated with increased PON1 activity and decreased inflammation in LDL receptor deficient mice [22].

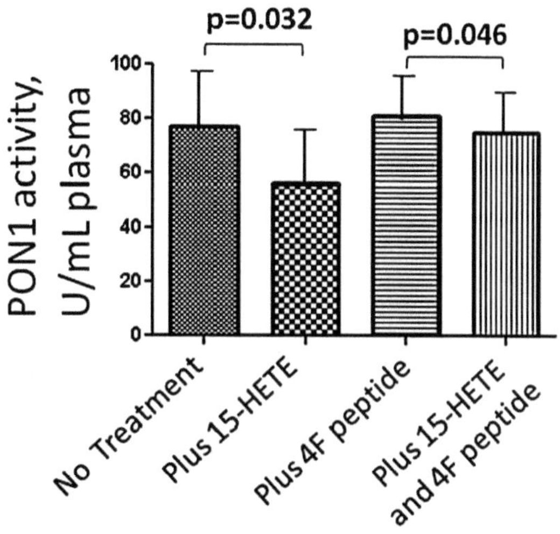

Fig. 8.2 Feeding LDLR$^{-/-}$ mice 15-HETE administered in mouse chow (without the WD) decreased plasma HDL-cholesterol levels, increased plasma triglyceride levels, decreased PON activity, and increased plasma SAA levels, all of which was ameliorated by D-4F treatment. Female LDLR$^{-/-}$ mice (10–11 months of age; n = 20 per group) were fed laboratory rodent chow (Ralston Purina) prepared and presented to the mice as described in Materials and Methods. The chow did or did not contain 15-HETE at a concentration of 1 ug per gram diet to provide each mouse on average with 5 ug of 15-HETE per day. The mice received daily SQ injections of saline or SQ saline containing D-4F at a dosage of 900 ug/mouse/day. After 2 weeks on the diet, 2–3 h after the last SQ dose, the mice were bled, and their plasma was analyzed for PON1 activity and other analytes. The data shown are mean ± SD

8.7 Summary

Under conditions of excess inflammatory pressure the ability of HDL to protect itself and other lipid-containing molecules and structures might be reduced, HDL be damaged, antioxidant enzymes such as PON1 be inactivated, and even HDL itself act as a proinflammatory molecule. Fortunately PON1 has the ability to prevent lipid oxidation and may even inactivate oxidized lipids once formed, and thus protect HDL against the inflammatory pressure. Reduction of lipid and protein oxidation by agents such as HDL mimetic peptides may prove to be an effective way of supporting the protective role of HDL.

References

1. Badimon JA, Badimon L, Fuster V. Regression of atherosclerotic lesions by high density lipoprotein plasma fraction in the cholesterol-fed rabbit. J Clin Invest. 1990;85:1234–41.
2. Plump AS, Scott CJ, Breslow JL. Human apolipoprotein A-I gene expression increases high density lipoprotein and suppresses atherosclerosis in the apolipoprotein E-deficient mouse. Proc Natl Acad Sci U S A. 1994;91:9607–11.
3. Nissen SE, Tsunoda T, Tuzcu EM, Schoenhagen P, Cooper CJ, Yasin M, et al. Effect of recombinant apoA-I Milano on coronary atherosclerosis in patients with acute coronary syndromes: a randomized controlled trial. JAMA. 2003;290:2292–300.
4. Tardif JC, Grégoire J, L'Allier PL, Ibrahim R, Lespérance J, Heinonen TM, Kouz S, Berry C, Basser R, Lavoie MA, Guertin MC, Rodés-Cabau J, Effect of rHDL on Atherosclerosis-Safety and Efficacy (ERASE) Investigators. Effects of reconstituted high-density lipoprotein infusions on coronary atherosclerosis: a randomized controlled trial. JAMA. 2007;297:1675–82.
5. Anantharamaiah GM, Jones JL, Brouillette CG, Schmidt CF, Chung BH, Hughes TA, et al. Studies of synthetic peptide analogs of amphipathic helix I: structure of peptide/DMPC complexes. J Biol Chem. 1985;260:10248–55.
6. Venkatachalapathi YV, Phillips MC, Epand RM, Epand RF, Tytler EM, Segrest JP, et al. Effect of end group blockage on the properties of a class A amphipathic helical peptide. Proteins Struct Funct Genet. 1993;15:349–59.
7. Yancey PG, Bielicki JK, Johnson WJ, Lund-Katz S, Palgunachari MN, Anantharamaiah GM, et al. Efflux of cellular cholesterol and phospholipid to lipid-free apolipoproteins and class A amphipathic peptides. Biochemistry. 1995;34:7955–65.
8. Datta G, Chaddha M, Hama S, Navab M, Fogelman AM, Garber DW, et al. Effects of increasing hydrophobicity on the physical–chemical and biological properties of a class A amphipathic helical peptide. J Lipid Res. 2001;42:1096–104.
9. Navab M, Anantharamaiah GM, Hama S, Garber DW, Chaddha M, Hough G, et al. Oral administration of an apoA-I mimetic peptide synthesized from D-amino acids dramatically reduces atherosclerosis in mice independent of plasma cholesterol. Circulation. 2002;105:290–2.
10. Li X, Chyu KY, Faria Neto JR, Yano J, Nathwani N, Ferreira C, et al. Differential effects of apolipoprotein A-I mimetic peptide on evolving and established atherosclerosis in apolipoprotein E-null mice. Circulation. 2004;110:1701–5.
11. Navab M, Anantharamaiah GM, Hama S, Hough G, Reddy ST, Frank JS, et al. D-4F and statins synergize to render HDL anti-inflammatory in mice and monkeys and cause lesion regression in old apolipoprotein E-null mice. Arterioscler Thromb Vasc Biol. 2005;25:1426–32.
12. Navab M, Anantharamaiah GM, Reddy ST, Van Lenten BJ, Wagner AC, Hama S, et al. An oral apoJ peptide renders HDL antiinflammatory in mice and monkeys and dramatically reduces atherosclerosis in apolipoprotein E-null mice. Arterioscler Thromb Vasc Biol. 2005;25:1932–7.
13. Navab M, Anantharamaiah GM, Reddy ST, Hama S, Hough G, Frank JS, et al. Oral small peptides render HDL antiinflammatory in mice and monkeys and reduce atherosclerosis in apoE null mice. Circ Res. 2005;97:524–32.
14. Van Lenten BJ, Wagner AC, Navab M, Anantharamaiah GM, Hama S, Reddy ST, et al. Lipoprotein inflammatory properties and serum amyloid A levels but not cholesterol levels predict lesion area in cholesterol-fed rabbits. J Lipid Res. 2007;48:2344–53.
15. Navab M, Anantharamaiah GM, Reddy ST, Hama S, Hough G, Grijalva VR, et al. Oral D-4F causes formation of pre-β high-density lipoprotein and improves high-density lipoprotein-mediated cholesterol efflux and reverse cholesterol transport from macrophages in apolipoprotein E-null mice. Circulation. 2004;109:3215–20.
16. Navab M, Anantharamaiah GM, Reddy ST, Hama S, Hough G, Grijalva VR, et al. Apolipoprotein A-I mimetic peptides. Arterioscler Thromb Vasc Biol. 2005;25:1325–31.
17. Datta G, Epand RF, Epand RM, Chaddha M, Kirksey MA, Garber DW, et al. Aromatic residue position on the non polar face of class A amphipathic helical peptides determines biological activity. J Biol Chem. 2004;18(279):26509–17.
18. Berliner JA, Watson AD. A role for oxidized phospholipids in atherosclerosis. N Engl J Med. 2005;353:9–12.
19. Gargalovic PS, Imura M, Zhang B, Gharavi NM, Clark MJ, Pagnon J, et al. Identification of inflammatory gene modules based on variations of human endothelial cell responses to oxidized lipids. Proc Natl Acad Sci U S A. 2006;103:12741–6.
20. Navab M, Hama-Levy S, Van Lenten BJ, Fonarow GC, Cardinez CJ, Castellani LW, et al. Mildly oxidized LDL induces an increased apolipoprotein J/paraoxonase ratio. J Clin Invest. 1997;99:2005–19.
21. Forte TM, Subbanagounder G, Berliner JA, Blanche PJ, Clermont AO, Jia Z, et al. Altered activities of anti-atherogenic enzymes LCAT, paraoxonase, and platelet-activating factor acetylhydrolase in atherosclerosis susceptible mice. J Lipid Res. 2002;43:477–85.
22. Navab M, Reddy ST, Anantharamaiah GM, Hough G, Buga GM, Danciger J. D-4F-mediated reduction in metabolites of arachidonic and linoleic acids in the small intestine is associated with decreased inflammation in low-density lipoprotein receptor-null mice. J Lipid Res. 2012;53:437–45.

Serotonin Modulation of Macrophage Polarization: Inflammation and Beyond

9

Mateo de las Casas-Engel and Angel L. Corbí

Abstract

Macrophages display a ample plethora of effector functions whose acquisition is promoted by the surrounding cytokine and cellular environment. Depending on the stimulus, macrophages become specialized ("polarized") for either pathogen elimination, tissue repair and wound healing or immunosuppression. This "polarization" versatility allows macrophages to critically contribute to tissue homeostasis, as they promote initiation and resolution of inflammatory responses. As a consequence, deregulation of the tissue macrophage polarization balance is an etiological agent of chronic inflammation, autoimmune diseases, cancer and even obesity and insulin resistance. In the present review we describe current concepts on the molecular basis and the patho-physiological implications of macrophage polarization, and describe its modulation by serotonin (5-HT), a neurotransmitter that regulates inflammation and tissue repair via a large set of receptors (5-HTR_{1-7}). 5-HT modulates the phenotypic and functional polarization of macrophages, and contributes to the maintenance of an anti-inflammatory state mainly via 5-HTR_{2B} and 5-HTR_7, whose activation has a great impact on macrophage gene expression profile. The identification of 5-HTR_{2B} and 5-HTR_7 as functionally-relevant polarization markers suggests their therapeutic value in inflammatory pathologies as well as their potential involvement in linking the immune and nervous systems.

Keywords

Inflammation • Macrophages • Psychiatric diseases • Serotonin • Serotonin receptor

M. de las Casas-Engel • A.L. Corbí (✉)
Centro de Investigaciones Biológicas, Consejo Superior de Investigaciones Científicas (CSIC), Ramiro de Maeztu, 9, Madrid 28040, Spain
e-mail: mcasas@cib.csic.es; acorbi@cib.csic.es

J. Camps (ed.), *Oxidative Stress and Inflammation in Non-communicable Diseases – Molecular Mechanisms and Perspectives in Therapeutics*, Advances in Experimental Medicine and Biology 824, DOI 10.1007/978-3-319-07320-0_9, © Springer International Publishing Switzerland 2014

9.1 Macrophages

Macrophages are bone marrow-derived cells that constitute the first line of defense of tissues against pathogens and potentially damaging and harmful stimuli. Macrophages provide a fast and unspecific response and, together with dendritic cells, are essential for the coordinated orchestration of innate and adaptive immune responses. Since the original description of "phagocytes" by Metchnikoff [1], numerous studies have addressed their developmental origin as well as their phenotypic and functional diversity. Although still a matter of debate, macrophages can differentiate from peripheral blood monocytes and become effector cells whose huge functional plasticity allows them to play essential roles in pathophysiological processes as diverse as pathogen clearance, tissue repair, angiogenesis and tumor progression and metastasis [2]. The functional plasticity of macrophages arises from their ability to respond to endogenous and non-self stimuli while adapting to the surrounding tissue environment [3–6], and explains both the existence of tissue-specific macrophages (microglia, osteoclasts, Kupffer cells) and the wide variety (*continuum*) of macrophage activation (or polarization) states. The huge plasticity of macrophage responses ultimately results from their ability to differentiate in a tissue-specific manner, and their constitutive expression of a large array of receptors for cytokines, chemokines, pathogen-associated molecular patterns (PAMP) and danger-associated molecular patterns (DAMP).

9.1.1 Macrophage Polarization

In response to microbe-derived factors, cytokines from Th1 cells (*e.g.* IFNγ) or other cytokines such as GM-CSF or TNFα, macrophages acquire pro-inflammatory, bactericidal, tumor suppressive and immunogenic activities, in a process commonly referred to as "classic" or M1 polarization, and whose hallmark is the ability to release large amounts of IL-12/IL-23, reactive nitrogen and oxygen intermediates, and expression of Th1-cell attracting chemokines [2, 4] (Fig. 9.1). M1 macrophages also drive the polarization and recruitment of Th1 cells through the expression of cytokines and chemokines like IL-12, CXCL9 and CXCL10, thereby amplifying Type-1 immune responses [2]. Conversely, Th2-derived cytokines like IL-4, IL-13, IL-10, TGFβ or M-CSF, as well as glucocorticoids, promote the acquisition of anti-inflammatory, scavenging, tumor-promoting, tissue repair and pro-angiogenic functions, all of which are grouped under the terms "alternative" or M2 polarization, that endows them with the ability to produce high levels of IL-10 [2, 5–8] (Fig. 9.1). It is now widely accepted that M1 and M2 macrophage polarization states are just two extreme examples of a wide range of functionally distinct macrophages activation states [3–6].

9.1.1.1 Macrophages in Homeostasis

Under homeostatic *in vivo* conditions, tissue-resident macrophages also exhibit a wide variety of polarization states, which are ultimately determined by the extracellular environment and the surrounding cell types. Thus, macrophages exhibit tissue-specific phenotypes and functions under homeostatic conditions. As representative examples, bone macrophages (osteoclasts) display potent bone-degrading functions, brain macrophages (microglia) [9] contribute to development of neural circuitry and modulation of angiogenesis and fluid balance in the brain [10], and liver macrophages (Kupffer cells) are primarily specialized in scavenging [11]. Due to their localization in liver sinusoids, Kupffer cells come in contact with antigens absorbed via the gastrointestinal tract and, therefore, play a crucial role in identifying and detoxifying bacteria, endotoxins [12], cell debris, apoptotic cells and immune complexes as well as toxic agents such as ethanol [13].

The existence of IL-10-producing (M2-polarized) macrophages has already been demonstrated in lungs [14] and gut [15, 16], tissues that are continuously exposed to exogenous and potentially damaging substances. Gut macrophages function in host defense through the recognition, phagocytosis, and killing of microorganisms

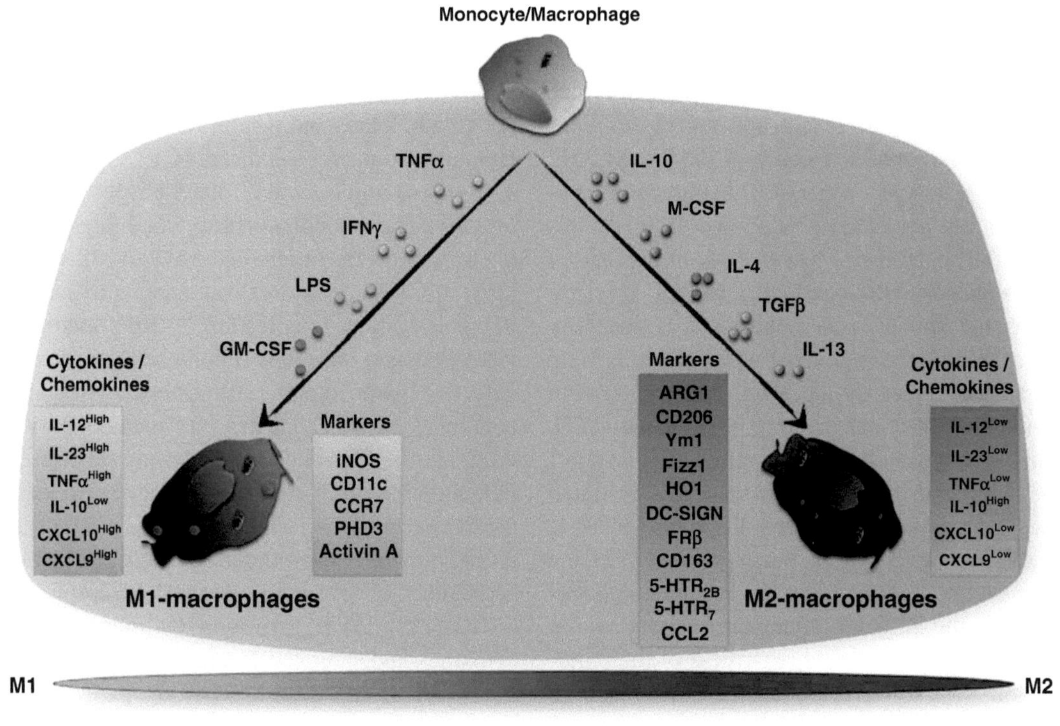

Fig. 9.1 Stimuli that trigger M1 or M2 macrophage polarization, with indication of the most representative chemokines, cytokines and markers expressed by each macrophage subtype

[17, 18], and display a weak pro-inflammatory cytokine profile. The same applies for lung macrophages, that constitutively secrete IL-10 and whose ability for pro-inflammatory cytokine production is low [14]. Other tissues that preferentially contain M2-skewed macrophages in homeostasis include brain and liver, since microglia [9] and most liver Kupffer cells (80–90 % of tissue macrophages of the body) [19] are M-CSF-dependent [20]. In the CNS, M2 polarized macrophages appear to regulate learning and memory [21]. Meningeal myeloid cells are polarized by Th2-derived IL-4 in response to a cognitive task, but acquire a pro-inflammatory phenotype in the absence of IL-4-producing Th2 cells [21]. Peritoneal fluid also contains large amounts of M-CSF [22], that ultimately results in peritoneal macrophages acquiring an M2-like phenotype [22, 23] characterized by a low capacity for T-cell activation and high IL-10 secretion levels after stimulation [23]. Therefore, macrophages display M2-associated effector functions in various tissues under basal conditions, implying that the control of macrophage polarization is of fundamental importance for tissue homeostasis.

9.1.1.2 Macrophages in Inflammation, Inflammation Resolution and Tissue Repair

It is during inflammatory responses when the pathological importance of fine-tuning macrophage polarization is most easily understood. Thus, M1-polarized macrophages predominate at the initial stages of an inflammatory response, when cytotoxic and tissue-damaging activities are more robust. Later, M2-type macrophages increase in number, as a means to promote resolution of inflammation [24]. Therefore, a critical issue for inflammation resolution is a change in the polarization state ("polarization switch") of the macrophages that are recruited towards inflamed tissues. Such a switch was elegantly demonstrated in the case of peritoneal infection by *L. monocytogenes*, where the initial influx of pro-inflammatory M1 monocytes is followed (after 6–8 h) by a potent increase in "resolving" M2 monocytes [25]. In fact, and in the resolution-phase, macrophages express a unique M1/M2 mixed phenotype, with cAMP being essential to restrain M1 activation [26]. However, it remains to be fully established whether pro-inflammatory macrophages within inflamed tissues are later converted into M2 macrophages. It is generally accepted that the capture of apoptotic cells is an important trigger for the M1-to-M2 polarization switch that takes place during inflammation [27]. This idea has gained considerable support after the studies of Brüne's group, who demonstrated that apoptotic cell-derived Sphingosine-1-Phosphate (S1P) triggers the expression of Heme Oxygenase 1 (HO1) and pushes macrophages towards the M2/anti-inflammatory polarization state [28]. However, and regardless of the polarization switch-triggering event, the sequential occurrence of both polarization states is required for adequate resolution of inflammation and return to tissue homeostasis. Some representative examples are briefly commented below.

In animal models of ischemic heart disease, the M1-to-M2 dynamic change in macrophage recruitment has been observed. After heart ischemia, macrophages primarily exhibit a "classical activation" (M1) phenotype, as exemplified by their high expression of TNF-α, whereas their phenotype shifts into an "alternatively activated" phenotype (M2) (high Arginase 1 and 2 expression) during the transition from inflammation to scar tissue formation [29]. Similarly, after spinal cord injury, activated microglia rapidly release pro-inflammatory cytokines, which contribute to the influx of neutrophiles and macrophages from the circulation [30]. However, the resolution phase is characterized by the influx of macrophages that exhibit anti-inflammatory and tissue-repair properties (M2), and whose injection promotes full recovery [31]. All current evidences indicate that the misbalance of the M1/M2 polarization equilibrium, or an inadequate switch in the macrophage polarization state within inflamed tissues, invariably leads to chronic inflammatory pathologies that include tumour development, autoimmune diseases (multiple sclerosis), modulation of T cell–mediated nervous system autoimmune disease [32], fat mass development, obesity-associated cardiovascular pathologies and insulin resistance [33–35]. Human chronic venous ulcers (CVUs) represent a good example where macrophages fail to resolve a chronic inflammatory condition probably because of the failure in the M1-to-M2 switch [36]. In CVUs, iron overload appears to sustain M1 polarization, thus leading to ROS-mediated DNA damage, fibroblast cellular senescence and defective tissue repair [36]. Obesity-associated insulin resistance, diabetes and metabolic syndrome are also sustained by a chronic subclinical inflammation. In obesity, adipocytes release mediators (CCL2, TNFα, free fatty acids) that promote the recruitment and subsequent M1-like activation of Adipose Tissue Macrophages (ATM) [27]. ATM activation by inflammatory cytokines and saturated fatty acids provokes the inhibition of the insulin signaling pathway by JNK, IKK and IRS, the activation of AP-1 and NF-κB (further increasing pro-inflammatory cytokine secretion) and, consequently, leads to insulin resistance [33, 37]. In fact, whereas normal ATMs express high levels of M2-associated genes, including IL-10 and Arg1, ATMs from obese mice and human exhibit an M1-like profile, with up-regulation of TNFα and NOS2 [34]. In brown adipose tissue, the IL-4-dependent polarization of ATM has been shown to orchestrate adaptive thermogenesis through the production of catecholamines [38].

Tumor progression probably constitutes the paradigmatic example of the pathological consequences of de-regulated macrophage polarization [39]. Macrophages play a key role in cancer-related inflammation, and their presence usually correlates with a poor outcome. The link between infiltration by Tumor Associated Macrophages (TAM) and a bad prognosis now extends to the case of Classic Hodgkin's Lymphoma [40], colon cancer [41], breast cancer [42], hepatocellular carcinoma [43], melanoma [44] and many other tumors. As an illustrative example, the antitumor activity of trabectedin has been recently shown to exert its activity by inducing apoptosis in TAM [45]. The tumor environment has a key role in determining the immune suppressive capacity of TAM. Monocytes from peripheral blood are recruited into the tumor by M-CSF and chemokines like CCL2 and CXCL12. At the early stages of tumor-promoted inflammation, macrophages secrete pro-inflammatory cytokines (TNFα, IL-12, IL-1 and IL-6), which might reduce tumour growth and progression and initiate an anti-tumor immune response [46, 47]. However, this anti-tumor behavior turns immunosuppressive and pro-tumoral as cancer progresses [48]. This is so because tumor and stroma cells contribute to the modulation of macrophage effector functions (polarization) by secreting M-CSF, IL-10, IL-6 and VEGF [49], all of which drive the acquisition of M2-associated properties. Moreover, and through the production of M2-skewing IL-4, IL-13 (mainly produced by tumor-infiltrating Th2 lymphocytes) [42], CCL2 and IL-6, tumor-recruited myeloid cells also favour tumor survival by committing TAM to a pro-tumoral setting [50]. In this scenario, TAM block cancer-induced immune responses, eliminate or switch off M1 macrophages, limit innate immune responses by impairing NK cells and T cells activation, display defective production of inflammatory cytokines and produce high levels of the Treg-inducing cytokine IL-10 [48, 51–54]. Therefore, macrophages are capable of destroying cancer cells and promoting anti-tumoral immune responses through their antigen-presentation capacity, but might also contribute to tumor progression via immune [39] and non-immune mechanisms [54] like promotion of angiogenesis [55], facilitation of tumor cell invasion and metastasis [56] and protection of tumor cells from chemotherapy-induced apoptosis [57]. As a consequence, TAM polarization represents a critical process in tumor development, and has become a target for the development of anti-tumoral therapeutic strategies.

9.1.2 Macrophage Polarization: Markers and Metabolism

A number of well-known markers are currently used to define the polarization state of mouse macrophages. Murine M1 "classically activated" macrophages (exposed to LPS and/or IFNγ) are usually identified by their high expression of NOS2, CD11c or CCR7 [58–60]. On the other hand, the IL-4-dependent M2 macrophage polarization is usually defined through the expression of Arginase-1, macrophage mannose receptor (CD206), the chitinase-like Ym1 molecule, and Fizz-1 (found in inflammatory zone-1) [60] (Fig. 9.1).

Cellular metabolism appears to lie at the basis of the opposite effector functions of M1 and M2 macrophages. The differential expression of NOS2 and Arginase-1 has become the best-established difference between classical and alternative mouse macrophages in terms of metabolism: while M1 macrophages use arginine to generate bactericidal nitric oxide through the stimulation of inducible nitric oxide synthase (NOS2), Arginase-1-expressing M2 macrophages produce the polyamine precursors urea and ornithine, needed for collagen synthesis and cell proliferation, respectively [61]. There are, however, other metabolic differences between M1 from M2 macrophages. Regarding lipid metabolism, M2 macrophages show significantly higher fatty-acid uptake and fatty acid oxidation than M1 macrophages [61]. This difference correlates with the opposite expression of COX-2 and COX-1, considered as M1 and M2 markers, respectively [62]. Iron metabolism also differs between both macrophage subsets. M1 macrophages are set to an iron-retention phenotype defined as $CD163^{Low}$, $Ferritin^{High}$ and $Ferroportin^{Low}$, a profile that agrees with their bacteriostatic and tumoricidal activity. Conversely, M2 macrophages are set to an iron-export mode that supports immune-regulation, matrix remodeling and cell proliferation [63, 64]. Molecules implicated in folate capture and metabolism are also differentially expressed in M1 and M2 macrophages [65, 66]. In the case of the glucose metabolism, M1 macrophages exhibit high expression of 6-Phospho-fructo-2-kinase (PKF2) and elevated glycolysis, probably to fulfill their high-energy requirements and to override potentially hypoxic microenvironments. Along this line, chronic activation of bone marrow derived-macrophages increases intracellular succinate, thus stabilizing HIF-1α and synergizing with Toll-like receptors for enhanced glycolysis [67]. On the contrary, M2-macrophages display a stronger oxidative-glucose metabolism and a potent beta-oxidation in lipid metabolism. The relevant contribution of the intrinsic glucose metabolism to macrophage polarization is illustrated by the ability of CARKL, an enzyme of the pentose phosphate pathway, to promote M2-polarization [68]. Therefore, glucose metabolism regulation is also linked and determines macrophage polarization [67, 68].

Numerous studies have now provided evidences of "difficulties of mouse-to-human extrapolation" regarding human macrophage polarization markers [4]. Although the functions attributed to M1 and M2 macrophages are usually common in both species, transcriptional comparison of monocyte-derived human macrophages and bone marrow-derived mouse macrophages demonstrated that only 26 % of the polarization-associated genes are conserved between both species [59]. In an attempt to identify human macrophage polarization markers with potential diagnostic and therapeutic value, numerous groups, including our own, have determined the transcriptional profiling of various *ex vivo*-isolated macrophage populations as well as of *in vitro* generated monocyte-derived macrophages exposed to M1 and M2-polarizing cytokines (IFNγ, GM-CSF, LPS, IL-4, M-CSF, IL-10, etc.) [3, 4, 39, 69]. The gene expression profiles of GM-CSF- and M-CSF-polarized macrophages indicates that *CCL1, CCL5, CCL22, CCR6, CSF1, FLT1* and *ADORA1* are more highly expressed in human M1 macrophages, while *IL-10, THBS1, ALK, DLL1, IGF-1, ADRB2* and *MSR1* are preferentially expressed by human M2 macrophages [59]. In addition, CD163, LXR, CD200R, CD36, c-Myc, FRβ, HTR2B, HO1 and DC-SIGN are preferentially expressed by human anti-inflammatory M2 macrophages as well as by tumour-associated macrophages (TAM) [2, 65, 69–74], whereas CD80, Activin A, PHD3, IRF5 and IRF4 are preferentially expressed in human M1 macrophages [59, 73, 75–77] (Fig. 9.1).

9.1.3 Factors and Signaling Pathways That Regulate Macrophage Polarization

A network of signaling molecules, transcription factors, epigenetic mechanisms and post-transcriptional regulators underlies the existence of the distinct M1 and M2 polarization/activation states. For M1 polarization, macrophages respond to TLR ligands (like LPS) by activating the NF-κB transcription factor family (RelA/p65, c-Rel, RelB, p50 and p52) [78] (Fig. 9.2). By contrast, the NFκB signaling pathway seems to be absent or defective in M2 polarization, where the presence of inhibitory NFκB p50 homodimers is augmented [79–81]. NFκB p50 homodimers enhance IL-10 gene transcription, inhibit the NFκB-dependent production of IFNγ and negatively regulate STAT1 activity, whereas their absence leads to increased levels of TNFα and IL-12 [81, 82]. Consequently, NFκB p50 homodimers impair M1 macrophage-mediated responses [79], and promote a defective inflammatory NFκB-dependent cytokine profile ($IL\text{-}10^{high}/IL\text{-}12^{low}$) [80, 81].

The interferon-related molecules of the IRF, STAT and SOCS families also regulate macrophage polarization, with the TRIF/IRF3/INFγ/STAT1 axis being involved in M1 polarization [27, 83] (Fig. 9.2). Stimulation of the IFNγ receptor triggers a JAK-mediated tyrosine phosphorylation and the subsequent dimerization of STAT1, which promotes the expression of "classical" M1 genes such as NOS2, MHC class II transactivator (CIITA), IL-12 and TNFα [84]. Although TAM display an M2-skewed phenotype with low expression of M1-associated genes, high levels of expression and activation of IRF3 and STAT1 has been found in some TAM [80], suggesting that the IRF3/STAT1 axis exerts a dual role in macrophage polarization. Several studies have also suggested that the expression of IL-10, which characterizes M2 macrophages and is required to avoid damage caused by an excess of pro-inflammatory cytokines, can be promoted in some settings by the TRIF/IRF3/INFγ/STAT1 pathway, thus highlighting the critical role of

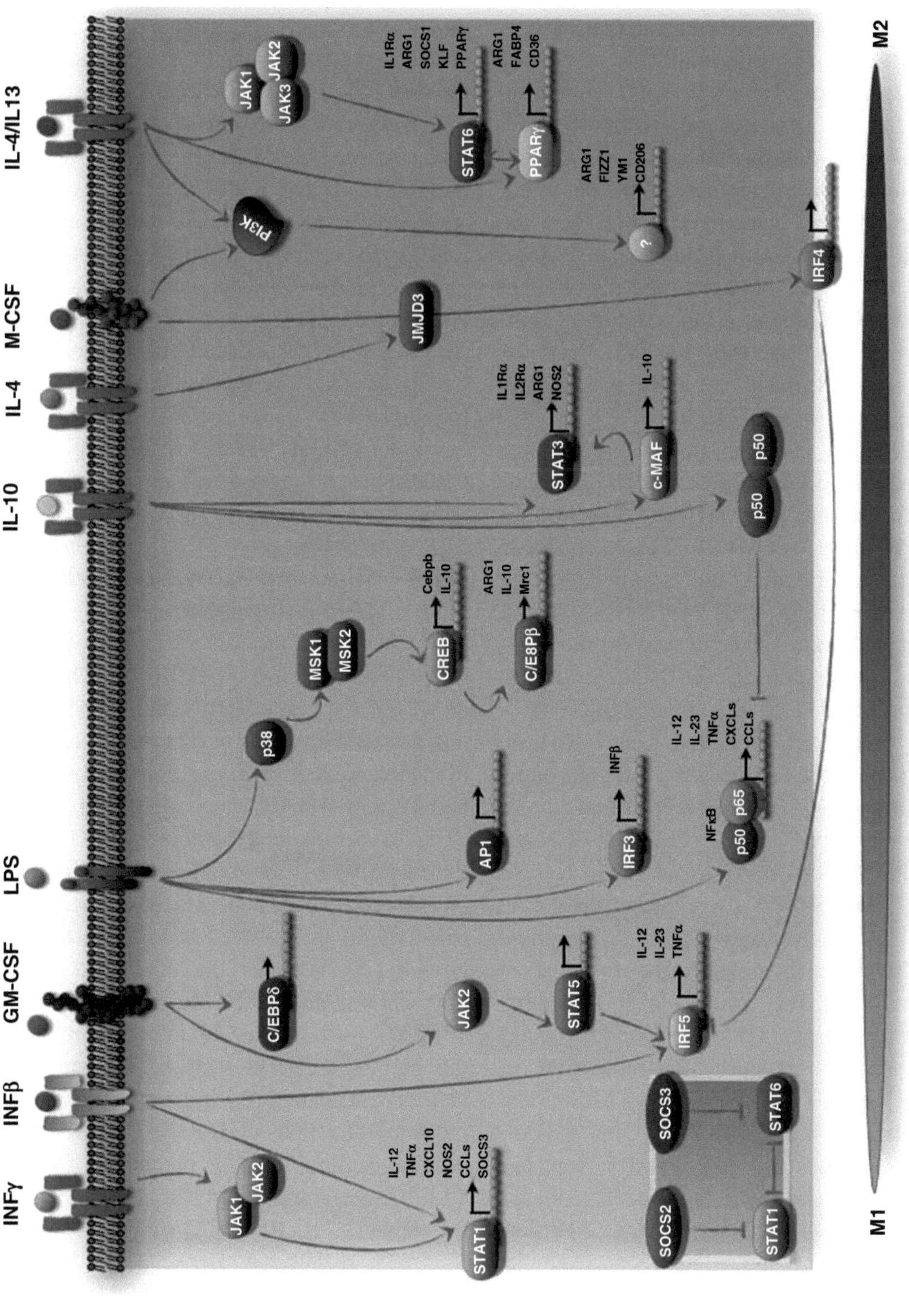

Fig. 9.2 Schematic representation of the signaling pathways and transcription factors that control macrophage polarization promoted by the indicated stimuli

STAT1 in both initiating and limiting M1 macrophage polarization [80, 85].

The contribution of IRF5 to human M1 polarization is somewhat controversial. Whereas Udalova's group provided evidence that IRF5 is required for the acquisition of M1-associated genes and functions [77], others have found no evidence for such a role and propose that, instead, IRF4 might play that function [59]. In fact, epigenetic studies have demonstrated the participation of the histone demethylase JMJD3 in the increased transcription of M2-associated genes triggered by M-CSF or chitin, and that JMJD3 inhibits the IRF4-mediated transcription of typical M1-associated genes [86] (Fig. 9.3). Recently, it has been demonstrated that SOCS also control macrophage polarization. The use of SOCS2- and SOCS3-deficient macrophages has led to the finding that SOCS2 depletion causes STAT1 activation and enrichment in M1-like macrophages. Conversely, the lack of SOCS3 leads to enhanced STAT6 phosphorylation and accumulation of M2-like macrophages [87] (Fig. 9.3). In fact, STAT6 is well-known to be critical for the IL-4-driven M2 polarization, and the IL-4Rα-triggered activation of STAT6 results in enhanced expression of many M2-associated gene markers, including *Arg1* and *Cd206* [88]. Along the same line, the antagonism between STAT1 and STAT6 has been described for the Th1 and Th2 cell polarization induced by IFNγ and IL-4, respectively [27, 84]. STAT6 also induces PPARδ in adipose tissue macrophages (ATM), driving them towards M2 polarization [89]. The transcription factor PPARγ, a master regulator of lipid metabolism in macrophages, inhibits pro-inflammatory gene expression through trans-repression of NF-κB [84], and is also required for M2 macrophage polarization [33, 90]. At present, the collaboration of STAT6 and PPARγ for M2 macrophage polarization in murine macrophages is widely accepted [91]. Similarly, STAT3 also contributes to M2 polarization. As an example, macrophage polarization in response to IL-10 is primarily dependent on STAT3, c-MAF, and NF-κB p50 homodimers, and the expression of IL-10, required for resolution of inflammation, is regulated by PU-1, STAT3 and c-MAF [90, 91].

Apart from NFB and STAT, the CREB and C/EBP families of transcription factors are also important factors involved in macrophage polarization (Fig. 9.2). Both STAT6 and C/EBPβ are essential for *Arg1* expression in macrophages in a stimulus-specific manner. C/EBPβ also regulates the expression of M2-associated genes, since the CREB-dependent activation of the CEBPβ promoter is needed for the expression of Arginase-1, IL-10 [92–94] and *SERPINB2* [95]. Recent studies have also demonstrated the implication of C/EBPδ in the GM-CSF-promoted M1-polarization on bone marrow derived macrophages, where it inhibits M2 polarization [96]. Conversely, in the case of the M-CSF-induced M2-like polarization, CREB has been proposed to exert a key role because it contributes to IL-10 expression, and because M-CSF-initiated signaling induces CREB phosphorylation [95].

9.1.4 Signaling Pathways Driving TAM Polarization

Like M2 macrophages, TAMs express high levels of M2-related genes [40, 97, 98]. TAM polarization has been dissected at the transcriptional level in several models, including murine fibrosarcoma and human ovarian carcinoma [79]. In those cases, TAMs displayed a defective M1 polarization caused by the nuclear accumulation of NF-κB p50 homodimers, which ultimately impair anti-tumoral responses and promote tumor growth [79, 80]. On the other hand, murine fibrosarcoma-derived TAM exhibited high levels of STAT1 [80]. In line with their role in M2 macrophage polarization fact, STAT transcription factors contribute to the immunosuppressive and pro-tumoral actions of TAM: STAT6 KO macrophages display an M1 phenotype [99], and STAT3 fosters tumor survival and dissemination by inducing IL-10, IL-23/p19 and inhibiting IL12/p35 expression [100, 101]. Altogether, these observations point to STAT3 as a potential therapeutic target in cancer, as demonstrated in the case of human squamous cell carcinoma [102] and animal models of glioma [103] and melanoma [104]. With all these signaling pathways

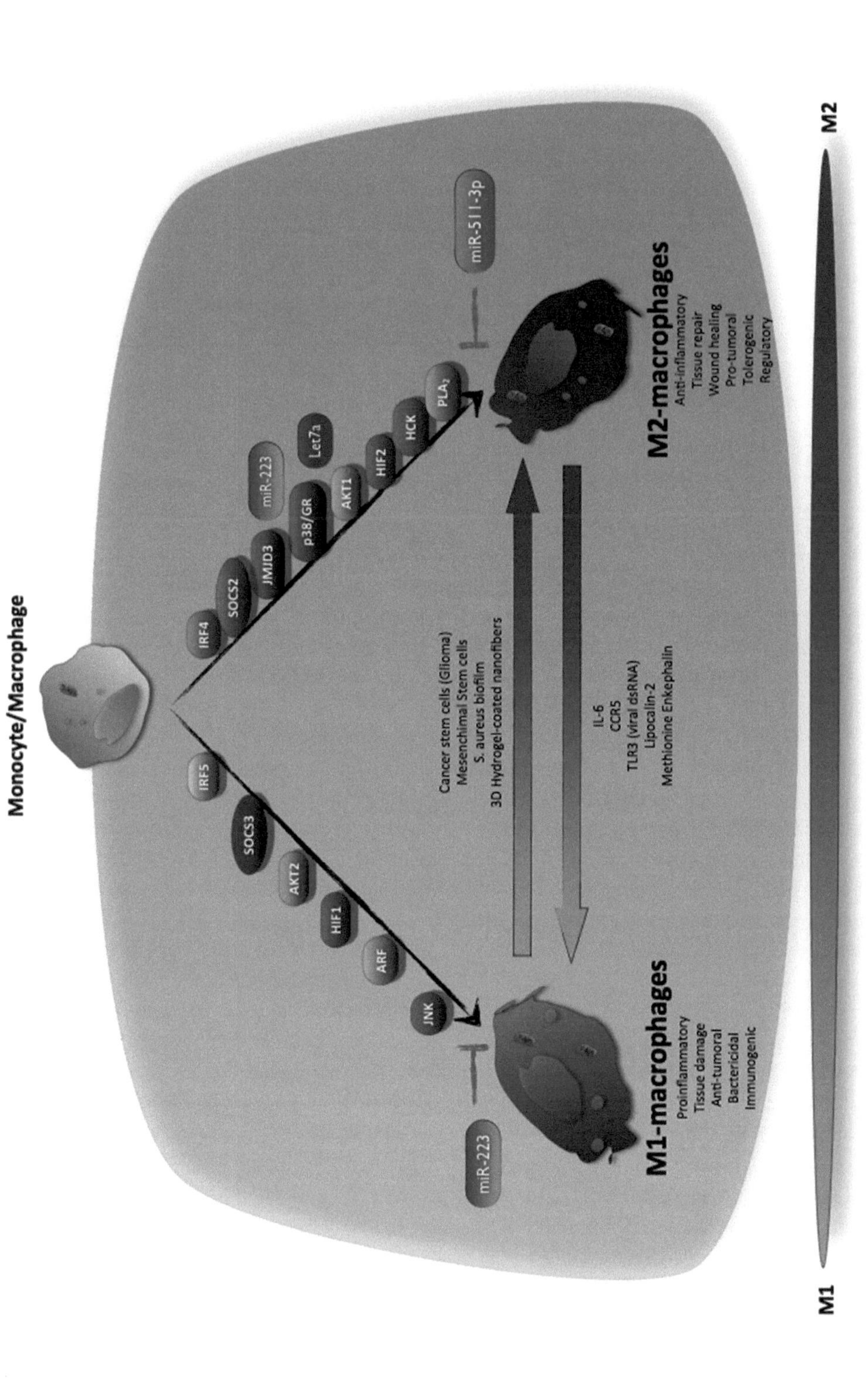

Fig. 9.3 Summary of factors known to modulate macrophage polarization

acting together, the stimulation of an M2-to-M1 switch in TAM appears as a desirable strategy to improve anti-tumoral therapies. Thus, it has been described that TLR9-mediated activation of NFκB in combination with anti-IL-10 treatment promotes a switch in the polarization state of TAMs in a mouse mammary carcinoma model [105]. Moreover, the fact that the M1-to-M2 polarization in TAMs might be driven by apoptotic-derived molecules (like S1P) within the tumor environment [106, 107] is in agreement with the reduced tumor progression detected after prevention of apoptotic cell recognition by macrophages [108]. A schematic representation of the factors and signaling pathways that regulate macrophage polarization is shown in Figs. 9.2 and 9.3.

9.2 Serotonin

The biogenic amine serotonin (5-hydroxytryptamine, 5-HT) has been known for more than a century. However, although its role as neurotransmitter is well established, evidences for its role in many other fundamental aspects of physiology and behavior (mood, aggression, sleep, appetite, pain sensation, bone mass, tissue regeneration, platelet coagulation, gastrointestinal function and thermo-regulation) [109–112] are still being gathered. Approximately 2 % of the tryptophan present in the diet is used for 5-HT synthesis. The rate-limiting enzyme involved in the synthesis of 5-HT is tryptophan hydroxylase (TPH), which converts L-tryptophan to 5-hydroxy-1-tryptophan through oxidation at position 5 of the pyrrole ring. A later step mediated by 5-OH-tryptamine decarboxylase leads to the generation of 5-HT. Two different isoforms of TPH exist: TPH-1 (expressed in the pineal gland and peripheral tissues) and TPH-2 (exclusively expressed in the dorsal raphe nucleus of the brain) [111]. TPH-1 provides 5-HT to non-neural cells whereas TPH-2 supplies 5-HT to the brain and mesenteric plexus, thus establishing the existence of two independent serotonin systems (brain and periphery), and whose independency is further supported by the hydrophobic nature of the molecule, unable to cross the blood–brain barrier [113]. It is still currently unclear, however, whether these two serotonin systems are completely independent as some evidences suggests a certain level of communication between them.

In the brain, 5-HT is one of the most widely distributed neurotransmitters. Serotonergic fibers originate in the brain raphe nuclei and their synaptic connections, where 5-HT mediates circadian rhythms and endocrine-related physiologic functions such as food intake, sleep, reproductive activity, cognition mood and anxiety [109]. However, almost 95 % of the whole body amounts of 5-HT is present outside the central nervous system, as it is produced by enterochromaffin cells (EC) of the gut [109]. Once released from EC, 5-HT is taken up and primarily stored by platelets, but also by other cells like lymphocytes, monocytes, macrophages, mast cells and pulmonary neuro-endocrine cells [111–113].

9.2.1 Serotonin Receptors

Seven families of 5-HT receptors (5-HTR$_1$ to 5-HTR$_7$) have been described to mediate the physiological and pathological functions of 5-HT in brain and periphery (Fig. 9.4). Up to 15 genes have been identified within these families, corresponding to a total of 20 subtypes with several alternative splicing variants [112]. Differences in the tissue distribution of each 5-HTR subclass explain the amplitude and tissue-specificity of 5-HT activities, and also serve to fine-tune physiological and cellular responses to 5-HT. This has led to the suggestion that each particular 5-HTR type is linked to a specific biological response to 5-HT [112].

At the molecular level, 5-HT receptors are G protein–coupled receptors (GPCRs) (except for the ligand-gated ion channel 5-HTR$_3$). 5-HT receptors possess seven transmembrane spanning helices, three intracellular and three extracellular loops, an extracellular amino-terminus, and an intracellular carboxy-terminal region [112] (Fig. 9.4). The intracellular domains couple these receptors to various intracellular signaling

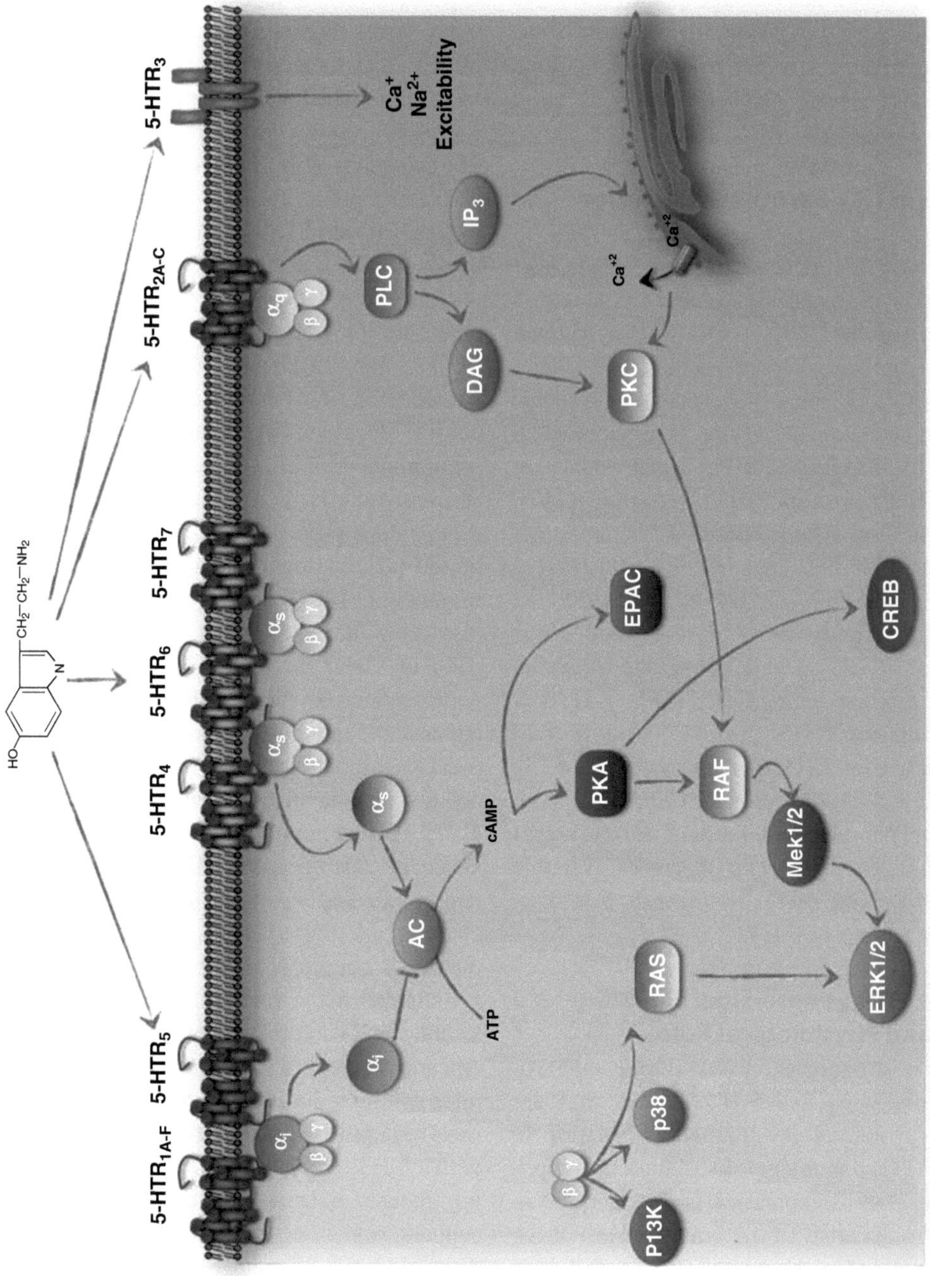

Fig. 9.4 Serotonin receptors and major intracellular signaling pathways initiated upon 5-HT binding to each specific receptor type

effectors, with the various receptors triggering different intracellular signaling cascades and distinct functional outcomes [112] (Fig. 9.4). Since 5-HTR$_{2B}$ and 5-HTR$_7$ receptors have been shown to regulate macrophage polarization and pro-inflammatory cytokine expression (*see below*), their most relevant roles in physiology and pathology are outlined below.

9.2.2 5-HTR$_{2B}$ Serotonin Receptor

5-HTR$_{2B}$ was first identified as the 5-HTR$_1$-like receptor that mediated 5HT-induced contraction of rat stomach fundus [114, 115]. Human 5-HTR$_{2B}$ is homologous to 5-HTR$_{2A}$ and 5-HTR$_{2C}$ (45 % and 42 % identity, respectively), shares with both receptors a similar pharmacological profile [116], and is encoded by a gene located at chromosomal position 2q36.3–2q37.1 [117]. 5-HTR$_{2B}$ has low affinity for ritanserin but higher affinity for yohimbine than 5-HTR$_{2A}$ or 5-HTR$_{2C}$. Along the same line, SB200646 and SB206553 have higher affinity for 5-HTR$_{2C/2B}$ and lower affinity for 5-HTR$_{2A}$, while spiperone shows an opposite binding properties [118]. 5-HTR$_{2B}$-specific antagonist (SB206741, 20-to-60-fold more selective for 5-HTR$_{2B}$ than for other 5-HTR$_2$ receptors) and agonist (BW723C86, with about 10-times higher selectivity for the 5-HTR$_{2B}$ *versus* 5-HTR$_{2A/2C}$) have recently become available [118] and proven useful to identify 5-HTR$_{2B}$-dependent actions of 5-HT.

9.2.2.1 5-HTR$_{2B}$ Expression, Signaling and Physiological Role

5-HTR$_{2B}$ is strongly expressed during embryogenesis, and mediates 5-HT actions that are essential for normal development [119, 120]. In adults, 5-HTR$_{2B}$ is mainly distributed in peripheral organs, but its expression is also detected in some restricted areas of the brain (at least at the mRNA level) [112]. A robust link exists between 5-HTR$_{2B}$ activation and cellular proliferation. The activation of 5-HTR$_{2B}$ leads to the stimulation of phospholipase C (PLC) pathway via the α subunit of the Gq GTP-binding protein. In 5-HTR$_{2B}$-expressing mouse fibroblasts, 5-HTR$_{2B}$ stimulates ERK1/2 (p42/44) in a G protein- and p21$_{Ras}$-dependent manner [121]. 5-HTR$_{2B}$ also activates the tyrosine kinase p60$_{Src}$, which induces both cyclin D1 and cyclin E via MAPKs-dependent or -independent pathways, respectively [122]. These cyclins lead to the induction of retinoblastoma protein and the transcription factor E2F, which subsequently activate the transcription of genes involved in DNA replication. The activation of the 5-HTR$_{2B}$ downstream targets p21$_{Ras}$ and p60$_{Src}$, together with the contribution of PDGFR and ErbB-2 growth factor receptors, seems to play a pivotal role in 5-HTR$_{2B}$-induced mitosis [123]. However, 5-HTR$_{2B}$ has also been proposed to exert anti-apoptotic activity via MAPKs and PI3-Kinase (PI3K). Consequently, it is accepted that 5-HTR$_{2B}$ receptors mediate cardiac development during embryogenesis by promoting proliferation and differentiation of cardiomyoblasts [123]. In adult heart, some groups have proposed that 5-HTR$_{2B}$ participates in the progression of myocardial hypertrophic remodeling [124], and activation of 5-HTR$_{2B}$ results in increased proliferation of interstitial cells of Cajal *in vivo* [125]. Small intestinal neuro-endocrine tumors also exhibit 5-HTR$_{2B}$-dependent mitosis [126], and blocking of 5-HTR$_{2B}$ signaling in pregnant mice limits beta cell expansion and causes glucose intolerance [127]. Further supporting the physiological relevance of this receptor, HTR$_{2B}^{-/-}$ mice present embryonic and neonatal lethality. Histological analysis of these HTR$_{2B}^{-/-}$ embryos reveals increased apoptosis in the heart and decreased cell number in the ventricular *trabeculae*, as well as abnormal sarcomeric organization in the sub-epicardial layer [128, 129]. Importantly, HTR$_{2B}^{-/-}$ mice exhibit reduced bone density, confirming its involvement in osteogenesis [130], and treatment with a 5-HT$_2$ receptor inverse-agonist of developing mouse embryos induces apoptosis in the cephalic region, neural tube and heart [123]. All these examples illustrate that 5-HTR$_{2B}$-dependent mitosis takes place in many physiological settings, and the crucial role of 5-HTR$_{2B}$ in development.

In the specific case of the nervous system, the role of 5-HTR$_{2B}$ has not been completely elucidated. Some studies implicate this receptor in the control of the respiratory network, because local administration of an 5-HTR$_{2B}$-specific agonist in the pre-Bötzinger complex increases respiratory frequency [131, 132]. However, HTR$_{2B}$$^{-/-}$ mice respiratory activity is unaltered, thus suggesting that 5-HTR$_{2B}$ might not be the only 5-HT receptor implicated in this process. In fact, 5-HTR$_{2B}$ is present in all respiratory nuclei and is co-expressed with low levels of 5-HTR$_{2A}$ in many cells. It has been postulated that, once circulating 5-HT concentration exceeds the levels required to activate 5-HTR$_{2A}$, 5-HTR$_{2B}$ receptor activation would modulate the respiratory rhythm in a dose-dependent manner [132].

9.2.2.2 5-HTR$_{2B}$ in Pathology

As indicated above, abnormal 5-HTR$_{2B}$ expression/function is strongly linked with the appearance of cardiac pathologies. In the 1980s and 1990s, the administration of the effective appetite suppressant fenfluramine was widely used until its use was linked to valvular heart disease (VHD) and Pulmonary arterial hypertension (PAH) [133, 134]. Patients taking the fenfluramine/phentermine (Fen-Phen) drug combination for 1–28 months developed heart valve abnormalities, with high myofibroblast proliferation [133]. Later studies reveled that fenfluramine and its metabolite norfenfluramine are potent agonists of 5-HTR$_{2B.}$ Therefore, activation of 5-HTR$_{2B}$ on heart valve interstitial cells leads to the formation of proliferative foci and subsequent changes that compromise tissue functions (e.g. increased extracellular matrix deposition and leukocyte infiltration) [135, 136].

Pulmonary arterial hypertension (PAH), a progressive and fatal disorder that results from an increase in pulmonary blood pressure associated with abnormal vascular proliferation [137], is also associated with abnormal 5-HTR$_{2B}$ activation. Analysis of a chronic-hypoxic-mouse model demonstrated that hypoxia-dependent increase in pulmonary blood pressure and lung remodeling is associated with a 5-HT- and 5-HTR$_{2B}$-dependent increase in vascular proliferation, elastase activity and TGFβ levels [138–140]. Very recently, bone marrow progenitor cells have been implicated in the etiology of PAH [141]. Thus, it seems that 5-HT contributes to pulmonary vascular remodeling and the pathogenesis of PAH by activating bone marrow progenitor through 5-HTR$_{2B}$ [142]. Altogether, all these observations point to 5-HTR$_{2B}$ as a key mediator of 5-HT-induced proliferation both under physiological and pathological conditions.

9.2.3 5-HTR$_7$ Serotonin Receptor

5-HTR$_7$ was identified as a serotonin receptor in 1993 [143–145] and, since then, has been described in numerous species. 5-HTR$_7$ is highly expressed in the brain, particularly in the neocortex, hippocampus, and hypothalamus, as well as in the suprachiasmatic nucleus [146]. In periphery, 5-HTR$_7$ is predominantly detected in smooth muscle cells of the cardiovascular [140], gastrointestinal [147] and reproductive system [148, 149], and in corneal epithelial cells [150]. The 5-HTR$_7$-encoding gene is located on human chromosome 10q21–q24, and gives rise to at least five splice variants in human, mouse, and rat. So far, three splice variants have been identified in human (5-HTR$_{7A}$, 5-HTR$_{7B}$, and 5-HTR$_{7D}$), three in mouse (5-HTR$_{7A}$, 5-HTR$_{7B}$, and 5-HTR$_{7C}$), and four in rat (5-HTR$_{7A}$, 5-HTR$_{7B}$, 5-HTR$_{7C}$, and 5-HTR$_{7E}$) [146]. The three human splice variants encode proteins of 448 (5-HTR$_{7A}$), 435 (5-HTR$_{7B}$), and 479 (5-HTR$_{7D}$) amino acids. The 5-HTR$_{7A}$ isoform is more widely expressed, followed by the 5-HTR$_{7B}$ variant, while the frequency of the 5-HTR$_{7C}$ and the 5-HTR$_{7D}$ isoforms is low. Splicing variants, however, do not seem to possess functional differences and are pharmacologically indistinguishable [146]. The availability of selective 5-HTR$_7$ ligands has been the limiting factor for elucidating the functions of this receptor. It is well established that 5-Carboxamidotryptamine (5-CT) and 8-OH-DPAT act as agonists for 5-HTR$_7$, but they can also activate other serotonin receptors. Fortunately, in the last years, 5-HTR$_7$-specific agonists have been described, including LP-12 [151] and AS-19 [152]. On the other hand, selective antagonists of 5-HTR$_7$, such as SB-258719 and SB-269970, have also helped in the analysis of the activity and functions of this receptor [112, 146].

9.2.3.1 5-HTR$_7$ Expression, Signaling and Physiological Role

5-HTR$_7$ is a GPCR that interacts with Gα_{12}. 5-HTR$_7$ activation leads to stimulation of adenylyl cyclase (AC), resulting in the conversion of ATP to cyclic AMP (cAMP) [146], an ubiquitous intracellular messenger that interacts with the phosphorylating enzyme protein-kinase-A (PKA) [153] and the exchange-protein-activated-by-cAMP (Epac) [154]. PKA phosphorylates cAMP-responsive transcription factors, such as the cAMP-response-element-binding-protein (CREB), thus affecting gene expression, whereas Epac activates Rap and Ras GTPases [112]. The activation of the 5-HTR$_7$-Gα_{12} signalling pathway also leads to stimulation of Cdc42 and RhoA, resulting in serum response element-mediated gene transcription, filopodia formation and cell rounding [155]. Ligation of 5-HTR$_7$ also activates ERK1/2 but in a cell-specific manner. 5-HT induces a rapid 5-HT$_7$-dependent phosphorylation of ERK and IκBα that enhances early T-cell activation and proliferation [156]. In the case of astrocytoma and microglia cell lines, the stimulation of 5-HT$_7$ triggers the expression of IL-6 via p38 and PKC activation [157, 158].

The activation of 5-HTR$_7$ in the central nervous system and the periphery modifies a number of cellular functions. In response to elevated concentrations of 5-HT in periphery, 5-HTR$_7$ mediates smooth muscle relaxation of the human colon [112, 159], and promotes IGF-1 synthesis in hepatocytes via cAMP/CREB/AKT [160]. Peripheral 5-HTR$_7$ also regulates the micturition reflex [161, 162], and modulates gut inflammation, since the blockade of 5-HTR$_7$ in dendritic cells improves the resolution of inflammation [163]. The role of 5-HTR$_7$ in the central nervous system has been addressed using *Htr7*$^{-/-}$ mice and selective antagonists (SB-269970). *Htr7*$^{-/-}$ mice exhibit a "antidepressant-like" phenotype [164, 165], whereas 5-HTR$_7$ antagonists increase the time to onset of REM sleep and reduce the time spent in REM [165, 166]. Antagonists of 5-HTR$_7$ also block 5-HT-induced hypothermia in both guinea pigs and rats. Interestingly, the role of 5-HTR$_7$ in thermoregulation has been confirmed in *Htr7*$^{-/-}$ mice, where 5-HT or 5-HTR$_7$ agonists fail to produce hypothermia [167]. Moreover, 5-HTR$_7$ has been implicated in anxiety, schizophrenia, nociception, epilepsy, and memory [146].

9.2.4 Serotonin and the Immune System

The determination of the role of 5-HT on immune cells has become a trendy objective when addressing the links between the nervous and immune systems. As mentioned above, 90 % of whole body serotonin is produced by enterochromaffin cells, and about 98 % of the remaining 5-HT is found in platelets, whereas only 2 % is located within CNS [110]. In platelets, 5-HT is stored in granules and constitutes a major secreted product upon platelet activation. Under physiological conditions, the plasmatic and vascular concentrations of 5-HT are maintained at low levels (nM) by mechanisms like uptake, storage or monoamine oxidase-mediated degradation [109]. Under inflammatory conditions, like thrombosis and ischemia, activated platelets release 5-HT and significantly increase its concentration around the inflamed area [110, 168]. Besides, pro-inflammatory stimuli (LPS, IFNγ) can directly induce platelet activation, causing a further elevation in peripheral blood 5-HT level [169, 170].

9.2.4.1 Cellular and Functional Effects of 5-HT on Immune Cells

5-HT has an immune-modulator role as it stimulates or inhibits numerous effector functions of B and T lymphocytes, NK cells and monocyte/macrophages/dendritic cells through interaction with distinct cell surface receptors (Fig. 9.5). 5-HT induces adhesion and chemotaxis in mouse and human mast cells, promoting their migration towards inflammatory sites through 5-HTR$_{1A}$ [171]. By contrast, the 5-HT-induced infiltration and migration of eosinophils is 5-HTR$_{2A}$-dependent [172, 173], as the lack of 5-HTR$_{2A}$

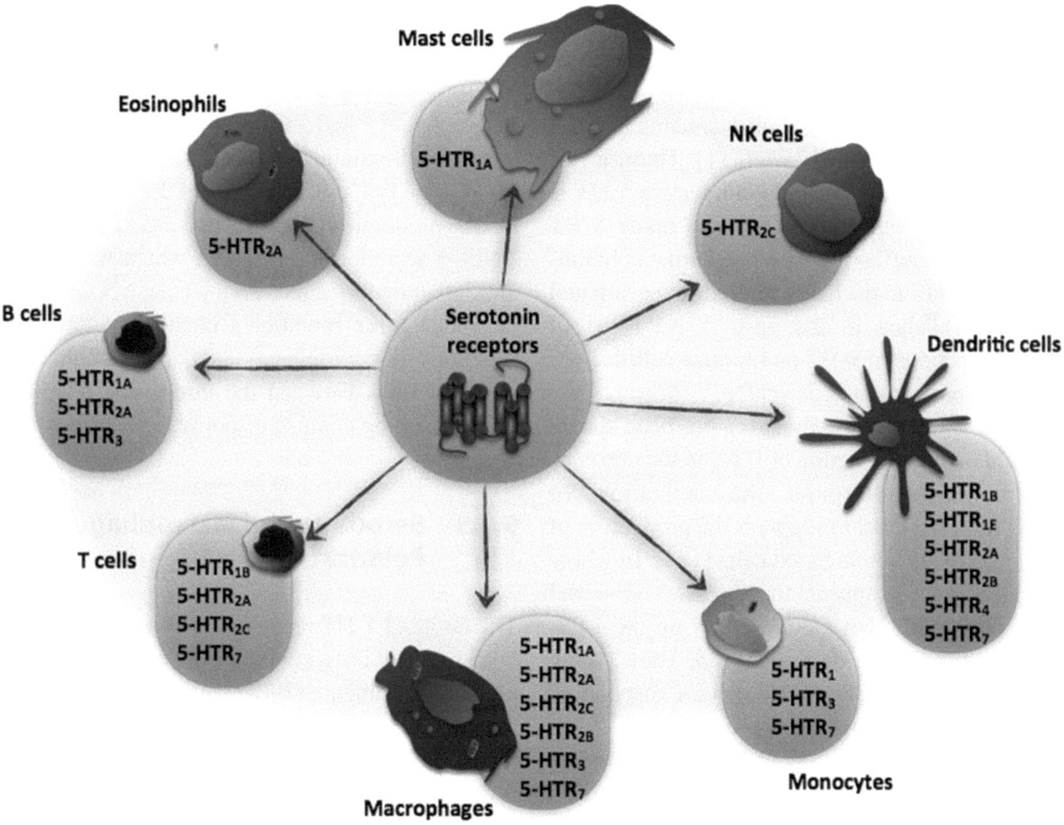

Fig. 9.5 Expression of serotonin receptors by immune cells

diminishes allergen-induced pulmonary eosinophilia in allergic asthma [173]. 5-HT also enhances the cytolytic function of NKs *in vitro* [174], and a long-term treatment with 5-HT reuptake inhibitors increases NK cell proliferation [175]. 5-HT can be shuttled from dendritic cells to T lymphocytes as a means to modulate activation, proliferation, and differentiation through 5-HTR_7 [156], and 5-HT appears to be necessary for optimal macrophage accessory function [176]. 5-HT also induces chemotaxis in immature human DC and enhances pulmonary DC migration to draining lymph nodes via the 5-HTR_1 and 5-HTR_2 receptor families [177]. In the case of macrophages, physiological concentrations of 5-HT (55 nM) suppress IFNγ–induced MHC class II expression and phagocytosis [178–180], regulate macrophage-mediated angiogenesis by reducing MMP12 expression in tumor-infiltrating macrophages [181], and modulate the production of chemotactic factors and cytokines from various immune cells [182]. Therefore, 5-HT participates in the control of numerous steps during immune and inflammatory responses.

9.2.4.2 5-HT and Cytokine Release from Immune Cells

The balance between pro-inflammatory and anti-inflammatory cytokines is crucial for the adequate control of inflammatory responses. As described above, the presence of 5-HT at inflammatory sites suggests its possible involvement in the control of the inflammation promotion/resolution equilibrium. The effect of 5-HT on immune cells has been studied in different cell models and most *in vitro* results have been obtained in the presence of foetal bovine serum. A potential experimental problem with previously published information, only recently appreciated, is the presence of considerable amounts of 5-HT in culture media, primarily derived from serum or from 5-HT producing cells. Indeed, 10 % of heat-inactivated foetal bovine serum contains 300 nM 5-HT as detected by ELISA [111]. These levels of 5-HT, assuming that immunoreactive 5-HT is bioactive, are sufficient to activate many 5-HT receptors. Thus, there is real danger that contaminating 5-HT may have altered experimental results published in the past. A brief list of reported actions of 5-HT on immune cell-derived cytokine production is presented.

In the case of NK cells, 5-HT has been shown to increase the production of IFNγ in the presence of monocytes through the activation of 5-HTR_{1A} [183], but to suppress the production of the cytokine in whole blood cells [184]. In whole blood, 5-HT also appears to decrease TNF-α and IL-6 production, but has no effect on the LPS-induced production of IL-10 [185, 186]. Along this line, human $CD14^+$ monocytes respond to 5-HT by increasing the production of LPS-stimulated IL-1β and IL-8, and by decreasing that of TNF-α [187]. Several studies have also revealed a role for 5-HT in DC cytokine secretion. Thus, 5-HT alters the cytokine profile of DC, enhancing IL-1β, IL-8, IL-6 and IL-10, and decreasing IL-12 and TNF- α [177, 188]. It has also been demonstrated that 5-HT impairs GM-CSF/IL-4–driven human monocyte-derived dendritic cell differentiation by reducing co-stimulatory molecule expression, CD1a levels and Mixed Lymphocyte Response stimulatory activity, and by increasing CD14 expression and IL-10 production through 5-HTR_1 or 5-HTR_7 [189]. In addition, 5-HT-treated DC show increased production of CCL22 (Th2 chemokine) and decreased levels of CXCL10 (Th1 chemokine) [177]. Consequently, DC treated with 5-HT induce Th2 polarization in naïve CD4 T cells [177]. However, recent studies have shown that gut DC produce IL-12 in response to 5-HT and LPS [190], and that the lack of 5-HTR_7 ameliorates mucosal inflammation [163].

Regarding macrophages, 5-HT is now known to modulate many of their effector functions, including phagocytosis [191]. 5-HT decreases the LPS-evoked production of TNF-α and IL-6 in murine peritoneal macrophages [192] and up-regulates the expression of CCL2 in 5-HTR_{2c}-expressing alveolar macrophages [193]. In human alveolar macrophages and macrophage-like synovial cells, 5-HT stimulation leads to overexpression of PGE2 [194, 195], enhances LPS-stimulated IL-10 production and decreases LPS-induced TNF- α secretion [195], all of which have a net anti-inflammatory effect. As a whole, it is therefore clear that 5-HT modifies a plethora of effectors functions of macrophages and dendritic cells, the essential links between the innate and adaptive branches of the immune response.

9.2.5 Serotonin and Macrophage Polarization

The range of 5-HT receptors in macrophages has been recently found to be dependent on the cytokine environment [196]. Both 5-HTR_{2B} and 5-HTR_7 are preferentially expressed by macrophages polarized by M-CSF, and their expression is considerably lower in macrophages exposed to GM-CSF [196]. In fact, *HTR2B* gene expression in macrophages is inhibited *in vitro* by stimuli inducing either M1 (LPS, GM-CSF) or M2 (IL-4, IL-10) polarization, whereas *HTR7* expression is downregulated by LPS, GM-CSF or IL-4. *In vivo*, expression of 5-HTR_{2B} can be primarily detected in M2-skewed macrophages, including human liver Kupffer cells, tissue-resident macrophages with anti-inflammatory ability (alveolar, colonic macrophages) and even TAM [196].

Since macrophage polarization is critically determined by the environment [197], it could be anticipated that 5-HT would influence both the macrophage phenotype and the range of tissue macrophage effector functions. Indeed, our results have revealed that 5-HT modulates macrophage functions and phenotype by acting through both 5-HTR_{2B} and 5-HTR_7 receptors [196] (Fig. 9.6). Regarding the phenotypic side of macrophage polarization, both 5-HT and the 5-HTR_{2B} agonist BW723C86 reduce the expression of GM-CSF-dependent M1 polarization markers *ALDH1A2*, *CD1B* and *MMP12*, whereas both agents enhance the expression of M-CSF-dependent M2 markers

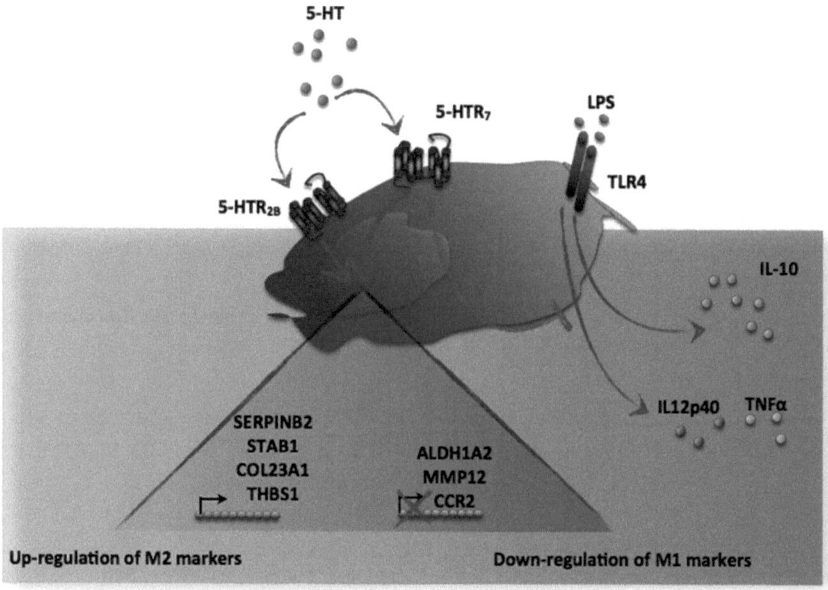

Fig. 9.6 Schematic representation of the influence of 5-HT on macrophage polarization, with indication of 5-HT-regulated cytokines and markers

SERPINB2, COL23A1, THBS1 and *STAB1*. Besides, 5-HT, but not the 5-HTR$_{2B}$ agonist, reduces the expression of GM-CSF-dependent M1 polarization markers *INHBA, CCR2* and *SERPINE1*, an effect that is hypothesized to be mediated by 5-HTR$_7$. In fact, 5-HTR$_7$ is also involved in the modulation of macrophage polarization by 5-HT since the 5-HTR$_7$-specific antagonist SB269970 prevents the 5-HT-induced upregulation of *STAB1, SERPINB2, THBS1* and *COL23A1* gene expression [196] (Fig. 9.6). Even more, the M-CSF-dependent macrophage polarization *in vitro* is inhibited by blockade of 5-HTR$_{2B}$ and 5-HTR$_7$. In this case, and regarding polarization-associated markers, *SERPINB2, HTR2B, ALDH1A2* and *CCR2* expression is controlled by 5-HTR$_{2B}$ activation, *STAB1* and *THBS1* expression is regulated by 5-HTR$_7$ activation, and both receptors influence *COL23A1, MMP12* and *CD1B* expression [196]. Therefore, 5-HT skews macrophages towards an M2 phenotype by inducing a set of M2-specific genes and repressing M1 markers [196]. Further transcriptomic analysis have now demonstrated that 5-HT exerts a fast effect on the macrophage gene expression profile, as two hours of exposure to 5-HT modify the mRNA levels of a significant number of genes. This fast effect is only partially mediated by 5-HTR$_{2B}$ and mostly mediated by 5-HTR$_7$ (*de las Casas-Engel and Corbí, unpublished*).

At the functional level, 5-HT inhibits the LPS-stimulated release of IL-12p40 and TNFα from M-CSF-dependent macrophages, an effect that is almost abrogated in the presence of the 5-HTR$_7$ antagonist SB269970. Therefore, 5-HT modulates the profile of pro-inflammatory cytokines through interaction with the 5-HTR$_7$ receptor [196]. Although 5-HT stimulation or 5-HTR$_{2B}$ activation activates ERK1/2 and its downstream targets MSK1/2, the precise intracellular signaling pathways underlying the modulator action of 5-HT on the cytokine production from M-CSF-dependent macrophages is currently unknown.

In summary, it can be concluded that 5-HTR$_{2B}$ and 5-HTR$_7$ are crucial receptors that mediate the skewing effect of 5-HT on human macrophage polarization, and that might also regulate the 5-HT-dependent pro-proliferative and tissue-repair activity of macrophages during physiological and pathological processes (*see below*).

9.2.6 Potential Implications of the Effect of Serotonin on Macrophage Polarization

9.2.6.1 5-HT and Cell Proliferation

The link between 5-HT and cell proliferation has been known for a long time, and is more evident in organs like heart and liver, where 5-HT levels have a clear pathophysiological influence. Both liver regeneration and hepatocyte proliferation are regulated by 5-HT. TPH1$^{-/-}$ hepatectomized mice (where lack of TPH1 impairs the 5-HT synthesis in peripheral tissues) display a robust reduction of several markers associated to hepatocyte proliferation [198], indicating that 5-HT is involved in the induction of liver regeneration. Interestingly, activation of 5-HT$_{2B}$ improves animal survival in small liver grafts transplantations [199] and age-associated impairments in regenerative capacity [200]. In fact, through its interaction with the 5-HTR$_{2B}$ receptor, 5-HT promotes cell proliferation in numerous cell types, including smooth muscle cells, fibroblasts, hepatocytes [201] and tumor cells [202]. Besides, 5-HT affects the expression of certain growth factors, as it upregulates TGFβ$_1$ synthesis in cardiac fibroblasts [203, 204] and potentiates TGFβ$_1$ signaling [205] in a 5-HTR$_{2B}$-dependent manner. In the case of cardiac fibroblasts, a cross-talk between EGF- and 5-HTR$_{2B}$-initiated signals has been proposed to explain the up-regulation of TGFβ$_1$ by 5-HT [204]. Along the same line, 5-HT stimulates IGF-1 and HGF production by hepatocytes in a 5-HTR$_7$-dependent manner [160].

The existence of the 5-HT/EGF, 5-HT/TGFβ$_1$ and 5-HT/IGF-1 axis in various cell types suggests that a similar link between 5-HT and growth factor production might be operative in macrophages. Macrophages are, in fact, capable of secreting a number of 5-HT-inducible proliferative factors such as EGF [206], VEGF [2, 207], PDGF, HGF [208], IGF-1 and TGFβ [2] a capacity that is specially prominent upon M2 polarization. Interestingly, preliminary profiling experiments have revealed that 5-HT triggers the expression of previously unnoticed growth factors in M-CSF-polarized M2 macrophages (*de las Casas-Engel and Corbí, unpublished*). The production of 5-HT-inducible growth factors by macrophages might explain the contribution of bone marrow-derived myeloid cells to the 5-HTR$_{2B}$-dependent development of pulmonary arterial hypertension [142]. Similarly, since Kupffer cells secrete factors that participate in liver regeneration and repair [209], it is tempting to speculate that 5-HTR$_{2B}$ on Kupffer cells [196] might have an impact on liver regeneration. Such hypothesis would be supported by the fact that 5-HT and the 5-HTR$_{2B}$ agonist BW723C86 favor the *in vitro* maintenance of the M2 macrophage polarization state (characterized by its tissue-repair and cell growth-promoting properties) [196].

Also related to the link between 5-HT and cell growth, the presence of 5-HTR$_{2B}$ on tumor-growth promoting TAM further suggests the putative contribution of M2 macrophages to the influence of 5-HT on tumor growth. 5-HT directly fosters tumor cell proliferation in some cases [210–212], and the growth of murine colon cancer allografts is dependent on the 5-HT-mediated reduction in MMP12 levels by TAM [181]. The fact 5-HT downregulates MMP12 and increases the expression of growth factors by M2 macrophages through 5-HTR$_{2B}$ [196] further suggests that the pro-tumoral actions of M2 macrophages might be explained, at least in part, through the 5-HT receptor-mediated modification of the macrophage gene expression program.

9.2.6.2 5-HT and Wound Healing

Wound healing is a second aspect that is worth mentioning regarding the influence of 5-HT on macrophage polarization. A subset of M2 macrophages are specialized and play an important role in wound healing, a process where platelets and 5-HT are well-defined players [213]. During wound healing, platelet aggregation and 5-HT release are rapidly followed by macrophage recruitment and release of pro-inflammatory cytokines in response to potentially damaging exogenous or altered endogenous products [213]. Therefore, it could be hypothesized that platelet-derived 5-HT might act on macrophages, limiting pro-inflammatory cytokine production and enhancing the release of growth factors that would enhance fibroblast/endothelial cell proliferation to restore tissue integrity and functionality.

9.2.6.3 5-HT and Psychiatric Diseases

The link between 5-HT and macrophages goes beyond inflammatory responses and extends to pathologies of psychiatric nature and even to the "sickness response". It is becoming increasing clear that some psychiatric disorders have an immune origin, especially those where sensory inputs result in altered behavior responses (e.g. schizophrenia, bipolar disorders, autism). Microglia cells are known to contribute to brain function regulation and play crucial role in psychiatric pathologies [214–217]. In this regard it is worth mentioning that 5-HTR$_{2B}$ is required for the antidepressant activity of 5-HT [218], and regulates severe impulsivity [219]. Therefore, the ability of 5-HT to modify the effector functions of myeloid cells (e.g., microglia) poses the question of whether 5-HT lies at the basis of these neurological pathologies by modifying macrophage polarization. A similar line of reasoning might apply to "sickness response", commonly used to refer to the changes in behaviour and physiology that take place during an infectious process, and whose occurrence is triggered by the activity of proinflammatory cytokines on brain cells [220]. During an infection, proinflammatory cytokines enhance the activity of the indoleamine 2,3 dioxygenase (IDO1) enzyme, thus increasing pro-depressive kynurenine levels and decreasing tryptophan (and 5-HT) levels [217]. The fact that IDO1 is preferentially expressed upon M1 polarization (its expression is greatly enhanced by TLR ligands and interferons), and that proinflammatory cytokines in the brain are downregulated by IL-10, IL-1Ra and IGF-1 (all of which are primarily produced by M2 macrophages), further strengthens the relevance of the physiological link between 5-HT and macrophage polarization.

9.2.6.4 5-HTR$_7$-Mediated Macrophage Polarization

A final, and purely speculative, putative link between 5-HT and macrophage polarization stems from the range of physiological processes where IL-4-polarized M2 macrophages are now known to participate. Recent reports have provided evidences that M2 ATM participate in adaptive thermogenesis in brown adipose tissue [38], and that M2 microglia cells regulate memory and learning in brain [21]. Intriguingly, both processes (memory/learning and thermogenesis) are under the control of 5-HT, and are known to be altered in *Htr7*$^{-/-}$ mice [221]. Given the presence of 5-HTR$_7$ in M-CSF-polarized human M2 macrophages and the ability of 5-HTR$_7$ antagonists to block the anti-inflammatory response to 5-HT [196], it might be worth determining the polarization state of *Htr7*$^{-/-}$ microglia cells, as well as assessing whether 5-HTR$_7$ expression in macrophages/microglia contributes to cognitive processes and adaptive thermogenesis.

Acknowledgements This work was supported by grants from the Ministerio de Ciencia e Innovación (SAF2011-23801), Instituto de Salud Carlos III (Red de Investigación en Enfermedades Reumáticas RIER RD12/009, Spanish Network for the Research in Infectious Diseases REIPI RD06/0008, and Red de Investigación en SIDA RIS RD06/0006/1016), and Programa de Actividades de I+D de la Comunidad de Madrid (RAPHYME-CM, S2010/BMD-2350), to ALC.

References

1. Karnovsky ML. Metchnikoff in Messina: a century of studies on phagocytosis. N Engl J Med. 1981;304:1178–80.
2. Biswas SK, Mantovani A. Macrophage plasticity and interaction with lymphocyte subsets: cancer as a paradigm. Nat Immunol. 2010;11:889–96.
3. Mosser DM, Edwards JP. Exploring the full spectrum of macrophage activation. Nat Rev Immunol. 2008;8:958–69.
4. Mantovani A, Sica A, Locati M. Macrophage polarization comes of age. Immunity. 2005;23:344–6.
5. Gordon S, Martinez FO. Alternative activation of macrophages: mechanism and functions. Immunity. 2010;32:593–604.
6. Mantovani A, Sozzani S, Locati M, Allavena P, Sica A. Macrophage polarization: tumor-associated macrophages as a paradigm for polarized M2 mononuclear phagocytes. Trends Immunol. 2002; 23:549–55.
7. Fleetwood AJ, Dinh H, Cook AD, Hertzog PJ, Hamilton JA. GM-CSF- and M-CSF-dependent macrophage phenotypes display differential dependence

on type I interferon signaling. J Leukoc Biol. 2009;86:411–21.
8. Fleetwood AJ, Lawrence T, Hamilton JA, Cook AD. Granulocyte-macrophage colony-stimulating factor (CSF) and macrophage CSF-dependent macrophage phenotypes display differences in cytokine profiles and transcription factor activities: implications for CSF blockade in inflammation. J Immunol. 2007;178:5245–52.
9. Erblich B, Zhu L, Etgen AM, Dobrenis K, Pollard JW. Absence of colony stimulation factor-1 receptor results in loss of microglia, disrupted brain development and olfactory deficits. PLoS One. 2011;6:e26317.
10. Wynn TA, Chawla A, Pollard JW. Macrophage biology in development, homeostasis and disease. Nature. 2013;496:445–55.
11. Erhardt A, Biburger M, Papadopoulos T, Tiegs G. IL-10, regulatory T cells, and Kupffer cells mediate tolerance in concanavalin A-induced liver injury in mice. Hepatology. 2007;45:475–85.
12. Ruiter DJ, van der Meulen J, Brouwer A, Hummel MJ, Mauw BJ, van der Ploeg JC, Wisse E. Uptake by liver cells of endotoxin following its intravenous injection. Lab Invest. 1981;45:38–45.
13. Bilzer M, Roggel F, Gerbes AL. Role of Kupffer cells in host defense and liver disease. Liver Int. 2006;26:1175–86.
14. Bedoret D, Wallemacq H, Marichal T, Desmet C, Quesada Calvo F, Henry E, et al. Lung interstitial macrophages alter dendritic cell functions to prevent airway allergy in mice. J Clin Invest. 2009;119:3723–38.
15. Denning TL, Wang YC, Patel SR, Williams IR, Pulendran B. Lamina propria macrophages and dendritic cells differentially induce regulatory and interleukin 17-producing T cell responses. Nat Immunol. 2007;8:1086–94.
16. Horsnell WG, Brombacher F. Genes associated with alternatively activated macrophages discretely regulate helminth infection and pathogenesis in experimental mouse models. Immunobiology. 2010;215:704–8.
17. Smythies LE, Sellers M, Clements RH, Mosteller-Barnum M, Meng G, Benjamin WH, et al. Human intestinal macrophages display profound inflammatory anergy despite avid phagocytic and bacteriocidal activity. J Clin Invest. 2005;115:66–75.
18. Smith PD, Ochsenbauer-Jambor C, Smythies LE. Intestinal macrophages: unique effector cells of the innate immune system. Immunol Rev. 2005;206:149–59.
19. Nemeth E, Baird AW, O'Farrelly C. Microanatomy of the liver immune system. Semin Immunopathol. 2009;31:333–43.
20. Yamamoto T, Kaizu C, Kawasaki T, Hasegawa G, Umezu H, Ohashi RJ, et al. Macrophage colony-stimulating factor is indispensable for repopulation and differentiation of Kupffer cells but not for splenic red pulp macrophages in osteopetrotic (op/op) mice after macrophage depletion. Cell Tissue Res. 2008;332:245–56.
21. Derecki NC, Cardani AN, Yang CH, Quinnies KM, Crihfield A, Lynch KR, Kipnis J. Regulation of learning and memory by meningeal immunity: a key role for IL-4. J Exp Med. 2010;207:1067–80.
22. Weinberg JB, Haney AF, Xu FJ, Ramakrishnan S. Peritoneal fluid and plasma levels of human macrophage colony-stimulating factor in relation to peritoneal fluid macrophage content. Blood. 1991;78:513–6.
23. Van Ginderachter JA, Movahedi K, Hassanzadeh Ghassabeh G, Meerschaut S, Beschin A, Raes G, De Baetselier P. Classical and alternative activation of mononuclear phagocytes: picking the best of both worlds for tumor promotion. Immunobiology. 2006;211:487–501.
24. Gordon S, Taylor PR. Monocyte and macrophage heterogeneity. Nat Rev Immunol. 2005;5:953–64.
25. Auffray C, Fogg D, Garfa M, Elain G, Join-Lambert O, Kayal S, et al. Monitoring of blood vessels and tissues by a population of monocytes with patrolling behavior. Science. 2007;317:666–70.
26. Bystrom J, Evans I, Newson J, Stables M, Toor I, van Rooijen N, et al. Resolution-phase macrophages possess a unique inflammatory phenotype that is controlled by cAMP. Blood. 2008;112:4117–27.
27. Sica A, Mantovani A. Macrophage plasticity and polarization: in vivo veritas. J Clin Invest. 2012;122:787–95.
28. Weigert A, Weichand B, Brune B. S1P regulation of macrophage functions in the context of cancer. Anticancer Agents Med Chem. 2011;11:818–29.
29. Troidl C, Mollmann H, Nef H, Masseli F, Voss S, Szardien S, et al. Classically and alternatively activated macrophages contribute to tissue remodelling after myocardial infarction. J Cell Mol Med. 2009;13:3485–96.
30. David S, Kroner A. Repertoire of microglial and macrophage responses after spinal cord injury. Nat Rev Neurosci. 2011;12:388–99.
31. Rapalino O, Lazarov-Spiegler O, Agranov E, Velan GJ, Yoles E, Fraidakis M, et al. Implantation of stimulated homologous macrophages results in partial recovery of paraplegic rats. Nat Med. 1998;4:814–21.
32. Weber MS, Prod'homme T, Youssef S, Dunn SE, Rundle CD, Lee L, et al. Type II monocytes modulate T cell-mediated central nervous system autoimmune disease. Nat Med. 2007;13:935–43.
33. Odegaard JI, Ricardo-Gonzalez RR, Goforth MH, Morel CR, Subramanian V, Mukundan L. Macrophage-specific PPARgamma controls alternative activation and improves insulin resistance. Nature. 2007;447:1116–20.
34. Lumeng CN, Bodzin JL, Saltiel AR. Obesity induces a phenotypic switch in adipose tissue macrophage polarization. J Clin Invest. 2007;117:175–84.
35. Bourlier V, Zakaroff-Girard A, Miranville A, De Barros S, Maumus M, Sengenes C, et al. Remodeling phenotype of human subcutaneous adipose tissue macrophages. Circulation. 2008;117:806–15.

36. Sindrilaru A, Peters T, Wieschalka S, Baican C, Baican A, Peter H, et al. An unrestrained proinflammatory M1 macrophage population induced by iron impairs wound healing in humans and mice. J Clin Invest. 2011;121:985–97.
37. Weisberg SP, Hunter D, Huber R, Lemieux J, Slaymaker S, Vaddi K, et al. CCR2 modulates inflammatory and metabolic effects of high-fat feeding. J Clin Invest. 2006;116:115–24.
38. Nguyen KD, Qiu Y, Cui X, Goh YP, Mwangi J, David T, et al. Alternatively activated macrophages produce catecholamines to sustain adaptive thermogenesis. Nature. 2011;480:104–8.
39. Allavena P, Sica A, Garlanda C, Mantovani A. The Yin-Yang of tumor-associated macrophages in neoplastic progression and immune surveillance. Immunol Rev. 2008;222:155–61.
40. Steidl C, Lee T, Shah SP, Farinha P, Han G, Nayar T, et al. A Tumor-associated macrophages and survival in classic Hodgkin's lymphoma. N Engl J Med. 2010;362:875–85.
41. Edin S, Wikberg ML, Dahlin AM, Rutegard J, Oberg A, Oldenborg PA, Palmqvist R. The distribution of macrophages with a M1 or M2 phenotype in relation to prognosis and the molecular characteristics of colorectal cancer. PLoS One. 2012;7:e47045.
42. DeNardo DG, Barreto JB, Andreu P, Vasquez L, Tawfik D, Kolhatkar N, Coussens LM. CD4(+) T cells regulate pulmonary metastasis of mammary carcinomas by enhancing protumor properties of macrophages. Cancer Cell. 2009;16:91–102.
43. Kuang DM, Zhao Q, Peng C, Xu J, Zhang JP, Wu C, Zheng L. Activated monocytes in peritumoral stroma of hepatocellular carcinoma foster immune privilege and disease progression through PD-L1. J Exp Med. 2009;206:1327–37.
44. Chen P, Huang Y, Bong R, Ding Y, Song N, Wang X, et al. Tumor-associated macrophages promote angiogenesis and melanoma growth via adrenomedullin in a paracrine and autocrine manner. Clin Cancer Res. 2011;17:7230–9.
45. Germano G, Frapolli R, Belgiovine C, Anselmo A, Pesce S, Liguori M, et al. Role of macrophage targeting in the antitumor activity of trabectedin. Cancer Cell. 2013;23:249–62.
46. Greten FR, Eckmann L, Greten TF, Park JM, Li ZW, Egan LJ, et al. IKKbeta links inflammation and tumorigenesis in a mouse model of colitis-associated cancer. Cell. 2004;118:285–96.
47. Maeda S, Kamata H, Luo JL, Leffert H, Karin M. IKKbeta couples hepatocyte death to cytokine-driven compensatory proliferation that promotes chemical hepatocarcinogenesis. Cell. 2005;121:977–90.
48. Sica A, Saccani A, Bottazzi B, Polentarutti N, Vecchi A, van Damme J, Mantovani A. Autocrine production of IL-10 mediates defective IL-12 production and NF-kappa B activation in tumor-associated macrophages. J Immunol. 2000;164:762–7.
49. Kusmartsev S, Gabrilovich DI. Effect of tumor-derived cytokines and growth factors on differentiation and immune suppressive features of myeloid cells in cancer. Cancer Metastasis Rev. 2006;25:323–31.
50. Roca H, Varsos ZS, Sud S, Craig MJ, Ying C, Pienta KJ. CCL2 and interleukin-6 promote survival of human CD11b+peripheral blood mononuclear cells and induce M2-type macrophage polarization. J Biol Chem. 2009;284:34342–54.
51. Sica A, Bronte V. Altered macrophage differentiation and immune dysfunction in tumor development. J Clin Invest. 2007;117:1155–66.
52. Curiel TJ, Coukos G, Zou L, Alvarez X, Cheng P, Mottram P, et al. Specific recruitment of regulatory T cells in ovarian carcinoma fosters immune privilege and predicts reduced survival. Nat Med. 2004;10:942–9.
53. Hagemann T, Wilson J, Burke F, Kulbe H, Li NF, Pluddemann A, et al. Ovarian cancer cells polarize macrophages toward a tumor-associated phenotype. J Immunol. 2006;176:5023–32.
54. Ruffell B, Affara NI, Coussens LM. Differential macrophage programming in the tumor microenvironment. Trends Immunol. 2012;33:119–26.
55. Lin EY, Li JF, Gnatovskiy L, Deng Y, Zhu L, Grzesik DA, et al. Macrophages regulate the angiogenic switch in a mouse model of breast cancer. Cancer Res. 2006;66:11238–46.
56. Qian B, Deng Y, Im JH, Muschel RJ, Zou Y, Li J, et al. A distinct macrophage population mediates metastatic breast cancer cell extravasation, establishment and growth. PLoS One. 2009;4:e6562.
57. Zheng Y, Cai Z, Wang S, Zhang X, Qian J, Hong S, et al. Macrophages are an abundant component of myeloma microenvironment and protect myeloma cells from chemotherapy drug-induced apoptosis. Blood. 2009;114:3625–8.
58. Han MS, Jung DY, Morel C, Lakhani SA, Kim JK, Flavell RA, Davis RJ. JNK expression by macrophages promotes obesity-induced insulin resistance and inflammation. Science. 2013;339:218–22.
59. Lacey DC, Achuthan A, Fleetwood AJ, Dinh H, Roiniotis J, Scholz GM, Chang MW, et al. Defining GM-CSF- and macrophage-CSF-dependent macrophage responses by in vitro models. J Immunol. 2012;188:5752–65.
60. Gordon S. Alternative activation of macrophages. Nat Rev Immunol. 2003;3:23–35.
61. Odegaard JI, Chawla A. Alternative macrophage activation and metabolism. Annu Rev Pathol. 2011;6:275–97.
62. Martinez FO, Gordon S, Locati M, Mantovani A. Transcriptional profiling of the human monocyte-to-macrophage differentiation and polarization: new molecules and patterns of gene expression. J Immunol. 2006;177:7303–11.
63. Biswas SK, Mantovani A. Orchestration of metabolism by macrophages. Cell Metab. 2012;15:432–7.
64. Cairo G, Recalcati S, Mantovani A, Locati M. Iron trafficking and metabolism in macrophages: contribution to the polarized phenotype. Trends Immunol. 2011;32:241–7.

65. Puig-Kroger A, Sierra-Filardi E, Dominguez-Soto A, Samaniego R, Corcuera MT, Gomez-Aguado F, et al. Folate receptor beta is expressed by tumor-associated macrophages and constitutes a marker for M2 anti-inflammatory/regulatory macrophages. Cancer Res. 2009;69:9395–403.
66. Samaniego R, Palacios BS, Domiguez-Soto A, Vidal C, Salas A, Matsuyama T, et al. Macrophage uptake and accumulation of folates are polarization-dependent in vitro and in vivo and are regulated by activin A. J Leukoc Biol. 2014;95:797–808.
67. Tannahill GM, Curtis AM, Adamik J, Palsson-McDermott EM, McGettrick AF, Goel G, et al. Succinate is an inflammatory signal that induces IL-1beta through HIF-1alpha. Nature. 2013;496:238–42.
68. Haschemi A, Kosma P, Gille L, Evans CR, Burant CF, Starkl P, et al. The sedoheptulose kinase CARKL directs macrophage polarization through control of glucose metabolism. Cell Metab. 2012;15:813–26.
69. Bories G, Colin S, Vanhoutte J, Derudas B, Copin C, Fanchon M, et al. Liver X receptor (LXR) activation stimulates iron export in human alternative macrophages. Circ Res. 2013;113:1196–205.
70. Sierra-Filardi E, Vega MA, Sanchez-Mateos P, Corbi AL, Puig-Kroger A. Heme oxygenase-1 expression in M-CSF-polarized M2 macrophages contributes to LPS-induced IL-10 release. Immunobiology. 2010;215:788–95.
71. Dominguez-Soto A, Sierra-Filardi E, Puig-Kroger A, Perez-Maceda B, Gomez-Aguado F, Corcuera MT, et al. Dendritic cell-specific ICAM-3-grabbing nonintegrin expression on M2-polarized and tumor-associated macrophages is macrophage-CSF dependent and enhanced by tumor-derived IL-6 and IL-10. J Immunol. 2011;186:2192–200.
72. Chaitidis P, Billett E, Kuban RJ, Ungethuem U, Kuhn H. Expression regulation of MAO isoforms in monocytic cells in response to Th2 cytokines. Med Sci Monit. 2005;11:BR259–65.
73. Jaguin M, Houlbert N, Fardel O, Lecureur V. Polarization profiles of human M-CSF-generated macrophages and comparison of M1-markers in classically activated macrophages from GM-CSF and M-CSF origin. Cell Immunol. 2013;281:51–61.
74. Ghassabeh GH, De Baetselier P, Brys L, Noel W, Van Ginderachter JA, Meerschaut S, et al. Identification of a common gene signature for type II cytokine-associated myeloid cells elicited in vivo in different pathologic conditions. Blood. 2006;108:575–83.
75. Sierra-Filardi E, Puig-Kroger A, Blanco FJ, Nieto C, Bragado R, Palomero MI, et al. Activin A skews macrophage polarization by promoting a proinflammatory phenotype and inhibiting the acquisition of anti-inflammatory macrophage markers. Blood. 2011;117:5092–101.
76. Escribese MM, Sierra-Filardi E, Nieto C, Samaniego R, Sanchez-Torres C, Matsuyama T, et al. The prolyl hydroxylase PHD3 identifies proinflammatory macrophages and its expression is regulated by activin A. J Immunol. 2012;189:1946–54.
77. Krausgruber T, Blazek K, Smallie T, Alzabin S, Lockstone H, Sahgal N, et al. TIRF5 promotes inflammatory macrophage polarization and TH1-TH17 responses. Nat Immunol. 2011;12:231–8.
78. Olefsky JM, Glass CK. Macrophages, inflammation, and insulin resistance. Annu Rev Physiol. 2010;72:219–46.
79. Saccani A, Schioppa T, Porta C, Biswas SK, Nebuloni M, Vago L, et al. p50 nuclear factor-kappaB overexpression in tumor-associated macrophages inhibits M1 inflammatory responses and antitumor resistance. Cancer Res. 2006;66:11432–40.
80. Biswas SK, Gangi L, Paul S, Schioppa T, Saccani A, Sironi M, et al. A distinct and unique transcriptional program expressed by tumor-associated macrophages (defective NF-kappaB and enhanced IRF-3/STAT1 activation). Blood. 2006;107:2112–22.
81. Porta C, Rimoldi M, Raes G, Brys L, Ghezzi P, Di Liberto D, et al. Tolerance and M2 (alternative) macrophage polarization are related processes orchestrated by p50 nuclear factor kappaB. Proc Natl Acad Sci U S A. 2009;106:14978–83.
82. Cao S, Zhang X, Edwards JP, Mosser DM. NF-kappaB1 (p50) homodimers differentially regulate pro- and anti-inflammatory cytokines in macrophages. J Biol Chem. 2006;281:26041–50.
83. Kawai T, Akira S. The role of pattern-recognition receptors in innate immunity: update on Toll-like receptors. Nat Immunol. 2010;11:373–84.
84. Lawrence T, Natoli G. Transcriptional regulation of macrophage polarization: enabling diversity with identity. Nat Rev Immunol. 2011;11:750–61.
85. Kimura A, Naka T, Nakahama T, Chinen I, Masuda K, Nohara K, et al. Aryl hydrocarbon receptor in combination with Stat1 regulates LPS-induced inflammatory responses. J Exp Med. 2009;206:2027–35.
86. Satoh T, Takeuchi O, Vandenbon A, Yasuda K, Tanaka Y, Kumagai Y, et al. The Jmjd3-Irf4 axis regulates M2 macrophage polarization and host responses against helminth infection. Nat Immunol. 2010;11:936–44.
87. Spence S, Fitzsimons A, Boyd CR, Kessler J, Fitzgerald D, Elliott J, et al. Suppressors of cytokine signaling 2 and 3 diametrically control macrophage polarization. Immunity. 2013;38:66–78.
88. Martinez FO, Helming L, Gordon S. Alternative activation of macrophages: an immunologic functional perspective. Annu Rev Immunol. 2009;27:451–83.
89. Kang K, Reilly SM, Karabacak V, Gangl MR, Fitzgerald K, Hatano B, et al. Adipocyte-derived Th2 cytokines and myeloid PPARdelta regulate macrophage polarization and insulin sensitivity. Cell Metab. 2008;7:485–95.
90. Hevener AL, Olefsky JM, Reichart D, Nguyen MT, Bandyopadyhay G, Leung HY, et al. Macrophage PPAR gamma is required for normal skeletal muscle and hepatic insulin sensitivity and full antidiabetic effects of thiazolidinediones. J Clin Invest. 2007;117:1658–69.

91. Szanto A, Balint BL, Nagy ZS, Barta E, Dezso B, Pap A, et al. STAT6 transcription factor is a facilitator of the nuclear receptor PPARgamma-regulated gene expression in macrophages and dendritic cells. Immunity. 2010;33:699–712.
92. Ruffell D, Mourkioti F, Gambardella A, Kirstetter P, Lopez RG, Rosenthal N, et al. A CREB-C/EBPbeta cascade induces M2 macrophage-specific gene expression and promotes muscle injury repair. Proc Natl Acad Sci U S A. 2009;106:17475–80.
93. Liu YW, Tseng HP, Chen LC, Chen BK, Chang WC. Functional cooperation of simian virus 40 promoter factor 1 and CCAAT/enhancer-binding protein beta and delta in lipopolysaccharide-induced gene activation of IL-10 in mouse macrophages. J Immunol. 2003;171:821–8.
94. Gray MJ, Poljakovic M, Kepka-Lenhart D, Morris Jr SM. Induction of arginase I transcription by IL-4 requires a composite DNA response element for STAT6 and C/EBPbeta. Gene. 2005;353:98–106.
95. Park JM, Greten FR, Wong A, Westrick RJ, Arthur JS, Otsu K, et al. Signaling pathways and genes that inhibit pathogen-induced macrophage apoptosis–CREB and NF-kappaB as key regulators. Immunity. 2005;23:319–29.
96. Banerjee S, Xie N, Cui H, Tan Z, Yang S, Icyuz M, et al. MicroRNA let-7c regulates macrophage polarization. J Immunol. 2013;190:6542–9.
97. Solinas G, Schiarea S, Liguori M, Fabbri M, Pesce S, Zammataro L, et al. Tumor-conditioned macrophages secrete migration-stimulating factor: a new marker for M2-polarization, influencing tumor cell motility. J Immunol. 2010;185:642–52.
98. Sakai Y, Honda M, Fujinaga H, Tatsumi I, Mizukoshi E, Nakamoto Y, et al. Common transcriptional signature of tumor-infiltrating mononuclear inflammatory cells and peripheral blood mononuclear cells in hepatocellular carcinoma patients. Cancer Res. 2008;68:10267–79.
99. Sinha P, Clements VK, Miller S, Ostrand-Rosenberg S. Tumor immunity: a balancing act between T cell activation, macrophage activation and tumor-induced immune suppression. Cancer Immunol Immunother. 2005;54:1137–42.
100. Kortylewski M, Kujawski M, Wang T, Wei S, Zhang S, Pilon-Thomas S, et al. Inhibiting Stat3 signaling in the hematopoietic system elicits multicomponent antitumor immunity. Nat Med. 2005;11:1314–21.
101. Kortylewski M, Xin H, Kujawski M, Lee H, Liu Y, Harris T, et al. Regulation of the IL-23 and IL-12 balance by Stat3 signaling in the tumor microenvironment. Cancer Cell. 2009;15:114–23.
102. Albesiano E, Davis M, See AP, Han JE, Lim M, Pardoll DM, et al. Immunologic consequences of signal transducers and activators of transcription 3 activation in human squamous cell carcinoma. Cancer Res. 2010;70:6467–76.
103. Zhang L, Alizadeh D, Van Handel M, Kortylewski M, Yu H, Badie B. Stat3 inhibition activates tumor macrophages and abrogates glioma growth in mice. Glia. 2009;57:1458–67.
104. Herrmann A, Kortylewski M, Kujawski M, Zhang C, Reckamp K, Armstrong B, et al. Targeting Stat3 in the myeloid compartment drastically improves the in vivo antitumor functions of adoptively transferred T cells. Cancer Res. 2010;70:7455–64.
105. Guiducci C, Vicari AP, Sangaletti S, Trinchieri G, Colombo MP. Redirecting in vivo elicited tumor infiltrating macrophages and dendritic cells towards tumor rejection. Cancer Res. 2005;65: 3437–46.
106. Weigert A, Tzieply N, von Knethen A, Johann AM, Schmidt H, Geisslinger G, et al. Tumor cell apoptosis polarizes macrophages role of sphingosine-1-phosphate. Mol Biol Cell. 2007;18:3810–9.
107. Weigert A, Johann AM, von Knethen A, Schmidt H, Geisslinger G, Brune B. Apoptotic cells promote macrophage survival by releasing the antiapoptotic mediator sphingosine-1-phosphate. Blood. 2006; 108:1635–42.
108. Bondanza A, Zimmermann VS, Rovere-Querini P, Turnay J, Dumitriu IE, Stach CM, et al. Inhibition of phosphatidylserine recognition heightens the immunogenicity of irradiated lymphoma cells in vivo. J Exp Med. 2004;200:1157–65.
109. Cloez-Tayarani I, Changeux JP. Nicotine and serotonin in immune regulation and inflammatory processes: a perspective. J Leukoc Biol. 2007;81:599–606.
110. Mossner R, Lesch KP. Role of serotonin in the immune system and in neuroimmune interactions. Brain Behav Immun. 1998;12:249–71.
111. Ahern GP. 5-HT and the immune system. Curr Opin Pharmacol. 2011;11:29–33.
112. Nichols DE, Nichols CD. Serotonin receptors. Chem Rev. 2008;108:1614–41.
113. Lesurtel M, Soll C, Graf R, Clavien PA. Role of serotonin in the hepato-gastroIntestinal tract: an old molecule for new perspectives. Cell Mol Life Sci. 2008;65:940–52.
114. Vane JR. The relative activities of some tryptamine analogues on the isolated rat stomach strip preparation. Br J Pharmacol Chemother. 1959;14:87–98.
115. Bradley PB, Humphrey PP, Williams RH. Evidence for the existence of 5-hydroxytryptamine receptors, which are not of the 5-HT2 type, mediating contraction of rabbit isolated basilar artery. Br J Pharmacol. 1986;87:3–4.
116. Bonhaus DW, Bach C, DeSouza A, Salazar FH, Matsuoka BD, Zuppan P, et al. The pharmacology and distribution of human 5-hydroxytryptamine2B (5-HT2B) receptor gene products: comparison with 5-HT2A and 5-HT2C receptors. Br J Pharmacol. 1995;115:622–8.
117. Foguet M, Hoyer D, Pardo LA, Parekh A, Kluxen FW, Kalkman HO, et al. Cloning and functional characterization of the rat stomach fundus serotonin receptor. EMBO J. 1992;11:3481–7.
118. Baxter GS. Novel discriminatory ligands for 5-HT2B receptors. Behav Brain Res. 1996;73:149–52.
119. Choi DS, Ward SJ, Messaddeq N, Launay JM, Maroteaux L. 5-HT2B receptor-mediated serotonin morphogenetic functions in mouse cranial

neural crest and myocardiac cells. Development. 1997;124:1745–5.
120. Nebigil CG, Etienne N, Schaerlinger B, Hickel P, Launay JM, Maroteaux L. Developmentally regulated serotonin 5-HT2B receptors. Int J Dev Neurosci. 2001;19:365–72.
121. Launay JM, Birraux G, Bondoux D, Callebert J, Choi DS, Loric S, et al. Ras involvement in signal transduction by the serotonin 5-HT2B receptor. J Biol Chem. 1996;271:3141–7.
122. Nebigil CG, Launay JM, Hickel P, Tournois C, Maroteaux L. 5-hydroxytryptamine 2B receptor regulates cell-cycle progression: cross-talk with tyrosine kinase pathways. Proc Natl Acad Sci U S A. 2000;97:2591–6.
123. Nebigil CG, Choi DS, Dierich A, Hickel P, Le Meur M, Messaddeq N, et al. Serotonin 2B receptor is required for heart development. Proc Natl Acad Sci U S A. 2000;97:9508–13.
124. Nebigil CG, Maroteaux L. Functional consequence of serotonin/5-HT2B receptor signaling in heart: role of mitochondria in transition between hypertrophy and heart failure? Circulation. 2003;108:902–8.
125. Tharayil VS, Wouters MM, Stanich JE, Roeder JL, Lei S, Beyder A, et al. Lack of serotonin 5-HT2B receptor alters proliferation and network volume of interstitial cells of Cajal in vivo. Neurogastroenterol Motil. 2010;22:462–9.
126. Svejda B, Kidd M, Giovinazzo F, Eltawil K, Gustafsson BI, Pfragner R, et al. The 5-HT(2B) receptor plays a key regulatory role in both neuroendocrine tumor cell proliferation and the modulation of the fibroblast component of the neoplastic microenvironment. Cancer. 2010;116:2902–12.
127. Kim H, Toyofuku Y, Lynn FC, Chak E, Uchida T, Mizukami H, et al. Serotonin regulates pancreatic beta cell mass during pregnancy. Nat Med. 2010;16:804–8.
128. Nebigil CG, Jaffre F, Messaddeq N, Hickel P, Monassier L, Launay JM, Maroteaux L. Overexpression of the serotonin 5-HT2B receptor in heart leads to abnormal mitochondrial function and cardiac hypertrophy. Circulation. 2003;107:3223–9.
129. Nebigil CG, Maroteaux L. A novel role for serotonin in heart. Trends Cardiovasc Med. 2001;11:329–35.
130. Baudry A, Bitard J, Mouillet-Richard S, Locker M, Poliard A, Launay JM, et al. Serotonergic 5-HT(2B) receptor controls tissue-nonspecific alkaline phosphatase activity in osteoblasts via eicosanoids and phosphatidylinositol-specific phospholipase C. J Biol Chem. 2010;285:26066–73.
131. Gunther S, Maroteaux L, Schwarzacher SW. Endogenous 5-HT2B receptor activation regulates neonatal respiratory activity in vitro. J Neurobiol. 2006;66:949–61.
132. Niebert M, Vogelgesang S, Koch UR, Bischoff AM, Kron M, Bock N, et al. Expression and function of serotonin 2A and 2B receptors in the mammalian respiratory network. PLoS One. 2011;6:e21395.
133. Connolly HM, Crary JL, McGoon MD, Hensrud DD, Edwards BS, Edwards WD, et al. Valvular heart disease associated with fenfluramine-phentermine. N Engl J Med. 1997;337:581–8.
134. Abenhaim L, Moride Y, Brenot F, Rich S, Benichou J, Kurz X, et al. Appetite-suppressant drugs and the risk of primary pulmonary hypertension. International Primary Pulmonary Hypertension Study Group. N Engl J Med. 1996;335:609–16.
135. Rothman RB, Baumann MH, Savage JE, Rauser L, McBride A, Hufeisen SJ, et al. Evidence for possible involvement of 5-HT(2B) receptors in the cardiac valvulopathy associated with fenfluramine and other serotonergic medications. Circulation. 2000;102:2836–41.
136. Fitzgerald LW, Burn TC, Brown BS, Patterson JP, Corjay MH, Valentine PA, et al. Possible role of valvular serotonin 5-HT(2B) receptors in the cardiopathy associated with fenfluramine. Mol Pharmacol. 2000;57:75–81.
137. Archer SL, Weir EK, Wilkins MR. Basic science of pulmonary arterial hypertension for clinicians: new concepts and experimental therapies. Circulation. 2010;121:2045–66.
138. Launay JM, Herve P, Peoc'h K, Tournois C, Callebert J, Nebigil CG, et al. Function of the serotonin 5-hydroxytryptamine 2B receptor in pulmonary hypertension. Nat Med. 2002;8:1129–35.
139. Esteve JM, Launay JM, Kellermann O, Maroteaux L. Functions of serotonin in hypoxic pulmonary vascular remodeling. Cell Biochem Biophys. 2007;47:33–44.
140. Jahnichen S, Glusa E, Pertz H. Evidence for 5-HT2B and 5-HT7 receptor-mediated relaxation in pulmonary arteries of weaned pigs. Naunyn Schmiedebergs Arch Pharmacol. 2005;371:89–98.
141. Diller GP, Thum T, Wilkins MR, Wharton J. Endothelial progenitor cells in pulmonary arterial hypertension. Trends Cardiovasc Med. 2010;20:22–9.
142. Launay JM, Herve P, Callebert J, Mallat Z, Collet C, Doly S, et al. Serotonin 5-HT2B receptors are required for bone-marrow contribution to pulmonary arterial hypertension. Blood. 2012;119:1772–80.
143. Lovenberg TW, Baron BM, de Lecea L, Miller JD, Prosser RA, Rea MA, et al. A novel adenylyl cyclase-activating serotonin receptor (5-HT7) implicated in the regulation of mammalian circadian rhythms. Neuron. 1993;11:449–58.
144. Ruat M, Traiffort E, Leurs R, Tardivel-Lacombe J, Diaz J, Arrang JM, et al. Molecular cloning, characterization, and localization of a high-affinity serotonin receptor (5-HT7) activating cAMP formation. Proc Natl Acad Sci U S A. 1993;90:8547–51.
145. Bard JA, Zgombick J, Adham N, Vaysse P, Branchek TA, Weinshank RL. Cloning of a novel human serotonin receptor (5-HT7) positively linked to adenylate cyclase. J Biol Chem. 1993;268:23422–6.
146. Matthys A, Haegeman G, Van Craenenbroeck K, Vanhoenacker P. Role of the 5-HT7 receptor in the

central nervous system: from current status to future perspectives. Mol Neurobiol. 2011;43:228–53.
147. Janssen P, Prins NH, Moreaux B, Meulemans AL, Lefebvre RA. Characterization of 5-HT7-receptor-mediated gastric relaxation in conscious dogs. Am J Physiol Gastrointest Liver Physiol. 2005;289: G108–15.
148. Inoue M, Kitazawa T, Cao J, Taneike T. 5-HT7 receptor-mediated relaxation of the oviduct in non-pregnant proestrus pigs. Eur J Pharmacol. 2003; 461:207–18.
149. Kitazawa T, Nakagoshi K, Teraoka H, Taneike T. 5-HT(7) receptor and beta(2)-adrenoceptor share in the inhibition of porcine uterine contractility in a muscle layer-dependent manner. Eur J Pharmacol. 2011;433:187–97.
150. Crider JY, Williams GW, Drace CD, Katoli P, Senchyna M, Sharif NA. Pharmacological characterization of a serotonin receptor (5-HT7) stimulating cAMP production in human corneal epithelial cells. Invest Ophthalmol Vis Sci. 2003;44:4837–44.
151. Leopoldo M, Lacivita E, Contino M, Colabufo NA, Berardi F, Perrone R. Structure-activity relationship study on N-(1,2,3,4-tetrahydronaphthalen-1-yl)-4-aryl-1-piperazinehexanamides, a class of 5-HT7 receptor agents. J Med Chem. 2007;50:4214–21.
152. Perez-Garcia GS, Meneses A. Effects of the potential 5-HT7 receptor agonist AS 19 in an autoshaping learning task. Behav Brain Res. 2005;163:136–40.
153. Norum JH, Hart K, Levy FO. Ras-dependent ERK activation by the human G(s)-coupled serotonin receptors 5-HT4(b) and 5-HT7(a). J Biol Chem. 2003;278:3098–104.
154. Lin SL, Johnson-Farley NN, Lubinsky DR, Cowen DS. Coupling of neuronal 5-HT7 receptors to activation of extracellular-regulated kinase through a protein kinase A-independent pathway that can utilize Epac. J Neurochem. 2003;87:1076–85.
155. Kvachnina E, Liu G, Dityatev A, Renner U, Dumuis A, Richter DW, et al. 5-HT7 receptor is coupled to G alpha subunits of heterotrimeric G12-protein to regulate gene transcription and neuronal morphology. J Neurosci. 2005;25:7821–30.
156. Leon-Ponte M, Ahern GP, O'Connell PJ. Serotonin provides an accessory signal to enhance T-cell activation by signaling through the 5-HT7 receptor. Blood. 2005;109:3139–46.
157. Mahe C, Loetscher E, Dev KK, Bobirnac I, Otten U, Schoeffter P. Serotonin 5-HT7 receptors coupled to induction of interleukin-6 in human microglial MC-3 cells. Neuropharmacology. 2005;49:40–7.
158. Lieb K, Biersack L, Waschbisch A, Orlikowski S, Akundi RS, Candelario-Jalil R, et al. Serotonin via 5-HT7 receptors activates p38 mitogen-activated protein kinase and protein kinase C epsilon resulting in interleukin-6 synthesis in human U373 MG astrocytoma cells. J Neurochem. 2005;93:549–59.
159. Irving HR, Tan YY, Tochon-Danguy N, Liu H, Chetty N, Desmond PV, et al. Comparison of 5-HT4 and 5-HT7 receptor expression and function in the circular muscle of the human colon. Life Sci. 2007;80:1198–205.
160. Svejda B, Kidd M, Timberlake A, Harry K, Kazberouk A, Schimmack S, et al. Serotonin and the 5-HT7 receptor: the link between Hepatocytes, IGF-1 and Small intestinal neuroendocrine tumors. Cancer Sci. 2013;104:844–55.
161. Read KE, Sanger GJ, Ramage AG. Evidence for the involvement of central 5-HT7 receptors in the micturition reflex in anaesthetized female rats. Br J Pharmacol. 2003;140:53–60.
162. Recio P, Barahona MV, Orensanz LM, Bustamante S, Martinez AC, Benedito S, et al. 5-hydroxytryptamine induced relaxation in the pig urinary bladder neck. Br J Pharmacol. 2009;157:271–80.
163. Kim JJ, Bridle BW, Ghia JE, Wang H, Syed SN, Manocha MM. Targeted inhibition of serotonin type 7 (5-HT7) receptor function modulates immune responses and reduces the severity of intestinal inflammation. J Immunol. 2013;190:4795–804.
164. Guscott M, Bristow LJ, Hadingham K, Rosahl TW, Beer MS, Stanton JA, et al. Genetic knockout and pharmacological blockade studies of the 5-HT7 receptor suggest therapeutic potential in depression. Neuropharmacology. 2005;48:492–502.
165. Thomas DR, Melotto S, Massagrande M, Gribble AD, Jeffrey P, Stevens AJ, et al. SB-656104-A, a novel selective 5-HT7 receptor antagonist, modulates REM sleep in rats. Br J Pharmacol. 2003;139:705–14.
166. Thomas DR, Hagan JJ. 5-HT7 receptors. Curr Drug Targets CNS Neurol Disord. 2004;3:81–90.
167. Hedlund PB, Danielson PE, Thomas EA, Slanina K, Carson MJ, Sutcliffe JG. No hypothermic response to serotonin in 5-HT7 receptor knockout mice. Proc Natl Acad Sci U S A. 2003;100:1375–80.
168. Benedict CR, Mathew B, Rex KA, Cartwright Jr J, Sordahl LA. Correlation of plasma serotonin changes with platelet aggregation in an in vivo dog model of spontaneous occlusive coronary thrombus formation. Circ Res. 1986;58:58–67.
169. Timmons S, Huzoor A, Grabarek J, Kloczewiak M, Hawiger J. Mechanism of human platelet activation by endotoxic glycolipid-bearing mutant Re595 of Salmonella minnesota. Blood. 1986;68:1015–23.
170. Finocchiaro LM, Arzt ES, Fernandez-Castelo S, Criscuolo M, Finkielman S, Nahmod VE. Serotonin and melatonin synthesis in peripheral blood mononuclear cells: stimulation by interferon-gamma as part of an immunomodulatory pathway. J Interferon Res. 1988;8:705–16.
171. Kushnir-Sukhov NM, Gilfillan AM, Coleman JW, Brown JM, Bruening S, Toth M, Metcalfe DD. 5-hydroxytryptamine induces mast cell adhesion and migration. J Immunol. 2006;177:6422–32.
172. Kang BN, Ha SG, Bahaie NS, Hosseinkhani MR, Ge XN, Blumenthal MN, Rao SP, Sriramarao P. Regulation of serotonin-induced trafficking and migration of eosinophils. PLoS One. 2013;8:e54840.
173. Boehme SA, Lio FM, Sikora L, Pandit TS, Lavrador K, Rao SP, Sriramarao P. Cutting edge: serotonin is a

chemotactic factor for eosinophils and functions additively with eotaxin. J Immunol. 2004;173:3599–603.
174. Evans DL, Lynch KG, Benton T, Dube B, Gettes DR, Tustin NB, et al. Selective serotonin reuptake inhibitor and substance P antagonist enhancement of natural killer cell innate immunity in human immunodeficiency virus/acquired immunodeficiency syndrome. Biol Psychiatry. 2008;63:899–905.
175. Hernandez ME, Martinez-Fong D, Perez-Tapia M, Estrada-Garcia I, Estrada-Parra S, Pavon L. Evaluation of the effect of selective serotonin-reuptake inhibitors on lymphocyte subsets in patients with a major depressive disorder. Eur Neuropsychopharmacol. 2010;20:88–95.
176. Young MR, Matthews JP. Serotonin regulation of T-cell subpopulations and of macrophage accessory function. Immunology. 1995;84:148–52.
177. Muller T, Durk T, Blumenthal B, Grimm M, Cicko S, Panther E, et al. 5-hydroxytryptamine modulates migration, cytokine and chemokine release and T-cell priming capacity of dendritic cells in vitro and in vivo. PLoS One. 2009;4:e6453.
178. Sternberg EM, Trial J, Parker CW. Effect of serotonin on murine macrophages: suppression of Ia expression by serotonin and its reversal by 5-HT2 serotonergic receptor antagonists. J Immunol. 1986;137:276–82.
179. Sternberg EM, Wedner HJ, Leung MK, Parker CW. Effect of serotonin (5-HT) and other monoamines on murine macrophages: modulation of interferon-gamma induced phagocytosis. J Immunol. 1987;138:4360–5.
180. Nakamura K, Sato T, Ohashi A, Tsurui H, Hasegawa H. Role of a serotonin precursor in development of gut microvilli. Am J Pathol. 2008;172:333–44.
181. Nocito A, Dahm F, Jochum W, Jang JH, Georgiev P, Bader M, et al. Serotonin regulates macrophage-mediated angiogenesis in a mouse model of colon cancer allografts. Cancer Res. 2008;68:5152–8.
182. Baganz NL, Blakely RD. A dialogue between the immune system and brain, spoken in the language of serotonin. ACS Chem Neurosci. 2013;4:48–63.
183. Hellstrand K, Czerkinsky C, Ricksten A, Jansson B, Asea A, Kylefjord H, et al. Role of serotonin in the regulation of interferon-gamma production by human natural killer cells. J Interferon Res. 1993;13:33–8.
184. Kubera M, Kenis G, Bosmans E, Scharpe S, Maes M. Effects of serotonin and serotonergic agonists and antagonists on the production of interferon-gamma and interleukin-10. Neuropsychopharmacology. 2000;23:89–98.
185. Cloez-Tayarani I, Petit-Bertron AF, Venters HD, Cavaillon JM. Differential effect of serotonin on cytokine production in lipopolysaccharide-stimulated human peripheral blood mononuclear cells: involvement of 5-hydroxytryptamine2A receptors. Int Immunol. 2003;15:233–40.
186. Kubera M, Maes M, Kenis G, Kim YK, Lason W. Effects of serotonin and serotonergic agonists and antagonists on the production of tumor necrosis factor alpha and interleukin-6. Psychiatry Res. 2005;134:251–8.
187. Durk T, Panther E, Muller T, Sorichter S, Ferrari D, Pizzirani C, et al. 5-Hydroxytryptamine modulates cytokine and chemokine production in LPS-primed human monocytes via stimulation of different 5-HTR subtypes. Int Immunol. 2005;17:599–606.
188. Idzko M, Panther E, Stratz C, Muller T, Bayer H, Zissel G, et al. The serotoninergic receptors of human dendritic cells: identification and coupling to cytokine release. J Immunol. 2004;172:6011–9.
189. Katoh N, Soga F, Nara T, Tamagawa-Mineoka R, Nin M, Kotani H, Masuda K, Kishimoto S. Effect of serotonin on the differentiation of human monocytes into dendritic cells. Clin Exp Immunol. 2006;146:354–61.
190. Li N, Ghia JE, Wang H, McClemens J, Cote F, Suehiro Y, et al. Serotonin activates dendritic cell function in the context of gut inflammation. Am J Pathol. 2011;178:662–71.
191. Freire-Garabal M, Nunez MJ, Balboa J, Lopez-Delgado P, Gallego R, Garcia-Caballero T, et al. Serotonin upregulates the activity of phagocytosis through 5-HT1A receptors. Br J Pharmacol. 2003;139:457–63.
192. Vollmar P, Nessler S, Kalluri SR, Hartung HP, Hemmer B. The antidepressant venlafaxine ameliorates murine experimental autoimmune encephalomyelitis by suppression of pro-inflammatory cytokines. Int J Neuropsychopharmacol. 2009;12:525–36.
193. Mikulski Z, Zaslona Z, Cakarova L, Hartmann P, Wilhelm J, Tecott LH, et al. Serotonin activates murine alveolar macrophages through 5-HT2C receptors. Am J Physiol Lung Cell Mol Physiol. 2010;299:L272–80.
194. Seidel MF, Fiebich BL, Ulrich-Merzenich G, Candelario-Jalil E, Koch FW, Vetter H. Serotonin mediates PGE2 overexpression through 5-HT2A and 5-HT3 receptor subtypes in serum-free tissue culture of macrophage-like synovial cells. Rheumatol Int. 2008;28:1017–22.
195. Menard G, Turmel V, Bissonnette EY. Serotonin modulates the cytokine network in the lung: involvement of prostaglandin E2. Clin Exp Immunol. 2007;150:340–8.
196. de las Casas-Engel M, Dominguez-Soto A, Sierra-Filardi E, Bragado R, Nieto C, Puig-Kroger A, et al. Serotonin skews human macrophage polarization through HTR2B and HTR7. J Immunol. 2013;190:2301–10.
197. Mantovani A, Biswas SK, Galdiero MR, Sica A, Locati M. Macrophage plasticity and polarization in tissue repair and remodelling. J Pathol. 2013;229:176–85.
198. Amireault P, Sibon D, Cote F. Life without peripheral serotonin: insights from tryptophan hydroxylase 1 knockout mice reveal the existence of paracrine/autocrine serotonergic networks. ACS Chem Neurosci. 2013;4:64–71.

199. Tian Y, Graf R, El-Badry AM, Lesurtel M, Furrer K, Moritz W, et al. Activation of serotonin receptor-2B rescues small-for-size liver graft failure in mice. Hepatology. 2013;53:253–62.
200. Furrer K, Rickenbacher A, Tian Y, Jochum W, Bittermann AG, Kach A, et al. Serotonin reverts age-related capillarization and failure of regeneration in the liver through a VEGF-dependent pathway. Proc Natl Acad Sci U S A. 2011;108:2945–50.
201. Lesurtel M, Graf R, Aleil B, Walther DJ, Tian Y, Jochum W, et al. Platelet-derived serotonin mediates liver regeneration. Science. 2006;312:104–7.
202. Soll C, Jang JH, Riener MO, Moritz W, Wild PJ, Graf R, et al. Serotonin promotes tumor growth in human hepatocellular cancer. Hepatology. 2010;51:1244–54.
203. Jian B, Xu J, Connolly J, Savani RC, Narula N, Liang B, Levy RJ. Serotonin mechanisms in heart valve disease I: serotonin-induced up-regulation of transforming growth factor-beta1 via G-protein signal transduction in aortic valve interstitial cells. Am J Pathol. 2002;161:2111–21.
204. Jaffre F, Bonnin P, Callebert J, Debbabi H, Setola V, Doly S, et al. Serotonin and angiotensin receptors in cardiac fibroblasts coregulate adrenergic-dependent cardiac hypertrophy. Circ Res. 2009;104:113–23.
205. Buskohl PR, Sun ML, Thompson RP, Butcher JT. Serotonin potentiates transforming growth factor-beta3 induced biomechanical remodeling in avian embryonic atrioventricular valves. PLoS One. 2012;7:e42527.
206. Rogers TL, Holen I. Tumour macrophages as potential targets of bisphosphonates. J Transl Med. 2011;9:177.
207. Kim R, Emi M, Tanabe K, Arihiro K. Tumor-driven evolution of immunosuppressive networks during malignant progression. Cancer Res. 2006;66: 5527–36.
208. D'Angelo F, Bernasconi E, Schafer M, Moyat M, Michetti P, Maillard MH, et al. Macrophages promote epithelial repair through hepatocyte growth factor secretion. Clin Exp Immunol. 2013;174:60–72.
209. Yamada Y, Kirillova I, Peschon JJ, Fausto N. Initiation of liver growth by tumor necrosis factor: deficient liver regeneration in mice lacking type I tumor necrosis factor receptor. Proc Natl Acad Sci U S A. 1997;94:1441–6.
210. Pai VP, Marshall AM, Hernandez LL, Buckley AR, Horseman ND. Altered serotonin physiology in human breast cancers favors paradoxical growth and cell survival. Breast Cancer Res. 2009;11:R81.
211. Soll C, Riener MO, Oberkofler CE, Hellerbrand C, Wild PJ, DeOliveira ML, et al. Expression of serotonin receptors in human hepatocellular cancer. Clin Cancer Res. 2012;18:5902–10.
212. Lesurtel M, Soll C, Humar B, Clavien PA. Serotonin: a double-edged sword for the liver? Surgeon. 2012;10:107–13.
213. Mann DA, Oakley F. Serotonin paracrine signaling in tissue fibrosis. Biochim Biophys Acta. 1832;2013:905–10.
214. Derecki NC, Cronk JC, Lu Z, Xu E, Abbott SB, Guyenet PG, et al. Wild-type microglia arrest pathology in a mouse model of Rett syndrome. Nature. 2012;484:105–9.
215. Chen SK, Tvrdik P, Peden E, Cho S, Wu S, Spangrude G, et al. Hematopoietic origin of pathological grooming in Hoxb8 mutant mice. Cell. 2010;141:775–85.
216. Blank T, Prinz M. Microglia as modulators of cognition and neuropsychiatric disorders. Glia. 2013;61:62–70.
217. Dantzer R, O'Connor JC, Freund GG, Johnson RW, Kelley KW. From inflammation to sickness and depression: when the immune system subjugates the brain. Nat Rev Neurosci. 2008;9:46–56.
218. Diaz SL, Doly S, Narboux-Neme N, Fernandez S, Mazot P, Banas SM, et al. 5-HT(2B) receptors are required for serotonin-selective antidepressant actions. Mol Psychiatry. 2012;17:154–63.
219. Bevilacqua L, Doly S, Kaprio J, Yuan Q, Tikkanen R, Paunio T, et al. A population-specific HTR2B stop codon predisposes to severe impulsivity. Nature. 2010;468:1061–6.
220. Dantzer R. Cytokine-induced sickness behaviour: a neuroimmune response to activation of innate immunity. Eur J Pharmacol. 2004;500:399–411.
221. Hedlund PB, Sutcliffe JG. Functional, molecular and pharmacological advances in 5-HT7 receptor research. Trends Pharmacol Sci. 2004;25:481–6.

Energy Metabolism and Metabolic Sensors in Stem Cells: The Metabostem Crossroads of Aging and Cancer

10

Javier A. Menendez and Jorge Joven

Abstract

We are as old as our adult stem cells are; therefore, stem cell exhaustion is considered a hallmark of aging. Our tumors are as aggressive as the number of cancer stem cells (CSCs) they bear because CSCs can survive treatments with hormones, radiation, chemotherapy, and molecularly targeted drugs, thus increasing the difficulty of curing cancer. Not surprisingly, interest in stem cell research has never been greater among members of the public, politicians, and scientists. But how can we slow the rate at which our adult stem cells decline over our lifetime, reducing the regenerative potential of tissues, while efficiently eliminating the aberrant, life-threatening activity of "selfish", immortal, and migrating CSCs? Frustrated by the gene-centric limitations of conventional approaches to aging diseases, our group and other groups have begun to appreciate that bioenergetic metabolism, *i.e.*, the production of fuel & building blocks for growth and division, and autophagy/mitophagy, *i.e.*, the quality-control, self-cannibalistic system responsible for "cleaning house" and "recycling the trash", can govern the genetic and epigenetic networks that facilitate stem cell behaviors. Indeed, it is reasonable to suggest the existence of a "metabostem" infrastructure that operates as a shared hallmark of aging and cancer, thus making it physiologically plausible to maintain or even

J.A. Menendez (✉)
Metabolism & Cancer Group, Translational Research Laboratory, Catalan Institute of Oncology, Girona, Spain

Molecular Oncology Group, Girona Biomedical Research Institute (IDIBGI), Girona, Spain

Catalan Institute of Oncology, Girona (ICO-Girona), Hospital Dr. Josep Trueta de Girona, Ctra. França s/n, Girona, Catalonia E-17007, Spain
e-mail: jmenendez@iconcologia.net; jmenendez@idibgi.org

J. Joven
Unitat de Recerca Biomèdica, Hospital Universitari Sant Joan and Hospital Universitari Joan XXIII, Institut d'Investigació Sanitària Pere Virgili, Universitat Rovira i Virgili, Reus, Spain
e-mail: jorge.joven@urv.cat

J. Camps (ed.), *Oxidative Stress and Inflammation in Non-communicable Diseases – Molecular Mechanisms and Perspectives in Therapeutics*, Advances in Experimental Medicine and Biology 824, DOI 10.1007/978-3-319-07320-0_10, © Springer International Publishing Switzerland 2014

increase the functionality of adult stem cells while reducing the incidence of cancer and extending the lifespan. This "metabostemness" property could lead to the discovery of new drugs that reprogram cell metabotypes to increase the structural and functional integrity of adult stem cells and positively influence their lineage determination, while preventing the development and aberrant function of stem cells in cancer tissues. While it is obvious that the antifungal antibiotic rapamycin, the polyphenol resveratrol, and the biguanide metformin already belong to this new family of metabostemness-targeting drugs, we can expect a rapid identification of new drug candidates (*e.g.,* polyphenolic xenohormetins) that reverse or postpone "geroncogenesis", *i.e.,* aging-induced metabolic decline as a driver of tumorigenesis, at the stem cell level.

Keywords
Aging • Autophagy • Cancer • Metabolism • Mitophagy • Stem cells • Stemness

10.1 Introduction

We are as old as our stem cells are. Adult somatic stem cells persist throughout our lifespan and are required to rejuvenate tissues. However, in most tissues, adult somatic stem cell function declines during the aging process. Because this decline in stem cell function with age contributes to tissue dysfunction and age-associated diseases by reducing the regenerative potential of tissues, stem cell exhaustion is considered a hallmark of aging [1–5]. When adult stem cells are damaged or mutated, they usually die. However, occasionally, these compromised stem cells become cancer stem cells (CSCs), special forms of malignant cells that have the frightening power to develop an entire tumor from a single cell [6–10]. Indeed, it would be reasonable to say that tumors are as aggressive as the number of CSCs they bear. Furthermore, we have recently learned that CSCs are *made* and not just *born*. That is, either normal or non-tumorigenic cancer cells can be endowed with stem cell-like abilities, including immortality and metastatic potential *via* the activation of largely unknown pathways [11–15]. CSCs increase the difficulty of curing cancer because they can survive treatment with hormones, radiation, chemotherapy, and molecularly targeted drugs. Thus, CSCs are ultimately responsible for the clinical failure of the majority of currently available oncology therapies. Developing techniques to reactivate and direct the activity of adult stem cells could create revolutionary opportunities to clinically manage age-related degenerative diseases. Moreover, strategies for eliminating drug-resistant CSCs and controlling their metastasis will create novel opportunities for the clinical management of deadly cancers. But how can we slow the rate at which adult stem cells decline over our lifetime while efficiently eliminating the aberrant, life-threatening activity of immortal CSCs? (Fig. 10.1). Learning how to maintain and/or increase adult stem cell function while reducing cancer rates will undoubtedly have important implications for human health.

One of the most commonly accepted paradigms in the field is that loss of function in aging stem cells correlates with reduced potential for adult stem cell transformation (*i.e.,* CSC generation). In other words, there is a balance between functional senescence and cancer risk. Therefore, the maximum potential lifespan of an organism may be limited by the increased risk of deadly cancer and the effectiveness of surveillance mechanisms that maintain stem cell function. Intriguingly, calorie restriction, which increases adult stem cell function to levels found in younger subjects, does not increase the incidence of cancer [16–20]. Regardless of the ultimate route

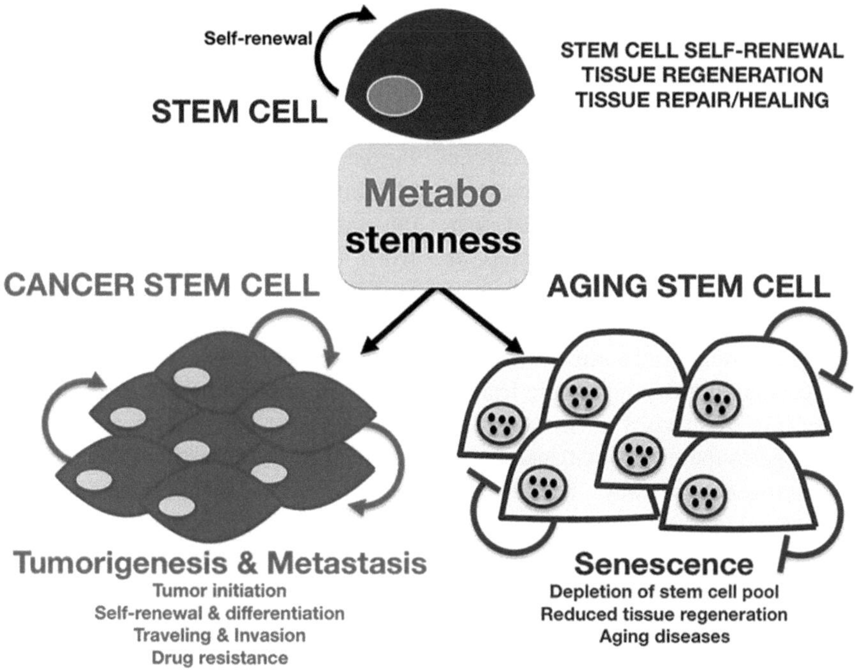

Fig. 10.1 Metabostemness: a new-dimensional hallmark shared by aging & cancer stem cells. Because the decline in stem cell function with age contributes to age-associated diseases by reducing the regenerative potential of tissues, stem cell exhaustion is considered a hallmark of aging. Occasionally, adult stem cells become cancer stem cells (CSCs), special forms of "selfish", immortal and migrating malignant cells with an intrinsically aberrant life-threatening stemness activity. Moreover, either normal or non-tumourigenic cancer cells can be endowed with "stemness powers", including immortality, drug resistance, and metastatic potential. We now know that the molecular circuitries that switch *on&off* bio-energetics metabolism, *i.e.*, the production of *fuel&building blocks* for growth and division, and autophagy/mitophagy, *i.e.*, the quality-control, *self-cannibalistic* system responsible for "cleaning house" and "recycling the trash", are wired differently in adult stem cells and cancer stem cells. The new cellular attribute at the intersection of aging and cancer has been called metabostemness. While the prolonged activation of critical metabostemness components leads to depletion of adult stem cell function and reduced health, the activation of analogous metabostemness machineries causally contributes to the aberrant acquisition of tumour-initiating and metastatic abilities in CSCs

by which calorie restriction circumvents the balance between reduced adult stem cell function (*via* activation of tumor suppressor-regulated cellular senescence) and increased risk of cancer with age, calorie restriction studies demonstrate that is physiologically plausible to maintain or even increase the number and/or function of adult stem cells while reducing the incidence of cancer and extend maximal lifespan, at least in rodents. Indeed, we are beginning to understand a metabolic hallmark of adult stem cells and CSCs that may create a therapeutic bridge between the fields of aging and cancer [21–31].

10.2 Aging & Cancer: A Metabolic View

Age-related diseases (*e.g.*, atherosclerosis, diabetes, cancer, and others) appear to reflect a synergistic interaction between our evolutionary path to sedentarism – which increases a number of gero-promoting factors (*e.g.*, nutrients, growth factors, cytokines, insulin) that overactivate key gerogenes (*e.g.*, the nutrient-sensing mammalian Target Of Rapamycin (mTOR)) – and the "defective design" of central energy metabolism sensors that function either as metabolic gerogenes (*e.g.*, mTOR) or metabolic gerosuppressors (*e.g.*, the AMP-activated protein kinase (AMPK), which antagonizes the gerogenic activity of mTOR) [32–43]. In this scenario, aging can be viewed as the ability of metabolic gerogenes to continue, in an aimless but harmful manner, a developmental program that was beneficial early in life but was not switched off upon its completion. In other words, overactivation of metabolic gerogenes limits the lifespan by accelerating age-related diseases. A crucial element in understanding this metabolic framework for aging-related diseases is the weakness of the metabolic gerosuppressors that antagonize the metabolo-gerogenic pathway. The responsiveness of metabolo-gerosuppressor signaling should decline with aging because robust, continuous activation of metabolic suppressors in response to metabolic stresses results in accelerated aging [44, 45]. In summary, the ability of metabolic gerogenes to drive aging can be triggered or accelerated by the loss of responsiveness to the activation of critical metabolic gerosuppressors. Understandably, if the overactivity of metabolic gerogenes limits lifespan by accelerating the progression of age-related diseases such as atherosclerosis or cancer, direct or indirect behavioral or pharmacological suppression of metabolic gerogene-driven aging (*e.g.*, *via* non-permanent activation of metabolic gerosuppressors) should increase the healthy lifespan.

10.3 Aging, Cancer, and Stemness: A Metabolic Intersection

Aging and cancer could be seen as opposite processes: whereas aging is characterized by a loss of fitness, cancer is the consequence of an aberrant gain of cellular fitness. From a metabolic perspective, however, aging and cancer can be regarded as two different manifestations of the same underlying process: cancer and most aging-associated pathologies such as atherosclerosis and inflammation involve uncontrolled cellular overgrowth or hyperactivity. Current evidence strongly supports the idea that anabolic metabolic signaling (*e.g.*, deregulated nutrient sensing and mitochondrial dysfunction, both hallmarks of aging) accelerates aging and that decreased nutrient signaling extends longevity. Further, treatment with drugs that mimic a state of limited nutrient availability, such as rapamycin, can extend longevity in mice [46–53]. It is worth noting that metabolism in cancer tissues is reorganized to increase the anabolic reactions linked to cell growth and proliferation, which are negatively impacted by energy states that mimic limited nutrient availability [54–60]. This apparent metabolic paradox can be easily resolved in the context of stemness as an integrative framework for understanding the antagonistic metabolic features that occur in aging and in aging-related diseases such as cancer (Fig. 10.2). In other words, metabolic gerosuppressors in adult stem cells are tumor suppressors in CSCs. Upon activation, these suppressors inhibit the geroconverting activity of gerogenes in adult stem cells and transformation of CSCs. This "metabostemness" property is the only scenario that explains how pharmacological inhibitors of gerogene activity (*e.g.*, mTOR inhibitors such as rapalogs) that significantly postpone aging by affecting energy sensing may also improve "normal" stem cell function in several tissues (*e.g.*, epidermis, hematopoietic system, intestine) and simultaneously blocking specific stemness-related functions in CSCs [61]. Similarly, the agonistic activity of molecules such as metformin toward metabolic gerosuppressors could significantly improve the structural and functional integrity of adult stem cells by preventing their entrance into potentially deleterious hyperproliferative cancer-like modes [62–80].

The most essential hallmark of aging and cancer (*i.e.*, stemness) is highly intertwined with a specific cell-intrinsic metabolism, either as consequence or as a cause. Therefore, an integrated "metabostemness" hallmark could be considered as ultimately responsible for both the depletion of adult stem cell function and the aberrant acquisition of tumor-initiating and metastatic abilities in CSCs.

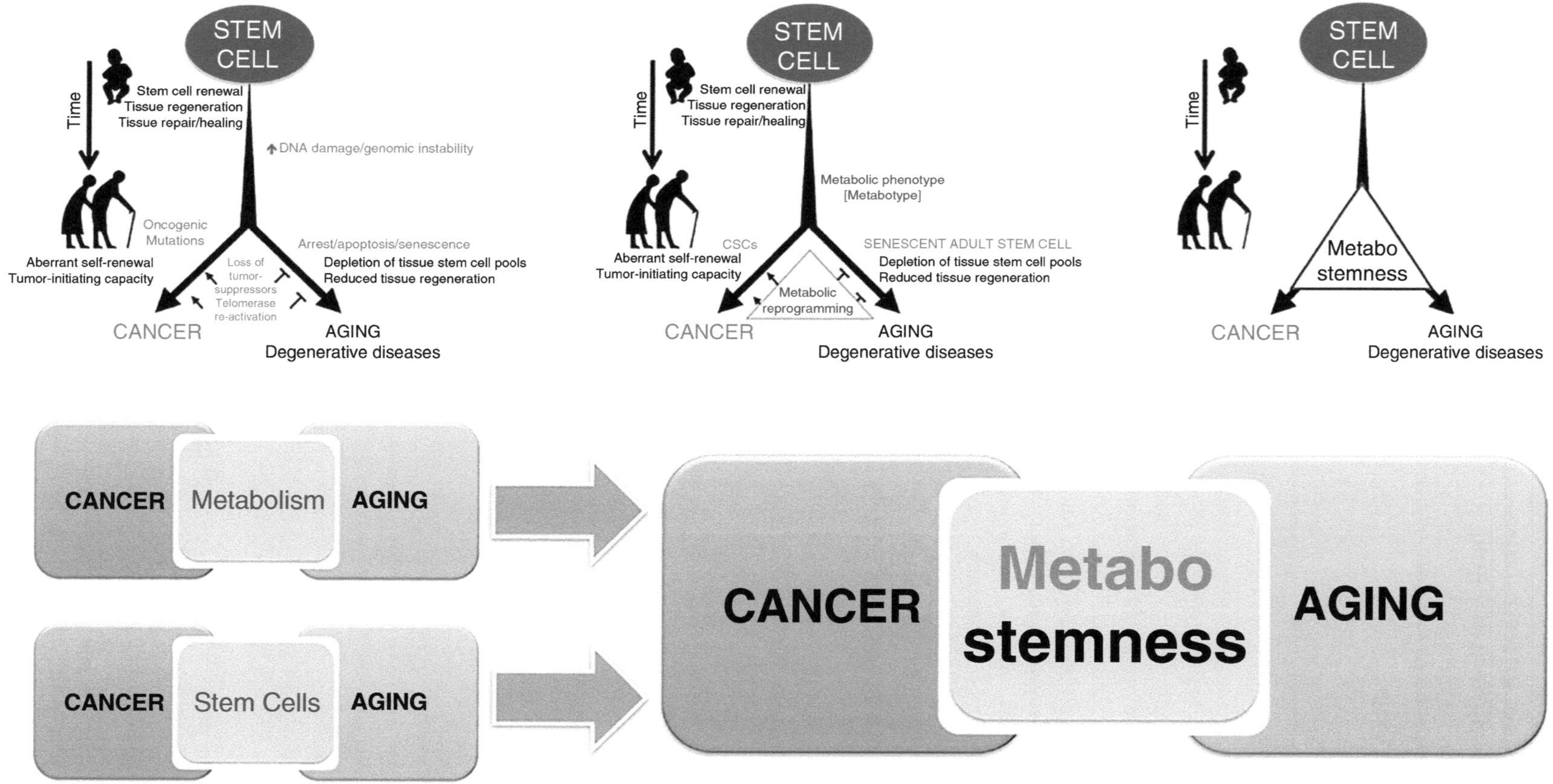

Fig. 10.2 Stemness: gene-centered *versus* metabolo-centered mechanistic views of aging and cancer. It is not simply the time taken to accumulate DNA damage (genomic mutations) and genome instability from intrinsic or external causes that accounts for the increased rate of CSCs-driven cancer and reduced adult stem cells-driven tissue regeneration with age. An alternative possibility is that the natural changes in energy utilization as we age could set a metabolic stage that might operate as a scavenger mechanism for fate determinants related to cell differentiation, thus causally contributing to both the depletion/exhaustion of adult stem cell pools in aging tissues and the gaining/retaining of the stem cell fate in malignant tissues

10.4 A Metabostem Infrastructure Is Shared Hallmark of Aging and Cancer: Lessons from Induced Pluripotent Stem (iPS) Cells

The discovery that somatic cells can be induced to enter a pluripotent stem state by the exogenous expression of reprogramming factors (*e.g.*, Oct4, Sox2, Klf4, c-Myc) has enormous potential for human disease modeling and therapeutics [81–89]. It is quite clear that reprogramming can reset an aged, somatic cell to a more youthful, pluripotent state. There are currently conflicting data regarding the ability of reprogramming to fully reset the aging clock by reversing the effects of damaged macromolecules, nuclear and mitochondrial mutations, telomere shortening, epigenomic changes, increased oxidative stress, and numerous other alterations that accrue with age. However, it is clear that reprogramming reverses many aspects of aging by resetting metabolic signatures, mitochondrial networks, and other factors to a youthful state. If the acquisition of stem cell-like properties in iPS cells is closely associated with the maintenance and/or enhancement of adult stem cell function, then determining the molecular mechanisms that *positively* regulate the efficiency and kinetics of reprogramming could provide proof-of-concept for novel mechanisms that regulate the number and function of adult stem cells (Fig. 10.3). Several reprogramming factors that can "reset" the epigenetic status of the somatic cells and allow them to adopt a plethora of possible fates were previously known for their oncogenic activity. A fundamental principle of cell biology is that stem cells with greater potential for self-renewal and pluripotency also have a higher probability of causing tumors [90–94]. Therefore, much research in the field has focused on the tumorigenic capacity traits of iPS cells to facilitate the development of safe, tumorigenesis-free iPS cell-based therapies. If the acquisition of stem cell-like properties in induced pluripotency is associated with the mechanisms underlying cancer stem cell-driven tumorigenesis, then determining the molecular mechanisms that *negatively* regulate tumorigenesis in iPS cells could provide proof-of-concept that novel self-renewing tumor-initiating

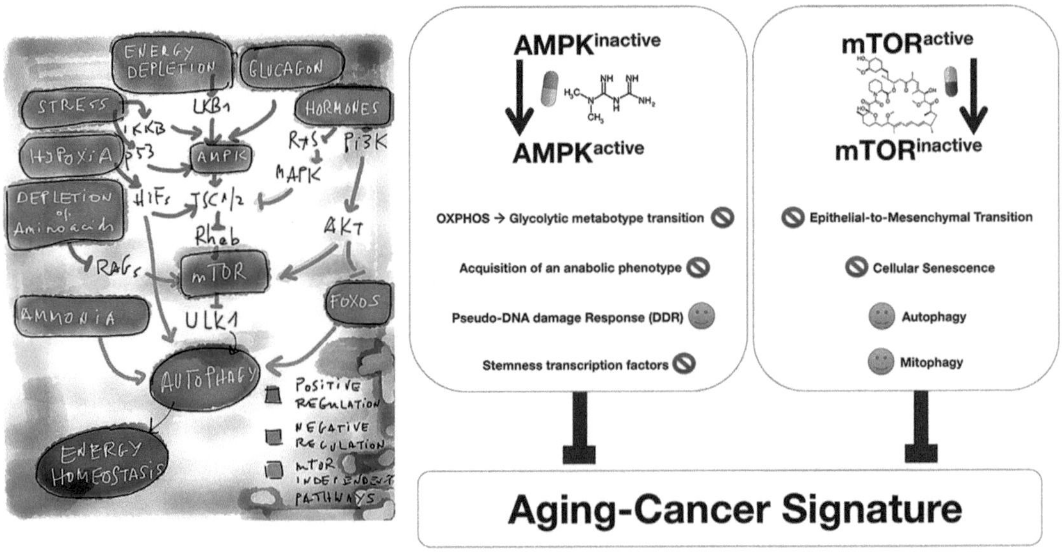

Fig. 10.3 The AMPK-mTOR axis: an example of the metabostemness machinery. Drugs able to inhibit the pro-aging activity of the mTOR gerogene (*e.g.*, rapamycin) and gerosuppressants of mTOR (*e.g.*, the AMPK agonist metformin) can simultaneously affect the aging-cancer signature in ways that significantly promote cell death in immortal CSCs and that weaken aging phenotypes in normal tissues

mechanisms regulate both the development and aberrant function of CSCs (Fig. 10.3). Remarkably, iPS cell-based studies have revealed that metabolic gerogenes (*e.g.*, mTOR) are involved in pivotal housekeeping pathways that may limit the decline in the number or function of adult stem cells. In contrast, deactivation of metabolic gerosuppressors (*e.g.*, AMPK) appears to regulate tumorigenesis in stem cells.

10.4.1 The Metabolic Gerogene mTOR and Stem Cells: Lesson from iPS Cells

Adult stem cells exist in most mammalian organs and tissues and are indispensable for normal tissue homeostasis and repair. The integrity, function, and proliferative capacity of adult stem cells is impaired by aging. Although stem cell function declines with age, it should be noted that there is not an age-related depletion of stem cells. Importantly, some age-related functional changes may be reversible, suggesting that aging stem cells could be rejuvenated. iPS cells are valuable tools that can be used to address stem cell aging and the metabolic pathways that are involved in this process. For example, if blockade of gerogene activity (*e.g.*, mTOR) significantly impacts both the speed and efficiency of iPS cell generation by alleviating aging-related roadblocks to somatic reprogramming (*e.g.*, cell senescence), then this would provide proof-of-concept that housekeeping metabolic pathways can drive aging (*e.g.*, mTOR-regulated cell senescence). Two inhibitors of mTOR activity (rapamycin and PP242) accelerate the early stages of the reprogramming process [95]. This is consistent with earlier studies that suggest that mTOR-regulated activation of the downstream effector p70S6K1 is a critical molecular mechanism for pluripotent cell differentiation [96]. mTOR inhibitors likely function in the very early stages of somatic cell reprogramming by helping to bypass at least two critical reprogramming barriers, namely the epithelial-to-mesenchymal transition (EMT) and cellular senescence, and by mediating the activity of transcription factors that lead to the acquisition of stemness.

10.4.1.1 mTOR and the Epithelial-to-Mesenchymal Transition (EMT)

Inhibition of the mTOR gerogene may facilitate the mesenchymal-to-epithelial (MET), as cells of mesenchymal origin such as fibroblasts are thought to undergo MET before reprogramming can be initiated [97–100]. TGFβ-induced EMT is regulated by activation of the mTOR pathway, and the stable inhibition of mTORC1 (Raptor) and mTORC2 (Rictor) induce MET in cancer cells [101, 102]. Therefore, it is plausible that mTOR blockade could circumvent a major barrier to reprogramming by suppressing microenvironmental pro-EMT signals and activating an epithelial program. EMT processes are a crucial link between cellular senescence and self-limiting mechanisms that limit fibrogenesis [103–105]. Fibrosis may lead to deleterious consequences and loss of tissue function, particularly in association with inflammation, if the rapid synthesis and deposition of extracellular matrix continues unchecked. Further studies should elucidate whether mTOR functioning is a pivotal regulator of the balance between EMT-related fibrosis and cellular senescence in aging stem cells.

10.4.1.2 mTOR and Cellular Senescence

Inhibition of mTOR may fine-tune its gerogenic activity to avoid cellular senescence, which is an initial barrier that limits the efficiency of somatic cells reprogramming. Indeed, the reprogramming process itself appears to trigger a stress response similar to senescence, known as reprogramming-induced senescence (RIS) [106, 107]. Senescence pathways including the $p19^{ARF}/p53/p21^{WAF1/CIP1}$ pathway and the $p16^{INK4a}/pRB$ pathway constitute critical barriers to the reprogramming process [108–114]. Because intrinsic senescence pathways may already be activated in cells from aging tissues (pre-senescent cells), the reprogramming process may push them into a fully senescent state, further hampering reprogramming. Notably, the pharmacological blockade of mTOR with rapamycin prevents the permanent loss of proliferative potential that occurs when cells enter senescence, and the cell cycle, but not cell growth, is arrested. Therefore, mTOR inhibition can induce "rejuvenated" cellular states that decelerate cellular senescence by turning irreversible arrest into a reversible condition. Indeed, when the cell cycle is arrested by inducing the expression cyclin dependent kinase inhibitors (CDKis), such as $p21^{WAF1/CIP1}$ and $p16^{INK4a}$ in the presence of rapamycin, the cells remain quiescent but not senescent because they can resume proliferation when the CDKis are removed [115–121]. This unique mechanism by which gerogenes such as mTOR link senescence to reprogramming is crucial for understanding how drugs that suppress the inappropriate inactivation of growth-promoting mTOR signaling without escaping from cell cycle arrest may be used to safely improve reprogramming without increasing the risk of cancer. Unlike the genetic abrogation of cell cycle arrest that is facilitated by losing *bona fide* tumor suppression (*e.g.*, p53) during reprogramming of somatic cells and malignant transformation, mTOR inhibition transiently inhibits senescence without permanently disabling tumor suppression because it does not reverse cell cycle arrest unless the initial cell cycle inhibitors (*e.g.*, $p21^{WAF1/CIP1}$ and $p16^{INK4a}$) are removed. The mTOR inhibitor rapamycin extends the maximal lifespan in cancer-prone mice but does not increase cancer incidence and/or the biological aggressiveness of the tumors that do arise.

10.4.1.3 mTOR and Autophagy

The ability of the mTOR gerogene to regulate the sensitivity of somatic cells to reprogramming can also be explained by mTOR regulation of autophagy. Autophagy is the homeostatic process of cytoplasmic degradation and recycling that evolved to respond to stress conditions. Cellular hypertrophy associated with senescence is driven by cell cycle arrest (a p53-regulated response to the overexpression of stemness factors that markedly decreases the efficiency of reprogramming), continuous protein synthesis, and insufficient autophagy [122–125]. It is therefore plausible that autophagy activation may balance cellular hypertrophy and proliferative potential to suppress cellular senescence and enhance reprogramming. Consistent with this notion, Wang et al. [126] recently reported that transient activation of autophagy is a key mechanism underlying the ability of the *Sox2* stemness factor to promote cellular reprogramming. First, the authors confirmed that depletion of autophagic proteins (Atg5, Atg3, Atg7) completely abrogates the formation of iPS cells from mouse embryonic fibroblasts. They did not observe any changes in the mRNA levels of various autophagy-related genes (Atg) during reprogramming; however, they did observe a dramatic decrease in mTOR mRNA on the first day of iPS cell generation. mTOR protein levels were also repressed for the first two days of the reprogramming process but were then restored to basal levels at day 3 of iPS cell generation. These results provide a mechanistic explanation for earlier observations by Chen et al. [95], who evaluated the dynamics of rapamycin-enhanced iPS cell generation. The authors observed that previous treatment with rapamycin correlated with enhanced reprogramming efficiency. He et al. [127] similarly reported that elaborate regulation of mTOR activity is required for somatic cell reprogramming induced by defined transcription factors. Taken together, these results strongly suggest that an initial short burst of mTOR suppression at an early stage followed by restoration of mTOR activity at a later stage are both required for successful reprogramming to occur. Perhaps more importantly, these results clearly indicate that proper autophagy driven by mTOR activation is indispensable for transcription factor-induced reprogramming, thus supporting the notion that stemness genes may have evolved as components of metabolic regulation. Thus, Sox2 initiates autophagy exclusively by suppressing the gerogene mTOR during reprogramming, including the cellular reprogramming that occurs in early embryogenesis [126].

10.4.1.4 mTOR and Mitophagy

mTOR-regulated autophagy may enable cells to overcome cellular senescence barriers by remodeling the cell cycle machinery or promoting the turnover of the "senescent" subcellular architecture, which may counterbalance the indefinite cellular growth of senescent cells. In addition, ever-growing genome-, proteome-, and metabolome-based studies have shown that during the first 3 days of reprogramming, multiple protein and metabolite profile changes occur in a highly coordinated fashion. Given this evidence, it is tempting to suggest that mTOR inhibition-induced (mitochondrial) autophagy may induce the early metabolic switch from mitochondrial respiration to glycolysis that underlies the acquisition of stemness.

Mitochondrial structure and function have been suggested as indicators of stem cell competence. Specifically, low mitochondrial activity and a relatively under-developed mitochondrial network are common features of stemness. Prigione et al. [128], Prigione and Adjaye [129] have pioneered the "metabolic state hypothesis" that links mitochondrial state and cellular metabolism to differentiation. According to this hypothesis, self-renewal of iPS cells is supported by glycolysis and mitochondrial properties that are similar to embryonic stem cells, including immature organelle shape with under-developed cristae and low levels of oxidative stress. The idea that alteration of mitochondrial homeostasis and cellular bioenergetics metabolism may be essential for the acquisition and maintenance of and exit from self-renewing pluripotent states in human cells is strongly supported by landmark studies by Folmes et al. [130], who showed that: (a) stemness factors, remarkably, induce the reversion of mitochondrial networks into cristae-poor structures, (b) functional metamorphosis of somatic-oxidative phosphorylation into acquired pluripotent-glycolytic metabolism corresponds to an embryonic-like pattern, (c) cell fate is determined by the upregulation of glycolytic enzymes and downregulation of electron transport chain complex I subunits, and (d) the *a priori* energy infrastructure of somatic cells is a crucial molecular feature for achieving stemness. Glycolytic gene potentiation occurs prior to the induction of pluripotent markers. Accordingly, stimulating glycolysis promotes reprogramming and inhibiting glycolytic enzyme activity inhibits reprogramming. Reprogramming to pluripotency is also more efficient the closer the glycolytic and OXPHOS energy metabolism profile of the starting somatic cells are to the pattern observed in embryonic stem cells [131–137]. Thus, it is tempting to suggest a type of "metabolic memory" that, similar to the "epigenetic memory" contained within specific chromatin modifications, may be partially retained through the reprogramming process. Metabolic reprogramming plays an early, active role in the acquisition of stemness and therefore presents a barrier that must be overcome to enhance reprogramming efficiency. Beyond this role, however, it is largely unknown how a retained metabolic memory could influence not only the differentiation potential and function of iPS cells but also resetting to a metabolic program that is compatible with stem cell function. This leads to the attractive hypothesis that gerogenes such as mTOR can regulate mitochondrial dynamics to segregate mitochondria that are destined for clearance through autophagy. Because this process results in compartmentalized cellular catabolism, loss of mitochondrial function, increased glucose uptake and, consequently, accelerated onset of glycolysis, mTOR-driven mitophagy may play a crucial role in regulating the number and quality of mitochondria to achieve stemness.

Mitochondrial homeostasis in pluripotency relies on mitochondrial biogenesis and dynamics (fission and fusion), as well as degradation through mitophagy [138–143]. Interestingly, although early studies suggested that iPS cells contain fewer mitochondria and lower copy numbers of mtDNA compared to somatic cells, recent studies have clarified that the ratio between mitochondrial mass and total cellular mass, is similar in iPS cells and differentiated cells [144, 145]. Intriguingly, we are beginning to accumulate evidence that mitochondrial dynamics are also actively involved in stem cell biology (*e.g.*, blockade of mitochondrial fission drastically decreases the efficiency of reprogramming to stemness) [143], indicating that not only functional but also structural changes in mitochondria are required to successfully acquire a stem cell-like state. Oncogene-induced mitophagy mediates mitochondrial functional loss (*i.e.*, decreased respiration that is not related to changes in mitochondrial biogenesis) during early tumorigenesis [146–148]. Similarly, mTOR activity-regulated mitophagy could be a crucial mechanism that regulates the transition to a glycolytic, rejuvenated metabotype that is able to bypass cellular senescence. An mTOR-regulated increase in mitochondrial fission during the reprogramming to stemness may upregulate mitophagy, which would therefore lead to significant reductions in both the number and the size of mitochondria, thus achieving the "mitochondrial phenotype" that is associated with stem cells. But how?

We know that the mTOR pathway regulates mitochondrial oxygen consumption and oxidative capacity. Indeed, mTOR activity plays an important role in determining the balance between mitochondrial and non-mitochondrial glycolytic sources of ATP [149, 150]. First, disrupting the mTOR-Raptor (mTORC1) complex with rapamycin lowers mitochondrial membrane potential, oxygen consumption and ATP synthesis. Second, altering mTOR-Raptor expression modulates the total oxidative capacity of cells and significantly alters the mitochondrial phosphoproteome. Third, mTOR inhibition modulates the relative contribution of mitochondrial metabolism to cytosolic glycolysis and therefore does not merely reflect a secondary response to changes in the energy demand. These metabolic peculiarities of the mTORC1 complex support a putative mitophagic role for mTOR during the acquisition of stemness. Impairing mitochondrial oxidation and increasing glycolysis-driven energy production may be necessary but not sufficient to convert somatic cells to a pluripotent state. The bioenergetic switch must be accompanied by alterations in the mitochondrial dynamics that facilitate bypass of the senescence checkpoint (*e.g.*, an increase in mitochondrial fusion could downregulate mitophagy, thus generating giant mitochondria that are associated with cell senescence). Notably, pharmacological blockade of mTOR activity with rapamycin promotes the colocalization of lysosomes with mitochondria [151–153], and the rapamycin-induced autophagosomal clearance of damaged mitochondria is partially blocked by specific suppression of autophagy-related genes (*i.e.*, ATG5). The fact that the mTOR-Raptor (mTORC1) complex directly associates with mitochondria and the recent discovery that inhibition of mTORC1 (but not mTOR-Rictor/mTORC2) is essential for the initiation of autophagy during iPS cell generation [126] strongly suggest that the dynamic (early) repression and (late) repression of mTOR activity during reprogramming to stemness is likely to be related to mitophagy induction.

Dai et al. [151] recently tested the hypothesis that enhancing mTOR-regulated mitophagy drives selection against dysfunctional mitochondria harboring high levels of mutations, thereby decreasing mutation levels over time. In their hands, pharmacological inhibition of mTOR activity with rapamycin induced colocalization of mitochondria with autophagosomes and increased the number of autophagic vacuoles containing mitochondria-like organelles, resulting in a striking decrease in heteroplasmic mtDNA mutations. The decreased mutational burden was not due to rapamycin-induced cell death or mtDNA depletion, but rather to the unique ability of mTOR inhibition to activate mitophagy as a way to select against heteroplasmic mtDNA mutations. This raises the exciting possibility that mTOR may have therapeutic potential as a target for the treatment of mitochondrial disorders associated with heteroplasmic mtDNA mutations. Reprogramming somatic cells from patients with mitochondrial dysfunction caused by mutations in heteroplasmic mtDNA would give rise to two populations of iPS cells: Mt-mutation-rich iPS cells with high levels of mutant mtDNA, and Mt-mutation-free iPS cells with undetectable levels of mutant mtDNA [154, 155]. Thus, it would be of interest to evaluate whether reprogramming in the presence of an early, transient blockade of mTOR activity drastically alters the above-mentioned balance of Mt-mutation-rich/Mt-mutation-free mtDNA and generates fully Mt-mutation-free iPS cell populations as a promising resource for the potential autologous cell therapy.

10.5 The Metabolic Gerosuppressor AMPK and Cancer Stem Cells: Lessons from iPS Cells

CSC-driven tumorigenesis is closely associated with the acquisition of stem cell-like properties in induced pluripotency. For example, the acquisition of both CSC and iPS cellular states requires a specific combination of oncogenes and tumor suppressor genes that leads to the loss of the cell's original identity and produces a less-differentiated cell that can proliferate and self-renew indefinitely [15]. Therefore, determining the mechanisms that negatively regulate the efficiency and kinetics of somatic reprogramming to iPS cells could provide proof-of-concept for the existence of pivotal self-renewing tumor-initiating mechanisms. These mechanisms could then be targeted to regulate both the number and aberrant function of CSCs. As discussed above, iPS cells appear to share all of the core components of metabolic reprogramming that have been observed in human tumors since the end of the 1920s, when Otto Warburg first described that the glycolytic pathway is favored over mitochondrial oxidative phosphorylation (OXPHOS) as the primary form of energy metabolism in tumor cells, even in the presence of oxygen. Indeed, it appears that the glycolytic phenotype is necessary for and promotes the acquisition of pluripotency [131–137]. Stimulation of aerobic glycolysis favors reprogramming, while blockade of glycolytic enzymes inhibits reprogramming. These findings strongly support cell bioenergetics as a novel prerequisite for acquired stemness in normal and malignant cells. Accordingly, an experimental model comparing oncogenic transformation and nuclear reprogramming confirmed that somatic cells must first acquire changes that lead to the downregulation of cell differentiation machinery as well as the activation of glycolysis and other glycolysis-related metabolic pathways. Only then do the oncogenic transformation/induced pluripotency pathways diverge, depending on other factors, such as the activity of pluripotency genes. Therefore, it is reasonable to suggest that the acquisition of, and exit from, malignant stemness in CSCs is governed not only by genetic and epigenetic factors but also by metabolic reprogramming. If energy metabolism plasticity facilitates the self-renewal and differentiation of CSCs in cancer tissues, then it is plausible that yet-to-be described cellular metabotypes defined by specific cellular energetic machineries and their quality control systems could regulate CSC-like states. Accordingly, only a few "permitted" cell metabotypes would possess the necessary plasticity to reprogram the tumor cell-of-origin so that it can become a CSC [30, 156]. The intrinsic and extrinsic genetic, epigenetic, and microenvironmental factors that regulate the transition to a CSC-like state could not operate in the presence of "protected" cell metabotypes. In this scenario, the activation status of metabolic gerosuppressors such AMPK, a master metabolic master that senses and decodes intracellular changes in the energy status [157–160], would play a crucial role in regulating the ability of oncogenes, transcription factors, or onco-microRNAs to promote the acquisition of a CSC-like state.

AMPK, whose ancestral role may have been related to the glucose starvation response, appears to have arisen very early during eukaryotic evolution [157–160]. Rapid cell growth requires the active synthesis of proteins, rRNA and lipids, all of which are switched off by activation of the gerosuppressor AMPK. Not surprisingly, expression of the catalytic subunit of AMPK is significantly downregulated in iPS cells [154], mimicking a mechanism employed by many tumor cells to escape the growth-restraining effects imposed by switching from anabolic to catabolic metabolism that may occur upon activation of AMPK. Loss of sensitivity to activation of the tumor-suppressor/gerosuppressant activity of AMPK in response to metabolic stresses may allow normal and non-CSC tumor cells to de-differentiate and acquire properties of CSCs. Therefore, characterizing the role of AMPK in inhibiting nuclear reprogramming during iPS cell generation could facilitate the development of clinically applicable AMPK activation strategies directed against stemness. In this regard, we have shown that AMPK activators can endow somatic cells with an energetic infrastructure that is protected against reprogramming to stemness [80]. Importantly, the metabolic barrier imposed by AMPK activation strategies cannot be bypassed even through p53 deficiency. p53 is the guardian of the genomic and metabolic checkpoint, and its loss greatly enhances the efficiency of stem cell production by facilitating immortalization [110–113]. How is the cell metabotype acquired in response to AMPK activation associated with a cellular state that is refractory to the induced acquisition of stemness?

10.5.1 AMPK and the OXPHOS/ Glycolytic Metabotype

There is an overall correlation between the bioenergetic state of the cell-of-origin and the efficiency of reprogramming it into a stem cell. Cells with an oxidative: glycolytic energy production ratio closer to that of pluripotent cells are reprogrammed more quickly and efficiently [133]. Therefore, it is reasonable to suggest that AMPK activation may efficiently impede nuclear reprogramming by preventing a glycolytic metabotype. Although data directly linking AMPK activation status with glycolysis are scarce, recent evidence shows that the activation status of AMPK either promotes (AMPK *on*) or inhibits (AMPK *off*) tumor development by regulating the Warburg effect [161–163]. While earlier studies suggested that AMPK may suppress glycolysis in tumor cells by inhibiting mTOR, AMPK regulation of glycolysis is also affected by p53, the best-known tumor suppressor that regulates the expression of several genes intimately linked to OXPHOS and glycolysis. Because the net effect of p53 deficiency is a reduction in mitochondrial respiration and strong activation of glycolysis, leading to the Warburg effect [164–168], AMPK activation may promote p53-driven suppression of the glycolytic flux. AMPK antagonizes the pro-immortalizing glycolysis triggered by p53 loss. Importantly, a direct link between aberrant glucose metabolism and CSCs was been recently confirmed in glioblastoma. The antidiabetic drug metformin promoted differentiation of stem-like glioma-initiating cells into non-tumorigenic cells *via* activation of AMPK, which is sensitive to glucose availability [169].

The activation status of the metabolic gerosuppressor AMPK can also affect reprogramming to stemness by regulating *primum movens* that establish the Warburg effect regardless the presence or absence of driver genetic alterations. First, the activity and protein expression level of H^+-ATPase synthase, a reversible engine in the inner mitochondrial membrane that regulates energy conservation by synthesizing or hydrolyzing ATP in response to changes in metabolic cellular conditions, are repressed in human carcinomas [170–177]. Overexpression of the ATPase inhibitor factor 1 (IF1) limits the activity of H^+-ATPase and triggers the metabolic switch to enhanced aerobic glycolysis, while silencing IF1 has the opposite metabolic effect. IF1 expression is generally negligible in normal tissues, and IF1 is highly overexpressed in numerous carcinomas. IF1 overexpression is sufficient to limit H^+-ATPase activity and promote the acquisition of the Warburg phenotype without any genetic changes. IF1 is highly expressed in iPS cells, whereas expression of the catalytic β-F1-ATPase subunit, which is the rate-limiting component of mitochondrial OXPHOS, is specifically repressed in iPS cells [156]. Notably, exposure to the AMPK agonist metformin drastically decreases IF1 levels. Conversely, AMPK activation by metformin promoted a significant increase in β-F1-ATPase content in individual iPS cells. Remarkably, AMPK activation-driven reversal of the IF1/β-F1-ATPase expression status concomitantly induced a switch to an SSEA1-negative state in smaller iPS cell colonies. This strongly suggests that AMPK can operate as an upstream regulator of the H^+-ATPase synthase-geared metabolism switch and that a mitochondria-mediated energy adaptation is sufficient to promote the acquisition of a stemness-competent Warburg-like metabotype [156]. Further studies should more clearly elucidate whether the AMPK-driven IF1/β-F1-ATPase ratio and H^+/ATPase activation status play an instrumental role in establishing and/or maintaining CSC-like states.

10.5.1.1 AMPK and the Lipogenic Metabotype

One reason for the high glycolytic rate of rapidly proliferating cells is that the TCA cycle ceases to be a purely catabolic pathway and becomes at least partially anabolic, actively providing precursors for biosynthesis, particularly citrate for lipid synthesis [178–181]. Indeed, the Warburg effect has been correctly redefined in terms of the obligatory dependence of some cellular states, including stem cell-like states, on maximizing the production of macromolecules and organelles. This is because aerobic glycolysis, but not OXPHOS, facilitates the rapid and efficient diversion of key metabolites into the major cellular biosynthetic pathways (amino acid synthesis, nucleic acid synthesis, and *de novo* fatty acid biogenesis). We recently demonstrated that, similar to biologically aggressive subtypes of human carcinomas and CSCs, iPS cells supercharge lipogenesis by triggering regulatory circuits that activate and provide substrates for the key lipogenic enzymes, including acetyl-CoA carboxylase (ACACA) and fatty acid synthase (FASN) [156]. Coincidentally, when the activity of ACACA and FASN is inhibited, the reprogramming efficiency decreases significantly. Treatment with the AMPK agonist metformin at concentrations that suppress the early self-renewal marker SSEA-1 significantly and specifically blocked the increase in ACACA and FASN proteins expression in iPS cells compared to the parental MEF population. Therefore, the activation of an AMPK-regulated lipogenic-phenotypeis instrumental in reprogramming somatic cells to induced pluripotency [156], most likely by enabling a rapid regeneration of NAD^+ by consuming large amounts of NADPH and avoiding low NAD^+/NADH ratios that would eventually inhibit glycolysis through feedback mechanisms.

10.5.1.2 AMPK and DNA Damage

The metabolic gerosuppressor AMPK may also render cells metabolically protected against the acquisition of stemness in response to DNA damage. The p53-dependent counterselection of DNA-damaged cells during reprogramming [109–113] has been demonstrated by increased DNA damage foci and increased phosphorylation of the serine/threonine kinase ataxia telangiectasia mutated (ATM), a primary regulator of the cellular response to DNA double-strand breaks that operates as a central DNA damage checkpoint connecting cancer predisposition with cellular bioenergetics [182, 183]. Thus, it has been suggested that the DNA damage response (DDR) activated during the induction of reprogramming might be equivalent to the oncogene-induced DDR that occurs during oncogene-induced senescence [107, 113]. In this case, the cell proliferation and transformation induced during oncogene activation in early tumorigenesis are inhibited by cellular senescence. AMPK activation may lower the threshold for cellular senescence by activating an ATM-dependent pseudo-DDR [184], which can occur regardless of the p53 status. This "pre-activation" of AMPK may protect cells against further damage, thus mimicking the precancerous stimuli that induce an intrinsic barrier against carcinogenesis [67, 68, 72, 184], which, in turn, would accelerate the onset of cellular senescence (RIS/OIS). The AMPK agonist metformin establishes a DDR-dependent cell cycle arrest that interacts synergistically with hyperoxic culture-induced DNA damage and cellular senescence in human diploid fibroblasts and accelerates the DDR-related cellular senescence response induced by DNA-damaging drugs [71]. This strongly suggests that activating the metabolic gerosuppressor AMPK could favor a metabolic imbalance that would lower the threshold for stress-induced senescence in response to the overexpression of stemness factors. Therefore, if AMPK-activating drugs strengthen RIS to decrease the rate of conversion of somatic cells into iPS cells, then drug strategies that promote AMPK activation should lead to an efficient entrance to senescence in premalignant and malignant tissues, preventing the acquisition of stemness in cell populations with tumor-propagating capacity.

10.5.1.3 AMPK and Stemness Factors

AMPK activation may negatively impact acquisition of a stem cell-like state by preventing the activation of transcriptional regulators that link the genetic and epigenetic regulation of stem cell states. The AMPK agonist metformin negatively regulates *Oct4*, a well-known transcription factor (also known as *Pou5f1*) that plays a fundamental role in stem cell self-renewal, pluripotency and somatic cell reprogramming [80, 81, 185, 186]. Some poorly differentiated, biologically aggressive carcinomas appear to hijack the *Oct4*-driven self-renewal machinery to support aberrant proliferation and tumor initiation. *Oct4* overexpression was sufficient to promote the growth of tumor-initiating cells in a mouse model of breast cancer, and subpopulations of self-renewing breast and ovarian cancer cells overexpress *Oct4* [187–191]. Indeed, transducing *Oct4* into primary human mammary epithelial breast cells is sufficient to generate cell lines that have a gene signature comparable to CSC-like-enriched claudin-low breast carcinomas that possess tumor-initiating and colonization capacities. The anti-*Oct4* activity of AMPK activators could translate into anticancer effects if *Oct4*-driven transcriptional networks are specifically reactivated in CSCs. Our group performed a study in which iPS cells were transplanted into immunodeficient mice. We showed that pharmacological activation of AMPK using systemic metformin fully recapitulated the ability of an increased dose of tumor suppressors (like *p53* and *Ink4a/ARF*) to prevent the occurrence or drastically reduce the size and weight of teratoma-like masses in these mice [81]. Importantly, AMPK activation facilitated the specific elimination of the teratoma-initiating pluripotent stem cells that are intermixed with the desired, non-tumorigenic iPS derivatives, at least in part by suppressing *Oct4* expression [80, 81]. The specific, efficient elimination of the *Oct4*-positive malignant iPS cells that give rise to teratocarcinomas strongly supports the hypothesis that the metabolic infrastructure of stem cells is an indispensable component of the CSC machinery.

Another *Oct4*-related molecular candidate that may link activation of the metabolic gerosuppressor AMPK to the reprogramming blockade is the Lin28/let-7 axis [192–198]. Lin28 is a gatekeeper of the pluripotent state that binds to and inhibits the processing of let-7, a gatekeeper of the differentiated state, in an intricately designed auto-regulatory loop. Let-7 opposes the actions of cell cycle-regulating miRNAs that maintain self-renewal in embryonic stem cells. Thus, inhibiting let-7 in human cells promotes reprogramming as much as the oncogene *c-Myc* when combined with the stemness factors Oct4, Sox2, and Klf4. Conversely, persistence of let-7-based signaling counteracts the activity of stemness factors by promoting the expression of pro-differentiation genes. Lin28 can functionally replace the oncogene *c-Myc* in the original Yamanaka cocktail of stemness transcription factors, supporting the idea that Lin28 promotes reprogramming by preventing the production of mature, tumor-suppressive let-7 miRNAs. Indeed, we are accumulating evidence that Lin28 acts as a *bona fide* oncogene in the absence of canonical genetic alterations. Moreover, CSCs could arise through a reprogramming-like mechanism regulated by a double-negative feedback loop between the reprogramming factor Lin28 and the microRNA let-7, which regulates aldehyde dehydrogenase-1 (ALDH1)-positive CSCs [199]. In our hands, short-term exposure to metformin was sufficient to drastically upregulate the expression of let-7 and coincidentally reduces the stem cell-like features of breast cancer cells, including the formation of mammospheres in non-adherent/non-differentiating conditions [200]. Bao et al. [185] later confirmed that AMPK activation causes de-represses expression of let-7, which is typically loss in pancreatic cancer, especially in pancreatospheres enriched with tumor-propagating CSC-like cells. Further mechanistic studies are needed to explore whether AMPK activation antagonizes Lin28 expression and/or activity. However, the impact of metformin on let-7 expression raises the tantalizing possibility that activation of the metabolic gerosuppressor AMPK impedes the acquisition of stemness by altering the differentiation *vs.* pluripotency states driven by the Lin28/let-7 loop. It should be noted that Lin28 binds to and enhances translation of mRNAs for several metabolic enzymes, thereby regulating glycolysis and OXPHOS. Lin28-mediated enhancement of tissue repair was negated by OXPHOS inhibition, whereas a pharmacologically induced increase in OXPHOS enhanced repair. Therefore, it could be relevant to evaluate whether upstream activation of AMPK upstream regulates the ability of Lin28 to exert anti-stemness activities by reprogramming cellular bioenergetics [201–204]. Moreover, let-7 on its own markedly reduces the expression of key transcriptional inducers of stemness, including *Oct4*, whereas Lin28 directly upregulates *Oct4* expression [205, 206]. The Lin28/let-7 signaling pathway is a central regulator of glucose metabolism [201–204], so it is reasonable to suggest that the interconnected metabolic controllers AMPK and Lin28/let-7 fine-tune the activation status of core reprogramming factors such as *Oct4*, rather than acting as on/off reprogramming switches.

10.6 A Therapeutic Corollary

We are beginning to understand how the molecular networks that switch bioenergetic metabolism on & off, *i.e.*, the production of fuel and building blocks for growth and division, and autophagy/mitophagy, *i.e.*, the quality-control, self-cannibalistic system responsible for "cleaning house" and "recycling trash", are wired differently in adult stem cells and CSCs. We propose that the term "metabostemness" is an appropriate designation for this new cellular attribute at the intersection of aging and cancer (Figs. 10.1 and 10.2). While the prolonged activation of critical components of metabostemness (*e.g.*, the gerogene mTOR) could lead to depletion of adult stem cell function and reduced health, the activation of analogous metabostemness machinery (or the loss of their negative regulators, *e.g.*, the gerosuppressor AMPK) could promote the aberrant acquisition of tumor-initiating and metastatic abilities in CSCs. In addition to the involvement of key metabolic sensors, we predict that metabostemness will likely include yet-to-be discovered metabolic parameters at the cell-intrinsic, tissue-microenvironmental, and systemic levels enable stem cells to self-renew and differentiate.

The metabostemness hallmark may offer a unique opportunity for developing innovative, effective metabolic drugs that target the metabolic infrastructure of stem cells (Fig. 10.4). Moreover, the co-development of drugs directed against metabostemness and diagnostic devices able to detect and monitor metabostemness function would revolutionize our current perception of stem cell-based healthcare. On the one hand, drugs that target features of metabostemness will improve the health of normal tissues by ameliorating the expansion, cell-fate plasticity, and lifespan of their resident adult stem cells. Drugs that target features of metabostemness will concomitantly impede the generation and maintenance of CSCs that are addicted to and, consequently, dependent on aberrant metabostemness function. At the same time, the metabostemness property could help delineate the structural and functional changes that allow adult stem cells and CSCs to produce specific or enriched "gerometabolites" and "oncometabolites" (or "gerometabolomic" or "oncometabolomic" expression profiles), respectively. Coupling the testing of drugs that target metabostemness with the use of non-invasive devices (*e.g.*, RMN, circulating metabolome) to accurately monitor the spatio-temporal distribution and function of adult and cancer stem cells in real time could accelerate the discovery and development of metabostemness drugs. Consequently, "all-in-one" theranostic approaches could take pharmacometabolomics-based personalized medicine from the lab to the "point-of-care", the patient.

Metabostemness could help identify and develop a new generation of drugs that reprogram cell metabotypes to increase the structural and functional integrity of adult stem cells and positively influence their lineage determination, while preventing the acquisition of stemness in human carcinomas (Fig. 10.4). The antidiabetic biguanide metformin exemplifies the anti-aging and anticancer effects that can be expected from drugs that target adult stem cells and CSCs [207]. Freda Miller and her colleagues recently showed that metformin promoted the repair and regeneration of endogenous adult stem cells [69]. Their group showed that the gerosuppressant drug metformin harnessed endogenous repair mechanisms to promote regeneration in situation in where regeneration does not normally occur. These results support the idea that inducing self-renewal and proliferation of endogenous adult stem cells using non-invasive and non-toxic therapies may eventually constitute a legitimate alternative to stem cell transplantation. Indeed, the findings of Miller and her colleagues [69] unambiguously confirm the potential of exploring the gerosuppressant activity of AMPK activators from a stem cell-centered perspective. In a series of experiments in cell culture, metformin-induced activation of AMPK was found to promote neurogenesis in both mouse and

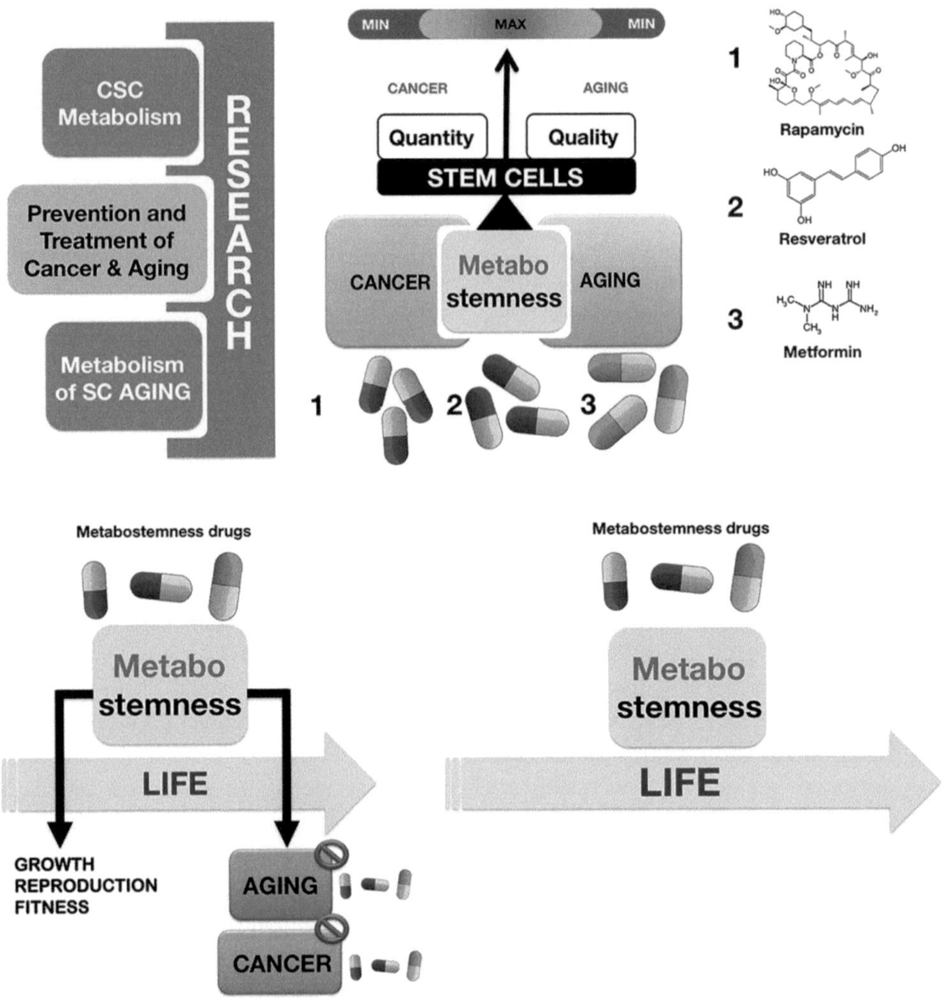

Fig. 10.4 Metabostemness drugs: decelerating aging and postponing cancer. Metabostemness targeting drugs (*e.g.*, rapamycin, resveratrol, metformin) will improve the health of normal tissues by improving the expansion capacity, cell-fate plasticity, and lifespan of their resident adult stem cells. Metabostemness targeting drugs will concomitantly impede the generation and maintenance of CSCs that are addicted to and, consequently, dependent on aberrant metabostemness function for their life-threatening activities. By metabolically impeding the generation and maintenance of CSCs while metabolically protecting adult stem cells from initiating cell deterioration programs, a new generation of metabostemness drugs will simultaneously permit the treatment of the underlying causes of cancer and metastasis while delaying aging. The metabostemness hallmark offers, for the first time, a combinatorial and interdisciplinary molecular scenario in which to acquire the power to manipulate a stem cell's destiny, and therefore life expectancy, using metabolic means

human neural stem cells. Compared to stem cells from control mice, stem cells from metformin-treated mice exhibited a nearly doubled capacity to produce new neurons. Notably, in living mice, metformin treatment induced an approximately 30 % increase in the number of new neurons in the hippocampus, a region of the brain that is closely involved in forming new memories. Perhaps more importantly, Miller and her colleagues [69] confirmed that mice treated with the gerosuppressant metformin formed new memories more rapidly than mice that were given a control substance. Notably, the enhanced spatial memory of metformin-treated mice was directly dependent on neurogenesis, as pharmacological killing of neural precursor cells efficiently blocked the effect of metformin on memory formation and reduced the number of new neurons.

The action of the gerosuppressant metformin as an AMPK agonist provides an efficient barrier to the reprogramming of somatic cells to stem cells, while uncoupling pluripotency from tumorigenesis [30, 156, 207]. The ability of individual somatic cells to enter reprogramming at different time points after induction with stemness factors and the length of time required to complete the reprogramming sequence are greatly affected by the bioenergetic and anabolic status of the cell, namely the activation status of the gerosuppressor AMPK. Activation of AMPK with metformin allows fewer cells to undergo the required stochastic epigenetic events and consequently become fully reprogrammed stem cells. Sustained mitochondrial OXPHOS and the suppression of cell anabolism in response to metformin-induced AMPK activation will therefore reduce the number of cells that activate the expression of essential transcription factors that regulate self-renewal and pluripotency. A model can be proposed in which the development of an AMPK-driven metabotype in response to metformin is a crucial stochastic event that imposes an *a priori* roadblock to dedifferentiation from somatic cells to pluripotent stem cells. This road block does not require the participation of oncogene-driven metabolic and non-metabolic changes. If metformin-dependent remodeling of the cell metabotype determines cell fate and the transition between non-CSC and CSC-like states, then metformin-like AMPK activating strategies could be used to pharmacologically manipulate the self-renewal and pluripotency that underlie CSC-driven tumorigenesis and metastasis. A growing number of studies have demonstrated that metformin selectively ablates CSCs, as evidenced by decreased expression of pluripotency-associated genes, CSC-associated surface markers, and other CSC-specific properties. We are thus beginning to delineate a new and complex scenario in which metformin-like gerosuppressant drugs specifically impact the expression of CSC-specific molecular networks to efficiently disrupt the stem cell compartment in multiple cancers, while also maintaining the precise balance between self-renewal and differentiation in adult stem cells. Although a fundamental principle of cell biology is that stem cells, which have a greater potential for self-renewal and pluripotency, are also more likely to cause tumors, the existing evidence strongly suggests that activation of AMPK *via* the systemic delivery of metformin may interfere with mechanisms that are important for stem cell-related tumorigenesis but are dispensable for adult stem cell development and function in mature tissues. Using iPS cells implanted into immunodeficient mice, we have provided proof-of-concept that systemic metformin efficiently suppresses the *Oct4*-driven compartment of malignant stem cells responsible for teratocarcinoma growth, while safeguarding the Oct4-independent ability to generate terminally differentiated tissues [81]. iPS cells implanted into metformin-treated mice produced the same number of distinct tissue types derived from the three embryonic germ layers as observed in untreated mice. If metformin can indeed uncouple tumorigenicity from pluripotency in stem cells, then new gerosuppressant approaches using AMPK-targeting drugs could potentially rejuvenate the tissue maintenance and repair processes driven by endogenous stem cells while decreasing the tumorigenic predisposition of aging tissues.

The recently proposed "geroncogenic" scenario suggests that metabolic changes during aging (*i.e.*, the normal decline in oxidative metabolism and the development of Warburg-like glycolytic metabolism in normal tissues) constitute an early and important "hit" that drives tumorigenesis. This scenario correlates well with our hypothesis that certain gerosuppressant agents that decelerate aging in turn can postpone cancer (anage-related disease), due to the "metabostemness" property described here. While it is clear that the antifungal antibiotic rapamycin, the polyphenol resveratrol, and the biguanide metformin already belong to this new family of metabostemness targeting drugs (Fig. 10.4), we expect a rapid identification of new drug candidates (*e.g.*, polyphenolic xenohormetins [32]) that may reverse or postpone geroncogenesis [208] at the stem cell level.

Acknowledgments This work was financially supported by the Ministerio de Ciencia e Innovación (SAF2012-38914), Plan Nacional de I+D+I, MICINN, Spain.

Conflict of Interest Statement The authors of this manuscript have no conflicts of interest to declare.

References

1. López-Otín C, Blasco MA, Partridge L, Serrano M, Kroemer G. The hallmarks of aging. Cell. 2013;153:1194–217.
2. Pollina EA, Brunet A. Epigenetic regulation of aging stem cells. Oncogene. 2011;30:3105–26.
3. Finkel T, Serrano M, Blasco MA. The common biology of cancer and ageing. Nature. 2007;448:767–74.
4. Rando TA. Stem cells, ageing and the quest for immortality. Nature. 2006;441:1080–6.
5. Signer RA, Morrison SJ. Mechanisms that regulate stem cell aging and life span. Cell Stem Cell. 2013;12:152–65.
6. Meacham CE, Morrison SJ. Tumour heterogeneity and cancer cell plasticity. Nature. 2013;501:328–37.
7. Marjanovic ND, Weinberg RA, Chaffer CL. Cell plasticity and heterogeneity in cancer. Clin Chem. 2013;59:168–79.
8. Shibata M, Shen MM. The roots of cancer: stemcells and the basis for tumor heterogeneity. Bioessays. 2013;35:253–60.
9. Visvader JE. Cells of origin in cancer. Nature. 2011;469:314–22.
10. Monteiro J, Fodde R. Cancer stemness and metastasis: therapeutic consequences and perspectives. Eur J Cancer. 2010;46:1198–203.
11. Chaffer CL, Brueckmann I, Scheel C, Kaestli AJ, Wiggins PA, Rodrigues LO, et al. Normal and neoplastic nonstem cells can spontaneously convert to a stem-like state. Proc Natl Acad Sci U S A. 2011;108:7950–5.
12. Gupta PB, Fillmore CM, Jiang G, Shapira SD, Tao K, Kuperwasser C, et al. Stochastic state transitions give rise to phenotypic equilibrium in populations of cancer cells. Cell. 2011;146:633–44.
13. Iliopoulos D, Hirsch HA, Wang G, Struhl K. Inducible formation of breast cancer stem cells and their dynamic equilibrium with non-stem cancer cells via IL6 secretion. Proc Natl Acad Sci U S A. 2011;108:1397–402.
14. Hirsch HA, Iliopoulos D, Struhl K. Metformin inhibits the inflammatory response associated with cellular transformation and cancer stem cell growth. Proc Natl Acad Sci U S A. 2013;110:972–7.
15. Menendez JA, Alarcón T, Corominas-Faja B, Cuyàs E, López-Bonet E, Martin AG, et al. Xenopatients 2.0: reprogramming the epigenetic landscapes of patient-derived cancer genomes. Cell Cycle. 2014; 13(3):358–70.
16. Mantel C, Broxmeyer HE. Sirtuin 1, stem cells, aging, and stem cell aging. Curr Opin Hematol. 2008;15:326–31.
17. Cerletti M, Jang YC, Finley LW, Haigis MC, Wagers AJ. Short-term calorie restriction enhances skeletal muscle stem cell function. Cell Stem Cell. 2012;10:515–9.
18. Yilmaz ÖH, Katajisto P, Lamming DW, Gültekin Y, Bauer-Rowe KE, Sengupta S, et al. mTORC1 in the Paneth cell niche couples intestinal stem-cell function to calorie intake. Nature. 2012;486:490–5.
19. Warr MR, Binnewies M, Flach J, Reynaud D, Garg T, Malhotra R, et al. FOXO3A directs a protective autophagy program in haematopoietic stem cells. Nature. 2013;494:323–7.
20. Park JH, Glass Z, Sayed K, Michurina TV, Lazutkin A, Mineyeva O, et al. Calorie restriction alleviates the age-related decrease in neural progenitor cell division in the aging brain. Eur J Neurosci. 2013;37:1987–93.
21. Rafalski VA, Mancini E, Brunet A. Energy metabolism and energy-sensing pathways in mammalian embryonic and adult stem cell fate. J Cell Sci. 2012;125:5597–608.
22. Shyh-Chang N, Daley GQ, Cantley LC. Stem cell metabolism in tissue development and aging. Development. 2013;140:2535–47.
23. Mahmoudi S, Brunet A. Aging and reprogramming: a two-way street. Curr Opin Cell Biol. 2012;24:744–56.
24. Ochocki JD, Simon MC. Nutrient-sensing pathways and metabolic regulation in stemcells. J Cell Biol. 2013;203:23–33.
25. Zhang G, Yang P, Guo P, Miele L, Sarkar FH, Wang Z, et al. Unraveling the mystery of cancer metabolism in the genesis of tumor-initiating cells and development of cancer. Biochim Biophys Acta. 2013;1836:49–59.
26. Menendez JA, Vellon L, Oliveras-Ferraros C, Cufí S, Vazquez-Martin A. mTOR-regulated senescence and autophagy during reprogramming of somatic cells to pluripotency: a roadmap from energy metabolism to stem cell renewal and aging. Cell Cycle. 2011;10:3658–77.
27. Guan JL, Simon AK, Prescott M, Menendez JA, Liu F, Wang F, et al. Autophagy in stemcells. Autophagy. 2013;9:830–49.
28. Pan H, Cai N, Li M, Liu GH, Izpisua Belmonte JC. Autophagic control of cell 'stemness'. EMBO Mol Med. 2013;5:327–31.
29. Del Barco S, Vazquez-Martin A, Cufí S, Oliveras-Ferraros C, Bosch-Barrera J, Joven J, et al. Metformin: multi-faceted protection against cancer. Oncotarget. 2011;2:896–917.
30. Menendez JA, Joven J, Cufí S, Corominas-Faja B, Oliveras-Ferraros C, Cuyàs E, et al. The Warburg effect version 2.0: metabolic reprogramming of cancer stem cells. Cell Cycle. 2013;12:1166–79.
31. Vazquez-Martin A, López-Bonet E, Cufí S, Oliveras-Ferraros C, Del Barco S, Martin-Castillo B, et al.

Repositioning chloroquine and metformin to eliminate cancer stem cell traits in pre-malignant lesions. Drug Resist Updat. 2011;14:212–23.
32. Menendez JA, Joven J, Aragonès G, Barrajón-Catalán E, Beltrán-Debón R, Borrás-Linares I, et al. Xenohormetic and anti-aging activity of secoiridoid polyphenols present in extra virgin olive oil: a new family of gerosuppressant agents. Cell Cycle. 2013;12:555–78.
33. Blagosklonny MV. Prevention of cancer by inhibiting aging. Cancer Biol Ther. 2008;7:1520–4.
34. Blagosklonny MV. Aging-suppressants: cellular senescence (hyperactivation) and its pharmacologic deceleration. Cell Cycle. 2009;8:1883–7.
35. Blagosklonny MV. TOR-driven aging: speeding car without brakes. Cell Cycle. 2009;8:4055–9.
36. Blagosklonny MV. Calorie restriction: decelerating mTOR-driven aging from cells to organisms (including humans). Cell Cycle. 2010;9:683–8.
37. Blagosklonny MV. Increasing healthy lifespan by suppressing aging in our lifetime: preliminary proposal. Cell Cycle. 2010;9:4788–94.
38. Blagosklonny MV. Molecular damage in cancer: an argument for mTOR-driven aging. Aging (Albany NY). 2011;3:1130–41.
39. Blagosklonny MV. Cell cycle arrest is not yet senescence, which is not just cell cycle arrest: terminology for TOR-driven aging. Aging (Albany NY). 2012;4:159–65.
40. Leontieva OV, Paszkiewicz GM, Blagosklonny MV. Mechanistic or mammalian target of rapamycin (mTOR) may determine robustness in young male mice at the cost of accelerated aging. Aging (Albany NY). 2012;4:899–916.
41. Blagosklonny MV. MTOR-driven quasi-programmed aging as a disposable soma theory: blind watchmaker vs. intelligent designer. Cell Cycle. 2013;12:1842–7.
42. Blagosklonny MV. Rapamycin extends life- and health span because it slows aging. Aging (Albany NY). 2013;5:592–8.
43. Blagosklonny MV. Aging is not programmed: Genetic pseudo-program is a shadow of developmental growth. Cell Cycle. 2013;12:3736–42.
44. Mariño G, Ugalde AP, Salvador-Montoliu N, Varela I, Quirós PM, Cadiñanos J, et al. Premature aging in mice activates a systemic metabolic response involving autophagy induction. Hum Mol Genet. 2008;17:2196–211.
45. Mariño G, López-Otín C. Autophagy and aging: new lessons from progeroid mice. Autophagy. 2008;4:807–9.
46. Harrison DE, Strong R, Sharp ZD, Nelson JF, Astle CM, Flurkey K, et al. Rapamycin fed late in life extends lifespan in genetically heterogeneous mice. Nature. 2009;460:392–5.
47. Cox LS, Mattison JA. Increasing longevity through caloric restriction or rapamycin feeding in mammals: common mechanisms for common outcomes? Aging Cell. 2009;8:607–13.
48. Johnson SC, Rabinovitch PS, Kaeberlein M. mTOR is a key modulator of ageing and age-related disease. Nature. 2013;493:338–45.
49. Anisimov VN, Zabezhinski MA, Popovich IG, Piskunova TS, Semenchenko AV, Tyndyk ML, et al. Rapamycin increases lifespan and inhibits spontaneous tumorigenesis in inbred female mice. Cell Cycle. 2011;10:4230–6.
50. Mercier I, Camacho J, Titchen K, Gonzales DM, Quann K, Bryant KG, et al. Caveolin-1 and accelerated host aging in the breast tumor microenvironment: chemoprevention with rapamycin, an mTOR inhibitor and anti-aging drug. Am J Pathol. 2012;181:278–93.
51. Comas M, Toshkov I, Kuropatwinski KK, Chernova OB, Polinsky A, Blagosklonny MV, et al. New nanoformulation of rapamycin Rapatar extends lifespan in homozygous p53−/− mice by delaying carcinogenesis. Aging (Albany NY). 2012;4:715–22.
52. Komarova EA, Antoch MP, Novototskaya LR, Chernova OB, Paszkiewicz G, Leontieva OV, et al. Rapamycin extends lifespan and delays tumorigenesis in heterozygous p53+/− mice. Aging (Albany NY). 2012;4:709–14.
53. Lamming DW, Ye L, Sabatini DM, Baur JA. Rapalogs and mTOR inhibitors as anti-aging therapeutics. J Clin Invest. 2013;123:980–9.
54. Jones RG, Thompson CB. Tumor suppressors and cell metabolism: a recipe for cancer growth. Genes Dev. 2009;23:537–48.
55. Vander Heiden MG, Cantley LC, Thompson CB. Understanding the Warburg effect: the metabolic requirements of cell proliferation. Science. 2009;324:1029–33.
56. Hanahan D, Weinberg RA. Hallmarks of cancer: the next generation. Cell. 2011;144:646–74.
57. Ward PS, Thompson CB. Metabolic reprogramming: a cancer hallmark even Warburg did not anticipate. Cancer Cell. 2012;21:297–308.
58. Cantor JR, Sabatini DM. Cancer cell metabolism: one hallmark, many faces. Cancer Discov. 2012;2:881–98.
59. Galluzzi L, Kepp O, Heiden MG, Kroemer G. Metabolic targets for cancer therapy. Nat Rev Drug Discov. 2013;12:829–46.
60. Cufí S, Corominas-Faja B, Lopez-Bonet E, Bonavia R, Pernas S, López IÁ, et al. Dietary restriction-resistant human tumors harboring the PIK3CA-activating mutation H1047R are sensitive to metformin. Oncotarget. 2013;4:1484–95.
61. Iglesias-Bartolome R, Gutkind SJ. Exploiting the mTOR paradox for disease prevention. Oncotarget. 2012;3:1061–3.
62. Anisimov VN. Metformin: do we finally have an anti-aging drug? Cell Cycle. 2013;12:3483–9.
63. Martin-Montalvo A, Mercken EM, Mitchell SJ, Palacios HH, Mote PL, Scheibye-Knudsen M, et al. Metformin improves health span and lifespan in mice. Nat Commun. 2013;4:2192.
64. Moiseeva O, Deschênes-Simard X, Pollak M, Ferbeyre G. Metformin, aging and cancer. Aging (Albany NY). 2013;5:330–1.

65. Anisimov VN. Metformin and rapamycin are master-keys for understanding the relationship between cell senescent, aging and cancer. Aging (Albany NY). 2013;5:337–8.
66. Cabreiro F, Au C, Leung KY, Vergara-Irigaray N, Cochemé HM, Noori T, et al. Metformin retards aging in C. elegans by altering microbial folate and methionine metabolism. Cell. 2013;153:228–39.
67. Menendez JA, Joven J. One-carbon metabolism: an aging-cancer crossroad for the gerosuppressant metformin. Aging (Albany NY). 2012;4:894–8.
68. Corominas-Faja B, Quirantes-Piné R, Oliveras-Ferraros C, Vazquez-Martin A, Cufí S, Martin-Castillo B, et al. Metabolomic fingerprint reveals that metformin impairs one-carbon metabolism in a manner similar to the antifolate class of chemotherapy drugs. Aging (Albany NY). 2012;4:480–98.
69. Wang J, Gallagher D, DeVito LM, Cancino GI, Tsui D, He L, et al. Metformin activates an atypical PKC-CBP pathway to promote neurogenesis and enhance spatial memory formation. Cell Stem Cell. 2012;11:23–35.
70. Berstein LM. Metformin in obesity, cancer and aging: addressing controversies. Aging (Albany NY). 2012;4:320–9.
71. Cufí S, Vazquez-Martin A, Oliveras-Ferraros C, Quirantes R, Segura-Carretero A, Micol V, et al. Metformin lowers the threshold for stress-induced senescence: a role for the microRNA-200 family and miR-205. Cell Cycle. 2012;11:1235–46.
72. Mercken EM, Carboneau BA, Krzysik-Walker SM, de Cabo R. Of mice and men: the benefits of caloric restriction, exercise, and mimetics. Ageing Res Rev. 2012;11:390–8.
73. Menendez JA, Cufí S, Oliveras-Ferraros C, Martin-Castillo B, Joven J, Vellon L, et al. Metformin and the ATM DNA damage response (DDR): accelerating the onset of stress-induced senescence to boost protection against cancer. Aging (Albany NY). 2011;3:1063–77.
74. Menendez JA, Cufí S, Oliveras-Ferraros C, Vellon L, Joven J, Vazquez-Martin A. Gerosuppressant metformin: less is more. Aging (Albany NY). 2011;3:348–62.
75. Anisimov VN, Berstein LM, Popovich IG, Zabezhinski MA, Egormin PA, Piskunova TS, et al. If started early in life, metformin treatment increases life span and postpones tumors in female SHR mice. Aging (Albany NY). 2011;3:148–57.
76. Anisimov VN, Piskunova TS, Popovich IG, Zabezhinski MA, Tyndyk ML, Egormin PA, et al. Gender differences in metformin effect on aging, life span and spontaneous tumorigenesis in 129/Sv mice. Aging (Albany NY). 2010;2:945–58.
77. Anisimov VN. Metformin for aging and cancer prevention. Aging (Albany NY). 2010;2:760–74.
78. Anisimov VN, Berstein LM, Egormin PA, Piskunova TS, Popovich IG, Zabezhinski MA, et al. Metformin slows down aging and extends life span of female SHR mice. Cell Cycle. 2008;7:2769–73.
79. Cufi S, Corominas-Faja B, Vazquez-Martin A, Oliveras-Ferraros C, Dorca J, Bosch-Barrera J, et al. Metformin-induced preferential killing of breast cancer initiating CD44+CD24-/low cells is sufficient to overcome primary resistance to trastuzumab in HER2+ human breast cancer xenografts. Oncotarget. 2012;3:395–8.
80. Vazquez-Martin A, Vellon L, Quirós PM, Cufí S, Ruiz de Galarreta E, Oliveras-Ferraros C, et al. Activation of AMP-activated protein kinase (AMPK) provides a metabolic barrier to reprogramming somatic cells into stem cells. Cell Cycle. 2012;11:974–89.
81. Vazquez-Martin A, Cufi S, Lopez-Bonet E, Corominas-Faja B, Oliveras-Ferraros C, Martin-Castillo B, et al. Metformin limits the tumourigenicity of iPS cells without affecting their pluripotency. Sci Rep. 2012;2:964.
82. Takahashi K, Yamanaka S. Induction of pluripotent stem cell from mouse embryonic and adult fibroblast cultures by defined factors. Cell. 2006;126:663–76.
83. Hanna JH, Saha K, Jaenisch R. Pluripotency and cellular reprogramming: facts, hypotheses, unresolved issues. Cell. 2010;143:508–25.
84. Drews K, Jozefczuk J, Prigione A, Adjaye J. Human induced pluripotent stem cells–from mechanisms to clinical applications. J Mol Med (Berl). 2012;90:735–45.
85. Ferreira LM, Mostajo-Radji MA. How induced pluripotent stem cells are redefining personalized medicine. Gene. 2013;520:1–6.
86. Cherry AB, Daley GQ. Reprogrammed cells for disease modeling and regenerative medicine. Annu Rev Med. 2013;64:277–90.
87. Liu GH, Ding Z, Izpisua Belmonte JC. iPSC technology to study human aging and aging-related disorders. Curr Opin Cell Biol. 2012;24:765–74.
88. Tiscornia G, Vivas EL, Izpisúa Belmonte JC. Diseases in a dish: modeling human genetic disorders using induced pluripotent cells. Nat Med. 2011;17:1570–6.
89. Rohani L, Johnson AA, Arnold A, Stolzing A. The aging signature: a hallmark of induced pluripotent stem cells? Aging Cell. 2014;13:2–7.
90. Blum B, Benvenisty N. The tumorigenicity of human embryonic stem cells. Adv Cancer Res. 2008;100:133–58.
91. Knoepfler PS. Deconstructing stem cell tumorigenicity: a roadmap to safe regenerative medicine. Stem Cells. 2009;27:1050–6.
92. Ben-David U, Benvenisty N. The tumorigenicity of human embryonic and induced pluripotent stem cells. Nat Rev Cancer. 2011;11:268–77.
93. Zhang Y, Yao L, Yu X, Ou J, Hui N, Liu S. A poor imitation of a natural process: a call to reconsider the iPSC engineering technique. Cell Cycle. 2012;11:4536–44.
94. Barrilleaux B, Knoepfler PS. Inducing iPSCs to escape the dish. Cell Stem Cell. 2011;9:103–11.
95. Chen T, Shen L, Yu J, Wan H, Guo A, Chen J, et al. Rapamycin and other longevity-promoting compounds

enhance the generation of mouse induced pluripotent stem cells. Aging Cell. 2011;10:908–11.
96. Easley 4th CA, Ben-Yehudah A, Redinger CJ, Oliver SL, Varum ST, Eisinger VM, et al. mTOR-mediated activation of p70 S6K induces differentiation of pluripotent human embryonic stem cells. Cell Reprogram. 2010;12:263–73.
97. Wang Y, Mah N, Prigione A, Wolfrum K, Andrade-Navarro MA, Adjaye J. A transcriptional roadmap to the induction of pluripotency in somatic cells. Stem Cell Rev. 2010;6:282–96.
98. Li R, Liang J, Ni S, Zhou T, Qing X, Li H, et al. A mesenchymal-to-epithelial transition initiates and is required for the nuclear reprogramming of mouse fibroblasts. Cell Stem Cell. 2010;7:51–63.
99. Esteban MA, Bao X, Zhuang Q, Zhou T, Qin B, Pei D. The mesenchymal-to-epithelial transition in somatic cell reprogramming. Curr Opin Genet Dev. 2012;22:423–8.
100. Liu X, Sun H, Qi J, Wang L, He S, Liu J, et al. Sequential introduction of reprogramming factors reveals a time-sensitive requirement for individual factors and a sequential EMT-MET mechanism for optimal reprogramming. Nat Cell Biol. 2013;15:829–38.
101. Lamouille S, Derynck R. Cell size and invasion in TGFbeta-induced epithelial to mesenchymal transition is regulated by activation of the mTOR pathway. J Cell Biol. 2007;178:437–51.
102. Gulhati P, Bowen KA, Liu J, Stevens PD, Rychahou PG, Chen M, et al. mTORC1 and mTORC2 regulate EMT, motility, and metastasis of colorectal cancer via RhoA and Rac1 signaling pathways. Cancer Res. 2011;71:3246–56.
103. Jun JI, Lau LF. The matricellular protein CCN1 induces fibroblast senescence and restricts fibrosis in cutaneous wound healing. Nat Cell Biol. 2010;12:676–85.
104. Kim KH, Chen CC, Monzon RI, Lau LF. Matricellular protein CCN1 promotes regression of liver fibrosis through induction of cellular senescence in hepatic myofibroblasts. Mol Cell Biol. 2013;33:2078–90.
105. Jun JI, Lau LF. Cellular senescence controls fibrosis in wound healing. Aging (Albany NY). 2010;2:627–31.
106. Banito A, Rashid ST, Acosta JC, Li S, Pereira CF, Geti I, et al. Senescence impairs successful reprogramming to pluripotent stem cells. Genes Dev. 2009;23:2134–9.
107. Banito A, Gil J. Induced pluripotent stem cells and senescence: learning the biology to improve the technology. EMBO Rep. 2010;11:353–9.
108. Li H, Collado M, Villasante A, Strati K, Ortega S, Cañamero M, et al. The Ink4/Arf locus is a barrier for iPS cell reprogramming. Nature. 2009;460:1136–9.
109. Utikal J, Polo JM, Stadtfeld M, Maherali N, Kulalert W, Walsh RM, et al. Immortalization eliminates a roadblock during cellular reprogramming into iPS cells. Nature. 2009;460:1145–8.
110. Liu Y, Hoya-Arias R, Nimer SD. The role of p53 in limiting somatic cell reprogramming. Cell Res. 2009;19:1227–8.
111. Kawamura T, Suzuki J, Wang YV, Menendez S, Morera LB, Raya A, et al. Linking the p53 tumour suppressor pathway to somatic cell reprogramming. Nature. 2009;460:1140–4.
112. Hanna J, Saha K, Pando B, van Zon J, Lengner CJ, Creyghton MP, et al. Direct cell reprogramming is a stochastic process amenable to acceleration. Nature. 2009;462:595–601.
113. Marión RM, Strati K, Li H, Murga M, Blanco R, Ortega S, et al. A p53-mediated DNA damage response limits reprogramming to ensure iPS cell genomic integrity. Nature. 2009;460:1149–53.
114. Son MJ, Son MY, Seol B, Kim MJ, Yoo CH, Han MK, et al. Nicotinamide overcomes pluripotency deficits and reprogramming barriers. Stem Cells. 2013;31:1121–35.
115. Blagosklonny MV. Aging, stem cells, and mammalian target of rapamycin: a prospect of pharmacologic rejuvenation of aging stem cells. Rejuvenation Res. 2008;11:801–8.
116. Demidenko ZN, Blagosklonny MV. Growth stimulation leads to cellular senescence when the cell cycle is blocked. Cell Cycle. 2008;7:3355–61.
117. Demidenko ZN, Zubova SG, Bukreeva EI, Pospelov VA, Pospelova TV, Blagosklonny MV. Rapamycin decelerates cellular senescence. Cell Cycle. 2009;8:1888–95.
118. Demidenko ZN, Blagosklonny MV. At concentrations that inhibit mTOR, resveratrol suppresses cellular senescence. Cell Cycle. 2009;8:1901–4.
119. Demidenko ZN, Korotchkina LG, Gudkov AV, Blagosklonny MV. Paradoxical suppression of cellular senescence by p53. Proc Natl Acad Sci U S A. 2010;107:9660–4.
120. Demidenko ZN, Blagosklonny MV. Quantifying pharmacologic suppression of cellular senescence: prevention of cellular hypertrophy versus preservation of proliferative potential. Aging (Albany NY). 2009;1:1008–16.
121. Leontieva OV, Blagosklonny MV. CDK4/6-inhibiting drug substitutes for p21 and p16 in senescence: duration of cell cycle arrest and MTOR activity determine geroconversion. Cell Cycle. 2013;12:3063–9.
122. Choi AM, Ryter SW, Levine B. Autophagy in human health and disease. N Engl J Med. 2013; 368:651–62.
123. Phadwal K, Watson AS, Simon AK. Tightrope act: autophagy in stem cell renewal, differentiation, proliferation, and aging. Cell Mol Life Sci. 2013; 70:89–103.
124. Rubinsztein DC, Mariño G, Kroemer G. Autophagy and aging. Cell. 2011;146:682–95.
125. Gewirtz DA. Autophagy and senescence in cancer therapy. J Cell Physiol. 2014;229:6–9.
126. Wang S, Xia P, Ye B, Huang G, Liu J, Fan Z. Transient activation of autophagy via Sox2-mediated suppression of mTOR is an important early step in reprogramming to pluripotency. Cell Stem Cell. 2013;13:617–25.

127. He J, Kang L, Wu T, Zhang J, Wang H, Gao H, et al. An elaborate regulation of Mammalian target of rapamycin activity is required for somatic cell reprogramming induced by defined transcription factors. Stem Cells Dev. 2012;21:2630–41.
128. Prigione A, Fauler B, Lurz R, Lehrach H, Adjaye J. The senescence-related mitochondrial/oxidative stress pathway is repressed in human induced pluripotent stem cells. Stem Cells. 2010;28:721–33.
129. Prigione A, Adjaye J. Modulation of mitochondrial biogenesis and bioenergetic metabolism upon in vitro and in vivo differentiation of human ES and iPS cells. Int J Dev Biol. 2010;54:1729–41.
130. Folmes CD, Nelson TJ, Martinez-Fernandez A, Arrell DK, Lindor JZ, Dzeja PP, et al. Somatic oxidative bioenergetics transitions into pluripotency-dependent glycolysis to facilitate nuclear reprogramming. Cell Metab. 2011;14:264–71.
131. Folmes CD, Nelson TJ, Terzic A. Energy metabolism in nuclear reprogramming. Biomark Med. 2011;5:715–29.
132. Folmes CD, Nelson TJ, Dzeja PP, Terzic A. Energy metabolism plasticity enables stemness programs. Ann N Y Acad Sci. 2012;1254:82–9.
133. Panopoulos AD, Yanes O, Ruiz S, Kida YS, Diep D, Tautenhahn R, et al. The metabolome of induced pluripotent stem cells reveals metabolic changes occurring in somatic cell reprogramming. Cell Res. 2012;22:168–77.
134. Folmes CD, Martinez-Fernandez A, Faustino RS, Yamada S, Perez-Terzic C, Nelson TJ, et al. Nuclear reprogramming with c-Myc potentiates glycolytic capacity of derived induced pluripotent stem cells. J Cardiovasc Transl Res. 2013;6:10–21.
135. Folmes CD, Arrell DK, Zlatkovic-Lindor J, Martinez-Fernandez A, Perez-Terzic C, Nelson TJ, et al. Metabolome and metaboproteome remodeling in nuclear reprogramming. Cell Cycle. 2013;12:2355–65.
136. Prigione A, Rohwer N, Hoffmann S, Mlody B, Drews K, Bukowiecki R, et al. HIF1α modulates cell fate reprogramming through early glycolytic shift and upregulation of PDK1-3 and PKM2. Stem Cells. 2014;32:364–76.
137. Bukowiecki R, Adjaye J, Prigione A. Mitochondrial function in pluripotent stemcells and cellular reprogramming. Gerontology. 2013;60:174–82.
138. Ding WX, Yin XM. Mitophagy: mechanisms, pathophysiological roles, and analysis. Biol Chem. 2012;393:547–64.
139. Liesa M, Shirihai OS. Mitochondrial dynamics in the regulation of nutrient utilization and energy expenditure. Cell Metab. 2013;17:491–506.
140. Michel S, Wanet A, De Pauw A, Rommelaere G, Arnould T, Renard P. Crosstalk between mitochondrial (dys) function and mitochondrial abundance. J Cell Physiol. 2012;227:2297–310.
141. Wang K, Klionsky DJ. Mitochondria removal by autophagy. Autophagy. 2011;7:297–300.
142. Todd LR, Damin MN, Gomathinayagam R, Horn SR, Means AR, Sankar U. Growth factor erv1-like modulates Drp1 to preserve mitochondrial dynamics and function in mouse embryonic stem cells. Mol Biol Cell. 2010;21:1225–36.
143. Vazquez-Martin A, Cufi S, Corominas-Faja B, Oliveras-Ferraros C, Vellon L, Menendez JA. Mitochondrial fusion by pharmacological manipulation impedes somatic cell reprogramming to pluripotency: new insight into the role of mitophagy in cell stemness. Aging (Albany NY). 2012;4:393–401.
144. Birket MJ, Orr AL, Gerencser AA, Madden DT, Vitelli C, Swistowski A, et al. A reduction in ATP demand and mitochondrial activity with neural differentiation of human embryonic stem cells. J Cell Sci. 2011;124:348–58.
145. Zhang J, Khvorostov I, Hong JS, Oktay Y, Vergnes L, Nuebel E, et al. UCP2 regulates energy metabolism and differentiation potential of human pluripotent stem cells. EMBO J. 2011;30:4860–73.
146. Kim JH, Kim HY, Lee YK, Yoon YS, Xu WG, Yoon JK, et al. Involvement of mitophagy in oncogenic K-Ras-induced transformation: overcoming a cellular energy deficit from glucose deficiency. Autophagy. 2011;7:1187–98.
147. Kim MJ, Woo SJ, Yoon CH, Lee JS, An S, Choi YH, et al. Involvement of autophagy in oncogenic K-Ras-induced malignant cell transformation. J Biol Chem. 2011;286:12924–32.
148. Mancias JD, Kimmelman AC. Targeting autophagy addiction in cancer. Oncotarget. 2011;2:1302–6.
149. Schieke SM, Phillips D, McCoy Jr JP, Aponte AM, Shen RF, Balaban RS, et al. The mammalian target of rapamycin (mTOR) pathway regulates mitochondrial oxygen consumption and oxidative capacity. J Biol Chem. 2006;281:27643–52.
150. Cunningham JT, Rodgers JT, Arlow DH, Vazquez F, Mootha VK, Puigserver P. mTOR controls mitochondrial oxidative function through a YY1-PGC-1alpha transcriptional complex. Nature. 2007;450:736–40.
151. Dai Y, Zheng K, Clark J, Swerdlow RH, Pulst SM, Sutton JP, et al. Rapamycin drives selection against a pathogenic heteroplasmic mitochondrial DNA mutation. Hum Mol Genet. 2014;23:637–47.
152. Groenewoud MJ, Zwartkruis FJ. Rheb and mammalian target of rapamycin in mitochondrial homoeostasis. Open Biol. 2013;3:130185.
153. Li Q, Zhang T, Wang J, Zhang Z, Zhai Y, Yang GY, Sun X. Rapamycin attenuates mitochondrial dysfunction via activation of mitophagy in experimental ischemic stroke. Biochem Biophys Res Commun. 2014;444:182–8.
154. Prigione A, Lichtner B, Kuhl H, Struys EA, Wamelink M, Lehrach H, et al. Human induced pluripotent stem cells harbor homoplasmic and heteroplasmic mitochondrial DNA mutations while maintaining human embryonic stem cell-like metabolic reprogramming. Stem Cells. 2011;29:1338–48.
155. Xu X, Duan S, Yi F, Ocampo A, Liu GH, Izpisua Belmonte JC. Mitochondrial regulation in pluripotent stem cells. Cell Metab. 2013;18:325–32.

156. Vazquez-Martin A, Corominas-Faja B, Cufi S, Vellon L, Oliveras-Ferraros C, Menendez OJ, et al. The mitochondrial H(+)-ATP synthase and the lipogenic switch: new core components of metabolic reprogramming in induced pluripotent stem (iPS) cells. Cell Cycle. 2013;12:207–18.
157. O'Neill LA, Hardie DG. Metabolism of inflammation limited by AMPK and pseudo-starvation. Nature. 2013;493:346–55.
158. Hardie DG, Ross FA, Hawley SA. AMPK: a nutrient and energy sensor that maintains energy homeostasis. Nat Rev Mol Cell Biol. 2012;13:251–62.
159. Hardie DG. AMP-activatedprotein kinase: an energy sensor that regulates all aspects of cell function. Genes Dev. 2011;25:1895–908.
160. Hardie DG. Sensing of energy and nutrients by AMP-activatedprotein kinase. Am J Clin Nutr. 2011;93:891S–6.
161. Shackelford DB, Vasquez DS, Corbeil J, Wu S, Leblanc M, Wu CL, et al. mTOR and HIF-1alpha-mediated tumor metabolism in an LKB1 mouse model of Peutz-Jeghers syndrome. Proc Natl Acad Sci U S A. 2009;106:11137–42.
162. Faubert B, Boily G, Izreig S, Griss T, Samborska B, Dong Z, et al. AMPK is a negative regulator of the Warburg effect and suppresses tumor growth in vivo. Cell Metab. 2013;17:113–24.
163. Dandapani M, Hardie DG. AMPK: opposing the metabolic changes in both tumour cells and inflammatory cells? Biochem Soc Trans. 2013; 41:687–93.
164. Matoba S, Kang JG, Patino WD, Wragg A, Boehm M, Gavrilova O, et al. p53 regulates mitochondrial respiration. Science. 2006;312:1650–3.
165. Assaily W, Benchimol S. Differential utilization of two ATP-generating pathways is regulated by p53. Cancer Cell. 2006;10:4–6.
166. Corcoran CA, Huang Y, Sheikh MS. The regulation of energy generating metabolic pathways by p53. Cancer Biol Ther. 2006;5:1610–3.
167. Bensaad K, Tsuruta A, Selak MA, Vidal MN, Nakano K, Bartrons R, et al. TIGAR, a p53-inducible regulator of glycolysis and apoptosis. Cell. 2006;126:107–20.
168. Zawacka-Pankau J, Grinkevich VV, Hünten S, Nikulenkov F, Gluch A, Li H, et al. Inhibition of glycolytic enzymes mediated by pharmacologically activated p53: targeting Warburg effect to fight cancer. J Biol Chem. 2011;286:41600–15.
169. Sato A, Sunayama J, Okada M, Watanabe E, Seino S, Shibuya K, et al. Glioma-initiating cell elimination by metformin activation of FOXO3 via AMPK. Stem Cells Transl Med. 2012;1:811–24.
170. Cuezva JM, Sánchez-Aragó M, Sala S, Blanco-Rivero A, Ortega AD. A message emerging from development: the repression of mitochondrial beta-F1-ATPase expression in cancer. J Bioenerg Biomembr. 2007;39:259–65.
171. López-Ríos F, Sánchez-Aragó M, García-García E, Ortega AD, Berrendero JR, Pozo-Rodríguez F, et al. Loss of the mitochondrial bioenergetic capacity underlies the glucose avidity of carcinomas. Cancer Res. 2007;67:9013–7.
172. Ortega AD, Sánchez-Aragó M, Giner-Sánchez D, Sánchez-Cenizo L, Willers I, Cuezva JM. Glucose avidity of carcinomas. Cancer Lett. 2009;276:125–35.
173. Sánchez-Aragó M, Chamorro M, Cuezva JM. Selection of cancer cells with repressed mitochondria triggers colon cancer progression. Carcinogenesis. 2010;31:567–76.
174. Sánchez-Cenizo L, Formentini L, Aldea M, Ortega AD, García-Huerta P, Sánchez-Aragó M, et al. Up-regulation of the ATPase inhibitory factor 1 (IF1) of the mitochondrial H+-ATP synthase in human tumors mediates the metabolic shift of cancer cells to a Warburg phenotype. J Biol Chem. 2010;285:25308–13.
175. Sánchez-Aragó M, Formentini L, Martínez-Reyes I, García-Bermudez J, Santacatterina F, Sánchez-Cenizo L, et al. Expression, regulation and clinical relevance of the ATPase inhibitory factor 1 in human cancers. Oncogenesis. 2013;2:e46.
176. Sánchez-Aragó M, Formentini L, García-Bermúdez J, Cuezva JM. IF1 reprograms energy metabolism and signals the oncogenic phenotype in cancer. Cell Cycle. 2012;11:2963–4.
177. Sánchez-Aragó M, Formentini L, Cuezva JM. Mitochondria-mediated energy adaption in cancer: the H(+)-ATP synthase-geared switch of metabolism in human tumors. Antioxid Redox Signal. 2013;19:285–98.
178. Menendez JA, Lupu R. Fatty acid synthase and the lipogenic phenotype in cancer pathogenesis. Nat Rev Cancer. 2007;7:763–77.
179. Menendez JA. Fine-tuning the lipogenic/lipolytic balance to optimize the metabolic requirements of cancer cell growth: molecular mechanisms and therapeutic perspectives. Biochim Biophys Acta. 1801;2010:381–91.
180. Pandey PR, Xing F, Sharma S, Watabe M, Pai SK, Iiizumi-Gairani M, et al. Elevated lipogenesis in epithelial stem-like cell confers survival advantage in ductal carcinoma in situ of breast cancer. Oncogene. 2013;32:5111–22.
181. Pandey PR, Okuda H, Watabe M, Pai SK, Liu W, Kobayashi A, et al. Resveratrol suppresses growth of cancer stem-like cells by inhibiting fatty acid synthase. Breast Cancer Res Treat. 2011;130: 387–98.
182. Cheema AK, Timofeeva O, Varghese R, Dimtchev A, Shiekh K, Shulaev V, et al. Integrated analysis of ATM mediated gene and protein expression impacting cellular metabolism. J Proteome Res. 2011;10:2651–7.
183. Ditch S, Paull TT. The ATM protein kinase and cellular redox signaling: beyond the DNA damage response. Trends Biochem Sci. 2012;37:15–22.
184. Vazquez-Martin A, Oliveras-Ferraros C, Cufí S, Martin-Castillo B, Menendez JA. Metformin activates an ataxia telangiectasia mutated (ATM)/

Chk2-regulated DNA damage-like response. Cell Cycle. 2011;10:1499–501.
185. Bao B, Wang Z, Ali S, Ahmad A, Azmi AS, Sarkar SH, et al. Metformin inhibits cell proliferation, migration and invasion by attenuating CSC function mediated by deregulating miRNAs in pancreatic cancer cells. Cancer Prev Res (Phila). 2012;5:355–64.
186. Jung JW, Park SB, Lee SJ, Seo MS, Trosko JE, Kang KS. Metformin represses self-renewal of the human breast carcinoma stem cells via inhibition of estrogen receptor-mediated OCT4 expression. PLoS One. 2011;6:e28068.
187. Hochedlinger K, Yamada Y, Beard C, Jaenisch R. Ectopic expression of Oct-4 blocks progenitor-cell differentiation and causes dysplasia in epithelial tissues. Cell. 2005;121:465–77.
188. Peng S, Maihle NJ, Huang Y. Pluripotency factors Lin28 and Oct4 identify a sub-population of stem cell-like cells in ovarian cancer. Oncogene. 2010;29:2153–9.
189. Hu T, Liu S, Breiter DR, Wang F, Tang Y, Sun S. Octamer 4 small interfering RNA results in cancer stem cell-like cell apoptosis. Cancer Res. 2008; 68:6533–40.
190. Kim RJ, Nam JS. OCT4 expression enhances features of cancer stem cells in a mouse model of breast cancer. Lab Anim Res. 2011;27:147–52.
191. Beltran AS, Rivenbark AG, Richardson BT, Yuan X, Quian H, Hunt JP, et al. Generation of tumor-initiating cells by exogenous delivery of OCT4 transcription factor. Breast Cancer Res. 2011;13:R94.
192. Shyh-Chang N, Daley GQ. Lin28: primal regulator of growth and metabolism in stem cells. Cell Stem Cell. 2013;12:395–406.
193. Zhou J, Ng SB, Chng WJ. LIN28/LIN28B: an emerging oncogenic driver in cancer stem cells. Int J Biochem Cell Biol. 2013;45:973–8.
194. Thornton JE, Gregory RI. How does Lin28 let-7 control development and disease? Trends Cell Biol. 2012;22:474–82.
195. Büssing I, Slack FJ, Grosshans H. let-7 microRNAs in development, stem cells and cancer. Trends Mol Med. 2008;14:400–9.
196. Gunaratne PH. Embryonic stem cell microRNAs: defining factors in induced pluripotent (iPS) and cancer (CSC) stem cells? Curr Stem Cell Res Ther. 2009;4:168–77.
197. Li MA, He L. microRNAs as novel regulators of stem cell pluripotency and somatic cell reprogramming. Bioessays. 2012;34:670–80.
198. Peter ME. Let-7 and miR-200 microRNAs: guardians against pluripotency and cancer progression. Cell Cycle. 2009;8:843–52.
199. Yang X, Lin X, Zhong X, Kaur S, Li N, Liang S, et al. Double-negative feedback loop between reprogramming factor LIN28 and microRNA let-7 regulates aldehyde dehydrogenase 1-positive cancer stem cells. Cancer Res. 2010;70:9463–72.
200. Oliveras-Ferraros C, Cufí S, Vazquez-Martin A, Torres-Garcia VZ, Del Barco S, Martin-Castillo B, et al. Micro(mi)RNA expression profile of breast cancer epithelial cells treated with the anti-diabetic drug metformin: induction of the tumor suppressor miRNA let-7a and suppression of the TGFβ-induced oncomiR miRNA-181a. Cell Cycle. 2011;10:1144–51.
201. Zhu H, Shyh-Chang N, Segrè AV, Shinoda G, Shah SP, Einhorn WS, et al. The Lin28/let-7 axis regulates glucose metabolism. Cell. 2011;147:81–94.
202. Shinoda G, Shyh-Chang N, Soysa TY, Zhu H, Seligson MT, Shah SP, et al. Fetal deficiency of lin28 programs life-long aberrations in growth and glucose metabolism. Stem Cells. 2013;31:1563–73.
203. Shyh-Chang N, Zhu H, Yvanka de Soysa T, Shinoda G, Seligson MT, Tsanov KM, et al. Lin28 enhances tissue repair by reprogramming cellular metabolism. Cell. 2013;155:778–92.
204. Frost RJ, Olson EN. Control of glucose homeostasis and insulin sensitivity by the Let-7 family of microRNAs. Proc Natl Acad Sci U S A. 2011;108:21075–80.
205. Qiu C, Ma Y, Wang J, Peng S, Huang Y. Lin28-mediated post-transcriptional regulation of Oct4 expression in human embryonic stem cells. Nucleic Acids Res. 2010;38:1240–8.
206. Ma W, Ma J, Xu J, Qiao C, Branscum A, Cardenas A, et al. Lin28 regulates BMP4 and functions with Oct4 to affect ovarian tumor microenvironment. Cell Cycle. 2013;12:88–97.
207. Menendez JA, Vazquez-Martin A. Rejuvenating regeneration: metformin activates endogenous adult stem cells. Cell Cycle. 2012;11:3521–2.
208. Wu LE, Gomes AP, Sinclair DA. Geroncogenesis: metabolic changes during aging as a driver of tumorigenesis. Cancer Cell. 2014;25:12–9.

11 Molecular Promiscuity of Plant Polyphenols in the Management of Age-Related Diseases: Far Beyond Their Antioxidant Properties

Enrique Barrajón-Catalán, María Herranz-López, Jorge Joven, Antonio Segura-Carretero, Carlos Alonso-Villaverde, Javier A. Menéndez, and Vicente Micol

Abstract

The use of plant-derived polyphenols for the management of diseases has been under debate in the last decades. Most studies have focused on the specific effects of polyphenols on particular targets, while ignoring their pleiotropic character. The multitargeted character of polyphenols, a plausible consequence of their molecular promiscuity, may suppose an opportunity to fight multifactorial diseases. Therefore, a wider perspective is urgently needed to elucidate whether their rational use as bioactive food components may be valid for the management of diseases. In this chapter, we discuss the most likely targets of polyphenols that may account for their salutary effects from a global perspective. Among these targets, the modulation of signalling and energy-sensitive pathways, oxidative stress and inflammation-related processes, mitochondrial functionality, epigenetic machinery, histone acetylation and membrane-dependent processes play central roles in polyphenols' mechanisms of action.

For the *Bioactive Food Components Platform*.

E. Barrajón-Catalán • M. Herranz-López
Instituto de Biología Molecular y Celular (IBMC), Universidad Miguel Hernández, Avenida de la Universidad s/n, Elche, Alicante E-03202, Spain
e-mail: e.barrajon@umh.es; mherranz@umh.es

J. Joven • C. Alonso-Villaverde
Unitat de Recerca Biomèdica, Hospital Universitari Sant Joan, Institut d'Investigació Sanitària Pere Virgili, Universitat Rovira i Virgili, Reus, Spain
e-mail: jorge.joven@urv.cat; calonvi@gmail.com

A. Segura-Carretero
Department of Analytical Chemistry, Faculty of Sciences, University of Granada, Granada, Spain
e-mail: ansegura@ugr.es

J.A. Menéndez
Head of the Traslation Research Unit, Catalan Institute of Oncology and Biomedical Research Institute, Girona, Spain
e-mail: jmenendez@iconcologia.net; jmenendez@idibgi.org

V. Micol (✉)
Instituto de Biología Molecular y Celular (IBMC), Universidad Miguel Hernández, Avenida de la Universidad s/n, Elche, Alicante E-03202, Spain

CIBER (CB12/03/30038, Fisiopatología de la Obesidad y la Nutrición, CIBERobn, Instituto de Salud Carlos III), Palma de Mallorca, Spain
e-mail: vmicol@umh.es

J. Camps (ed.), *Oxidative Stress and Inflammation in Non-communicable Diseases – Molecular Mechanisms and Perspectives in Therapeutics*, Advances in Experimental Medicine and Biology 824, DOI 10.1007/978-3-319-07320-0_11, © Springer International Publishing Switzerland 2014

Sufficient evidence on polyphenols has accumulated for them to be considered a serious option for the management of non-communicable diseases, such as cancer and obesity, as well as infectious diseases. The remaining unresolved issues that must be seriously addressed are their bioavailability, metabolite detection, specific molecular targets, interactions and toxicity. The Xenohormesis hypothesis, which postulates that polyphenols are the product of plant evolutive adaptation to stress and conferee their resistance to mammals, offers a reasonable explanation to justify the beneficial and non-toxic effects of plant mixtures, but do not fully meet expectations. Hence, future research must be supported by the use of complex polypharmacology approaches and synergic studies focused on the understanding of the pleiotropic effects of polyphenols. Revisiting polyphenol mechanisms of action with the help of these techniques may allow for the improvement of human health and wellness by using intelligent nutritional intervention.

Keywords

AMPK • Cancer • CCL2 • Inflammation • Obesity • Polyphenols • Synergy • Xenohormesis

11.1 Introduction

Eighty years ago, and most likely unwittingly, Linus Pauling made his first contribution to the field of the relationship between oxidative stress and ageing by predicting the existence of the superoxide radical, which was confirmed based on quantum mechanics a few decades later [1]. He also established that a quadratic function fit the relationship between weight and shorter lifespan [2] and predicted that the elimination of oxidative stress-related factors, such as smoking or medical radiology, would considerably increase life expectancy. Throughout his life, Pauling was convinced that vitamin C consumption would be the right approach to controlling age-related diseases such as cancer and cardiovascular diseases. Although he was not aware at that time of the importance of these contributions, he was a visionary that provided the basis for the current knowledge of the molecular biology of ageing.

For decades, plant polyphenols have been considered as "simple" antioxidant molecules, and most of their benefits have been related to their radical-scavenging properties. Several studies have recently questioned the correlation between their antioxidant activity and their efficacy in health promotion [3]. Nevertheless, there is increasing evidence that these compounds must have multiple targets, most of which remain to be discovered. First, it was believed that some plant polyphenols specifically targeted the nucleotide-binding site of protein kinases [4]. However, crystallographic studies on polyphenol-protein complexes revealed that polyphenols bind to several hydrophobic pockets in the protein structure and do not compete with nucleotides [5]. Many other studies have reported the capability of polyphenols to modulate pro-inflammatory gene expression, such as cyclooxygenase, lipoxygenase, nitric oxide synthases and several pivotal cytokines regulated by nuclear factor-kappa B (NF-κB) and mitogen-activated protein kinase (MAPK) signalling [6, 7]. Polyphenols also interact with phospholipid membranes, establishing hydrogen bonds and hydrophobic interactions through their hydroxyl groups and phenolic rings, which may lead to the modulation of many membrane-dependent proteins and processes [8–10]. Recent evidence also supports the capability of polyphenols to modulate pathways related to chronic inflammation and energy metabolism. Such is the case of the CCL2 (chemokine [C-C motif] ligand 2)/CCR2 (chemokine [C-C motif] receptor 2) pathway and the energy sensor AMP-activated protein kinase (AMPK), which are altered in obesity-related pathologies and may become attractive therapeutic targets as preventive management of disease [11, 12]. It has been postulated that polyphenols active the gerosuppressor AMPK, which leads to the inhibition of the mammalian target of rapamycin (mTOR) gerogene, which is overactivated in ageing and human chronic diseases. Some plant polyphenols also demonstrate significant anticancer activity that is concomitant with the activation of endoplasmic reticulum (ER) stress and the unfolded protein response signature (UPR), sirtuin-1 activation (SIRT1), AMPK up-regulation and epithelial-to-mesenchymal (EMT) transition inhibition [13]. In summary, considering the large diversity of this class of compounds (Fig. 11.1), there is sufficient evidence to believe that polyphenols possess a multitargeted mechanism of action. In this chapter, the potential molecular targets of plant polyphenols, their putative synergistic mechanism of action and their xenohormetic effects on human age-related diseases will be reviewed based on recent evidence.

11.2 Metabolic Stress in Non-communicable Diseases

11.2.1 The Relationship Between Oxidative Stress and Chronic Inflammation

Chronic inflammation is accepted as linked to oxidative stress via the immune response, especially through monocyte/macrophage cell activation. An altered red/ox state caused by an increase in oxidative stress is able to recruit and activate inflammatory cells. In most cases, these actions are mediated by CCL2, also called MCP-1 (monocyte chemotactic protein) [14]. CCL2 is produced when tissue cells are not able to control their red/ox state and attract monocytes to their surroundings. Once there, the monocytes differentiate to macrophages and initiate the immune response that ultimately will lead to an inflammatory state, which may develop into a chronic situation. In this scenario, a vicious cycle begins as the inflammation state increases oxidative stress and the latter leads to a subsequent inflammation state. In most tissues, there is an autoprotective system to counterbalance oxidative stress, including intracellular enzymes and specific antioxidant mechanisms. It seems reasonable, therefore, to postulate that when oxidative stress is reduced by polyphenols, there is reduced CCL2 production and secretion, and decreased monocyte recruitment and inflammation.

Dietary plant polyphenols can help to reduce this oxidative stress by acting as direct antioxidants and/or promoting endogenous defences against oxidative stress. The antioxidant defence system includes paraoxonases (PON), which are associated in serum with HDL, protecting serum lipids from oxidation, most likely as a result of their ability to hydrolyse specific oxidised lipids [15]. PONs are very important in atherosclerosis, and their failure causes circulating monocytes to be drawn into the subendothelium, where they differentiate into macrophages, become activated and secrete proinflammatory cytokines such as CCL2 [16], which accelerates atherosclerosis and increases the inflammatory state [17, 18].

Other members of the endogenous antioxidant arsenal are glutathione-related enzymes, such as glutathione reductase (GR). We have described a correlation between decreased oxidative damage in blood cell lipids and proteins, a decrease in blood pro-inflammatory cytokines and a concomitant activation of blood cell antioxidant enzymes, such as GR, in athletes supplemented with plant polyphenols who performed an aerobic training routine [19, 20]. All this evidence confirms that polyphenols, which were first known as antioxidants, also possess an anti-inflammatory activity that underlies their beneficial cardiovascular properties [21, 22]. Nonetheless, the capability of polyphenols to interact other molecular targets cannot be ruled out, as discussed below.

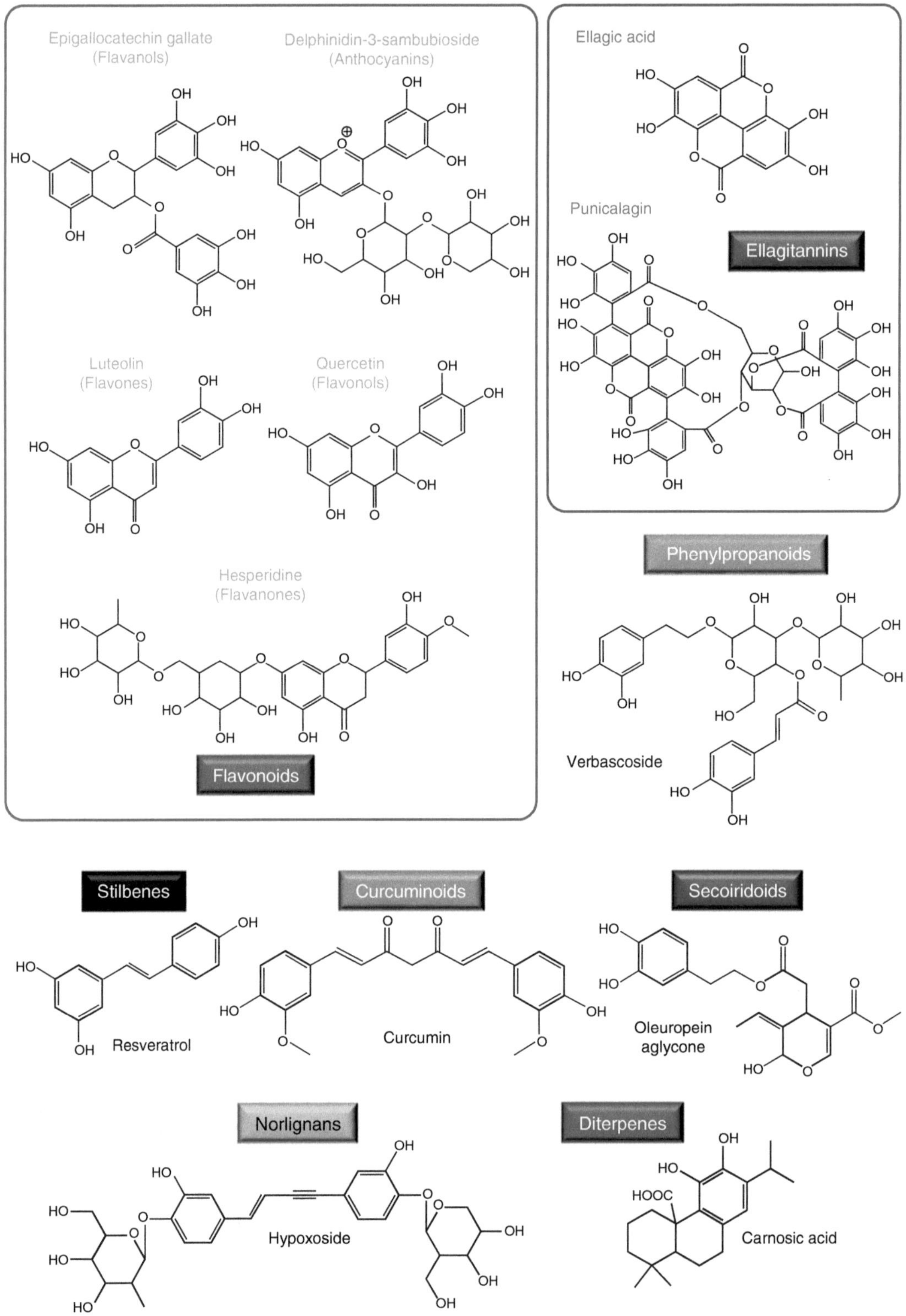

Fig. 11.1 Representative polyphenolic compounds with multitargeted action in human age-related diseases: main scaffolds are flavonoids (flavanols, anthocyanins, flavones, flavonols and flavanones), ellagitannins, phenylpropanoids (phenolic acid derivatives), stilbenes, curcuminoids, secoiridoids, norlignans and diterpenes. A large variability of structures is achieved by substitutions, interaction and polymerization to generate polypharmacological effects

11.2.2 Non-communicable Diseases Such as Obesity and Cancer Are Linked Through Inflammation

Obesity is commonly recognised as a chronic inflammation state. This situation is accompanied by altered metabolism, atherosclerosis and insulin resistance in correlation with macrophage activation and increased plasma levels of CCL2 [11]. In fact, abnormal plasma CCL2 levels and macrophage activation are associated with different manifestations of cardiovascular disease and heart injury conditions [17]. However, the relationship between cancer and inflammation is less evident. Chemokines such as CCL2 and macrophages also appear to be involved in tumourigenesis [11], and inflammatory infiltration is similar in both cancer and atherosclerosis [23]. Although the pro- and anti-tumour effects of CCL2 are controversial, it appears to be well supported that this chemokine is directly related to tumour cell physiology [24], and, in breast cancer, tumour-associated macrophages (TAMs) are present to a great extent in the tumour cell mass [25]. It is reported that, at least in breast tumours, CCL2 expression and macrophage infiltration are related to a poor prognosis and metastatic disease [26, 27]. Therefore, it is plausible to propose that searching for CCL2 inhibitors may represent an important therapeutic target, not only in obesity, but also in cancer [28].

In obesity, the links among CCL2, inflammation and their associated disturbances are generally accepted [29]. In obesity, metabolic homeostasis is altered in adipose tissue, and its regulation is coordinated by adipokines, among which the role of leptin and adiponectin must be stressed. Recently, the pro-carcinogenic effect of leptin and the anticarcinogenic effect of adiponectin have been reported [30, 31]. In addition, tumour necrosis factor-α (TNFα) produced by monocytes and released in the inflammatory milieu of adipose tissue may also contribute to the pathogenesis of tumours in the obese state [32]. It is widely recognised that the chemokines released by macrophages and other immune cells are closely related to inflammation-related tumourigenesis and atherogenesis in mice models [16, 29] but remains elusive for human diseases [33]. Obesity is usually accompanied by an increase in the production of pro-inflammatory cytokines, macrophage recruitment, an increase in circulating lipids and an aberrant insulin response. All this metabolic stress promotes phosphatidylinositol 3-kinase (PI3K) deregulation, leading to a decreased capacity to control the cell proliferation, growth, apoptosis and metabolism present in cancer, obesity and other human diseases. Therefore, pharmacological PI3K inhibitors have been proposed to play a role in the management of cancer and obesity [34, 35].

In summary, atherosclerosis and cancer share common pathways linked by adipokines, inflammation, CCL2, macrophages activation and proliferative signaling. Recent evidence on polyphenol research, in both cell and animal models, reveals that these compounds may be useful for managing such metabolic stress at different stages [13, 22, 36–38]. Finding polyphenols targets is likely the most challenging task in the future. Recently, we have shown that quercetin glucoside colocalises with activated macrophages using a novel antibody against this polyphenol and immunohistochemical detection [36]. Nevertheless, the *in vivo* metabolic targets of polyphenols are still an unresolved issue due to limitations on the absorption and metabolism of polyphenols and to undeveloped metabolite detection techniques.

11.2.3 The Putative Interconnection of AMPK and CCL2 in Metabolic Stress: The Mammalian Target of Rapamycin (mTOR)

Recently, the energy sensor enzyme AMPK has been proposed as a therapeutic target for the treatment of non-communicable diseases, such as cancer and obesity [39, 40]. This enzyme is a key regulator of lipid and glucose metabolism (sensitive to the ATP/ADP ratio) and is activated by drugs such as metformin and salicylate [39, 41] and also by glucose restriction [42]. Once activated, AMPK switches off biosynthetic pathways

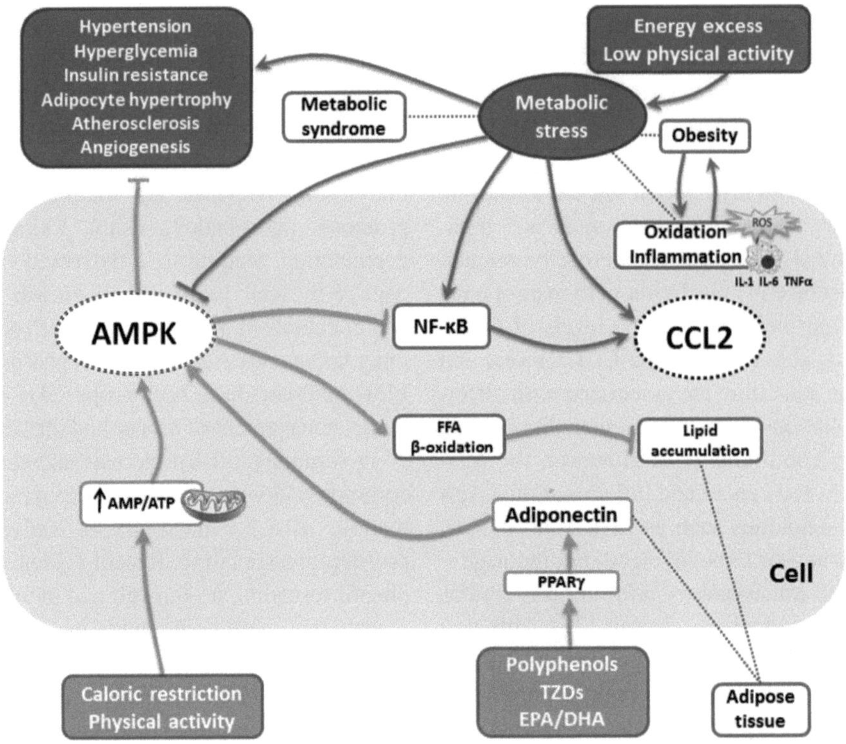

Fig. 11.2 AMPK and CCL2 pattern in obesity-related metabolic disorders: metabolic stress, as a result of low physical activity and energy excess, upregulates CCL2 and inhibits AMPK activity leading to chronic inflammation and oxidative stress that turn into most metabolic disturbances in obesity. Polyphenols are capable to invert this signature by adiponectin up-regulation

and cell cycle progression while switching on catabolic pathways that yield ATP and the NAD^+ biosynthetic enzyme Nampt, leading to SIRT1 activation and the deacetylation of a variety of substrates. In fact, many polyphenols, such as stilbenes, flavonols and secoiridoids (Fig. 11.1), have shown the capability to active AMPK [13, 36, 43].

There is a growing body of evidence showing that AMPK inhibition is linked to the inflammatory status and CCL2 signalling [44–46]. In this chapter, we propose that polyphenols act as AMPK activators counteracting the AMPK-suppressed signature present in some age-related chronic diseases based on our own and extensive experience and that of others [11, 13, 47, 48]. The relationship between CCL2 and AMPK is not evident and may be understood on the basis of energetic balance, inflammation and changes in mitochondrial function and morphology (Fig. 11.2). Metabolic stress induced by excess caloric intake and low physical activity may develop into the metabolic disturbances associated with obesity, i.e. oxidative stress, inflammation and insulin resistance. Adipose tissue is the major player in metabolic stress, in which a series of events occurs: oxidative stress, CCL2 release, macrophage infiltration, aerobic glycolysis enhancement ("Warburg effect"), insulin resistance, mitochondria rearrangement and inflammatory status. Later, this abnormal response and the systemic action of CCL2 and other chemokines extend further to the liver and vascular endothelium, eventually resulting in atherosclerosis.

One of the potential convergence points between CCL2 and AMPK may be mTOR (Fig. 11.3), a serine/threonine kinase that is located downstream in the PI3K/Akt signalling

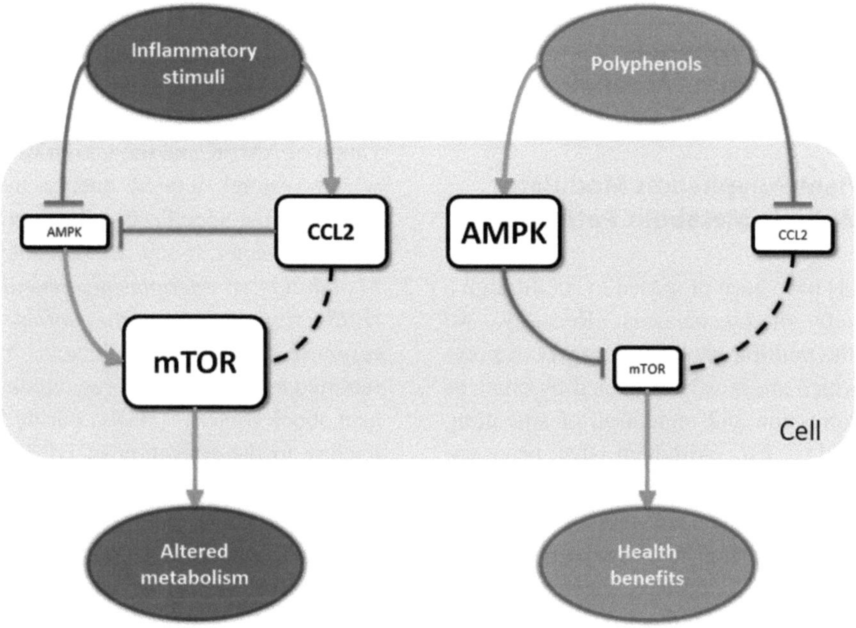

Fig. 11.3 Antagonic effects of inflammation vs. polyphenols on the regulation of the AMPK/mTOR energy sensing metabolic regulator: age-related diseases are associated to a loss of the AMPK activation capability by inflammation and subsequent overactivation of mTOR gerogene. The blockade of inflammation by polyphenols is proposed to restore activation of AMPK with the concomitant mTOR repression

pathway and has pleiotropic functions mainly related to the response to nutrients and energy, growth stimuli and cell progression. mTOR has been implicated in the pathogenesis of both cancer and diabetes. This protein is hyperactivated in almost 50 % of human cancers [49] and modulates mitochondrial biogenesis, hypoxia signalling, autophagy and cell cycle progression [50]. Autophagy attenuates inflammation by maintaining mitochondrial homeostasis, and it has been found to be defective in obesity and atherosclerosis models. Therefore, the activation of autophagy by mTOR inhibition would become a therapeutic target in such diseases [51]. Inflammatory factors (e.g., CCL2) and oxidative stress in adipose tissue progress into altered metabolism, resulting in mitochondrial failure and insulin resistance. In the opposite scenario, polyphenols and physiological conditions such as glucose restriction would lead to reduced inflammation (macrophage infiltration) and oxidative stress, triggering AMPK activation. This situation restores glucose homeostasis and cytokine regulation in several tissues and responds to hormones such as leptin and adiponectin [52]. Adiponectin has been proposed as a systemic functional link for the activation of AMPK in different tissues [11, 53, 54] (Fig. 11.2). Indeed, selected polyphenols have shown the capability of increasing adiponectin expression while decreasing that of leptin [37, 55, 56]. Therefore, we propose that dietary polyphenols act as AMPK activators and decrease CCL2 expression by reducing inflammation, oxidative stress and insulin resistance. Although these actions appear to be clearly related, polyphenols may not be direct effectors. Polyphenols are expected to increase insulin sensitivity in a mechanism mediated by the peroxisome proliferator-activated receptor gamma (PPAR-γ)-dependent transcriptional up-regulation of adiponectin, in a manner similar to thiazolidinediones or glitazones [57].

11.3 Plant Polyphenols: A Therapeutic Arsenal with Multiple Targets

11.3.1 Plant Polyphenols Modulate Multiple Metabolic Pathways

Polyphenols have been proposed to act through a wide variety of mechanisms. Recently, we reviewed the multiple molecular targets of polyphenols, which are mostly mediated by changes in gene expression and regulation of signalling pathways [11, 13]. Although the proposed molecular targets of polyphenols are individually considered, most of them are interconnected due to the pleiotropic character of the pathways and proteins involved. The putative mechanisms for polyphenol action and their proposed metabolic checkpoints are as follows:

1. *Direct antioxidant activity*: regulation of oxidative stress-sensitive factors and increased expression of antioxidant proteins: The *in vivo* antioxidant effects of a myriad of polyphenols have been proven in both animal models and human trials. In some cases, polyphenols have promoted a direct decrease in ROS, reactive carbonyls derived from proteins or malondialdehyde from lipid oxidation. Polyphenols (e.g., resveratrol, verbascoside, secoiridoids) (Fig. 11.1) have also demonstrated the capacity to increase the expression and/or activity of several antioxidant enzymes, such as catalase (CAT), PONs, GR, glutathione peroxidase (GPx), glutathione-S-transferase (GST) and γ-glutamyl cysteine synthetase (γ-GCS), [19, 20, 58, 59], mostly mediated by the activation of nuclear redox factor 2 (NRF2) [60–62]. As a consequence of the capacity of polyphenols to modulate redox-sensible networks, the transcription factor NF-κB and p38 MAPK kinase pathways, which modulate several of a large number of genes mediating immune and inflammatory responses and apoptosis, are also negatively modulated [63].
2. *Blockade of pro-inflammatory cytokines*: The regulation of the inflammatory response by polyphenols through their capacity to decrease CCL2 expression in macrophages and other tissues may be due, in part, to their radical-scavenging capacity, which has been fully reviewed [11, 14, 22]. This anti-inflammatory action appears to be clearly linked to the activation of AMPK and the inhibition of mTOR, which restores altered energy metabolism. However, the specific targets of polyphenols in macrophages, if any, are still unknown.
3. *Modulation of endoplasmic reticulum stress (ERS) signalling and the unfolded protein response (UPR)*: Polyphenols have been reported to induce the up-regulation of several heat shock proteins (HSPs) during ER stress, leading to the activation of UPR and subsequent cell cycle arrest. The prolonged exposure to ER stress can switch cell survival to cell death. This has been proposed as the putative mechanism of the antiproliferative and cytotoxic effects of polyphenols such as resveratrol and secoiridoids from extra virgin olive oil [13, 64] (Fig. 11.1).
4. *SIRT1 and histone acetylation inhibition*: Since the discovery of the effects of resveratrol [65] (Fig. 11.1), a growing list of polyphenols have shown the ability to up-regulate the expression of SIRT1, a class of histone deacetylases (HDACs) that play a major role in the "organ morphology" gene network inhibiting histone hyperacetylation [13, 66]. The changes in the gene expression machinery exerted by polyphenols have been associated with the silencing of genes responsible for inflammation or cell hyperproliferation [67]. Interestingly, polyphenols offer an allegedly contradictory response with regard to histone acetylation by promoting regeneration and lifespan in normal cells but inducing tumour-suppressive activity in cancer cells, which surely will lead to challenging future research.
5. *Modulation of genes that regulate metabolic and energy-sensing pathways*: the Warburg effect and AMPK/mTOR signalling: Major human diseases such as cancer and obesity have been closely related to the energetic signature consisting of the exacerbated induction of glycolysis with limited respiration and Krebs cycle in mitochondria, the so-called "Warburg effect". Many polyphenols have shown the capacity to restore mitochondrial

function in obesity by fighting inflammation and oxidative stress [11, 12]. The potential inhibitory role of polyphenols against the Warburg effect has also been documented in cancer cells [13], which is in agreement with their capacity to down-regulate lactate dehydrogenase (LDH). A high lactate level is a key feature of the "aerobic glycolysis" signature in cancer cells and is associated with a poor prognosis and metastasis in cancer [68]. Furthermore, several polyphenols (e.g., resveratrol, secoiridoids, flavonols) (Fig. 11.1) are activators of the energy sensor AMPK, mimicking the effects of caloric restriction. We have proposed that this action is closely related to the inhibition of the nutrient-sensing mTOR gerogene and that repairing the AMPK/mTOR-driven programme may lower the impact of ageing-related diseases [13, 47] (Fig. 11.3).

6. *Epigenetic regulation by modulating microRNA (miRNA) expression*: miRNAs are small, gene-silencing RNAs that regulate mRNA translation by blockage or degradation. A number of miRNAs appear to play important regulatory roles in cell differentiation, insulin action and fat metabolism in adipocytes. Recent studies have demonstrated that miRNA deregulation is involved in liver-associated metabolic disturbances in obese mice and humans [36]. Several polyphenols (e.g., curcumin, galloylated catechins, soy polyphenols and *Hibiscus sabdariffa* polyphenols) (Fig. 11.1) significantly modify the expression of several miRNAs in cell and animal models [67].

11.3.2 Biological Membranes Are Underexploited Molecular Targets of Polyphenols

A significant number of polyphenols have been shown to interact strongly with the lipid domains of cell membranes, altering the properties of the immediate lipidic environment in which a representative number of crucial protein receptors are embedded. In such a scenario, the goal is to discern whether the metabolic effects of some polyphenols are due to their direct modulation of receptors or to the modification of the lipid bilayer, which subsequently alters membrane protein activity. We have shown that a variety of phenolic compounds (Fig. 11.1), such as stilbenes [69], galloylated catechins [10], secoiridoids [70], phenylpropanoids [8], diterpenes [71, 72] and norlignans [73], promote significant perturbations in the physical properties of phospholipid bilayers (e.g., viscosity, lipid packing, phase transition temperature, lateral segregation, surface charge) and can be considered factors influencing cell metabolism.

These reports have established how the biological activity of polyphenols (i.e., antioxidant, anticancer, anti-inflammatory or antimicrobial effects) is strongly influenced by their interaction with lipid membranes. For instance, the strong antioxidant properties of rosemary diterpenes and green tea catechins have been demonstrated through different assays. Nevertheless, these compounds partition into phospholipid membranes and significantly increase lipid order, also modifying the phospholipid-water interface. These effects contribute to their antioxidant capacity and cooperate with their electron donating ability in protecting lipid membranes against oxidative damage [10, 71, 72]. Other phenolic compounds are known to target cell surface receptors and proteins that are localised in detergent-insoluble ordered membrane domains, so-called "lipid rafts". These structures participate in cellular signal transduction, endocytosis and the transmembrane translocation of different compounds [74]. The immunomodulatory and anticancer effects of galloylated catechins have also been related to their capacity to modulate the activity of the cell receptors that are located in lipid rafts [75–78]. Moreover, these compounds incorporate into the membrane of methicillin-resistant *Staphylococcus aureus* (MRSA), modifying the composition and fluidity of the bilayer that result in the modulation of the cell-surface properties necessary to maintain the beta-lactam-resistant phenotype [79–81]. The flavones quercetin and luteolin are also able to modulate cell surface receptors by altering membrane lipid rafts [82, 83]. The anti-inflammatory activity of anthocyanins has also been proposed to occur through the regulation of cholesterol distribution in lipid rafts, preventing CD40-induced proinflammatory signalling [84].

In addition, the lipid environment is also an important issue to consider with regard to increasing the bioavailability of certain polyphenols to cellular targets. In this sense, the polyphenols curcumin and resveratrol (Fig. 11.1), with reported anticancer activity in cell models, exhibit markedly increased cytotoxic activity in human breast cancer cells when incorporated into liposomes, indicating that these compounds preferentially partition in membranes and increase their uptake capacity in cells through a lipid pathway [85]. All these data strongly suggest that future research on polyphenols should be partially reoriented toward the study of the effects of these compounds on the lipid environment of protein receptors to fully understand their modes of action. Lipid nanoemulsions offer promising options to increase the bioavailability of many polyphenolic compounds [86]. Nevertheless, much work needs to be performed in relation to the stability and absorption of these formulations and, more importantly, to test their safety in human nutrition. Once all these aspects are solved, formulations to enhance the bioavailability and efficacy of polyphenols will become part of human nutrition.

11.3.3 Mitochondria Are Also a Therapeutic Target for Polyphenols

All the above-mentioned evidence clearly shows that the correct management of energy and food intake and the control of excessive oxidative stress are crucial to lowering the impact of atherosclerosis and cancer. In such a scenario, mitochondria play a central role, and severe mitochondrial dysfunction due to mitochondrial DNA mutations or mitochondrial proteins oxidation is a key process leading to a number of diseases (e.g., myopathies, neuropathies, diabetes). Overdeveloped mitochondrial dysfunction is the cause of the opening of the mitochondrial permeability transition pore (MPT) in the inner membrane, an end-stage that can lead to mitochondrial swelling and cell death (Fig. 11.4a). Mitochondria adapt to environmental factors and respond to energetic demand by producing effectors that activate multiple pathways related to oxidative stress and inflammation. Therefore, mitochondria are self-targets in the management of human diseases [87].

Animal studies using antioxidant vitamins have yielded positive effects in the management of diabetes but have revealed severe limitations in targeting oxidative stress at specific molecular sites. Paradoxical effects of green tea and vitamins C and E have been obtained in their administration to diabetic rats. Although these antioxidants improved several diabetes-related cellular dysfunctions, such as mitochondrial protein integrity and respiration, conversely, extracellular matrix glycoxidation worsened considerably [88]. These data suggest that targeting cellular oxidative stress should be addressed and demonstrate that some antioxidants are potentially double-edged swords; accordingly, their dosage and delivery route should be carefully studied [89].

Mitochondrially targeted antioxidants have been recently reviewed as a potential therapeutic option for a vast range of oxidative stress-related diseases [90, 91]. One of the strategies is to exploit the highly negative internal mitochondrial potential by targeting lipophilic cations such as MitoQ (ubiquinone derivative) or $MitoE_2$ (vitamin E analogue) (Fig. 11.4b) [92]. The second strategy is based on the utilisation of cell-permeable mitochondria-targeted cationic peptides (SS-peptides) (Fig. 11.4b), which are resistant to hydrolysis and contain antioxidant residues (Tyr). A third alternative employs the unique expression of enzymatic systems (i.e., monoamine oxidases, cysteine conjugation or β-oxidation) that catalyse the release of antioxidants from prodrugs in mitochondria. When administered orally to mice, Mito Q and $MitoE_2$ reached the liver, heart and skeletal muscle and were even able to cross the blood-brain barrier [93]. Some of these mitochondrially targeted antioxidants have shown promising results in preclinical models for hypertensive cardiovascular disease, cardiomyopathy and amyotrophic lateral sclerosis. Nevertheless, preclinical toxicological studies are mandatory to evaluate their potential for therapeutic development. At the moment, only MitoQ and SS-31 have reached human clinical trials. Although MitoQ exhibited the capacity to decrease serum alanine transaminase in patients suffering from

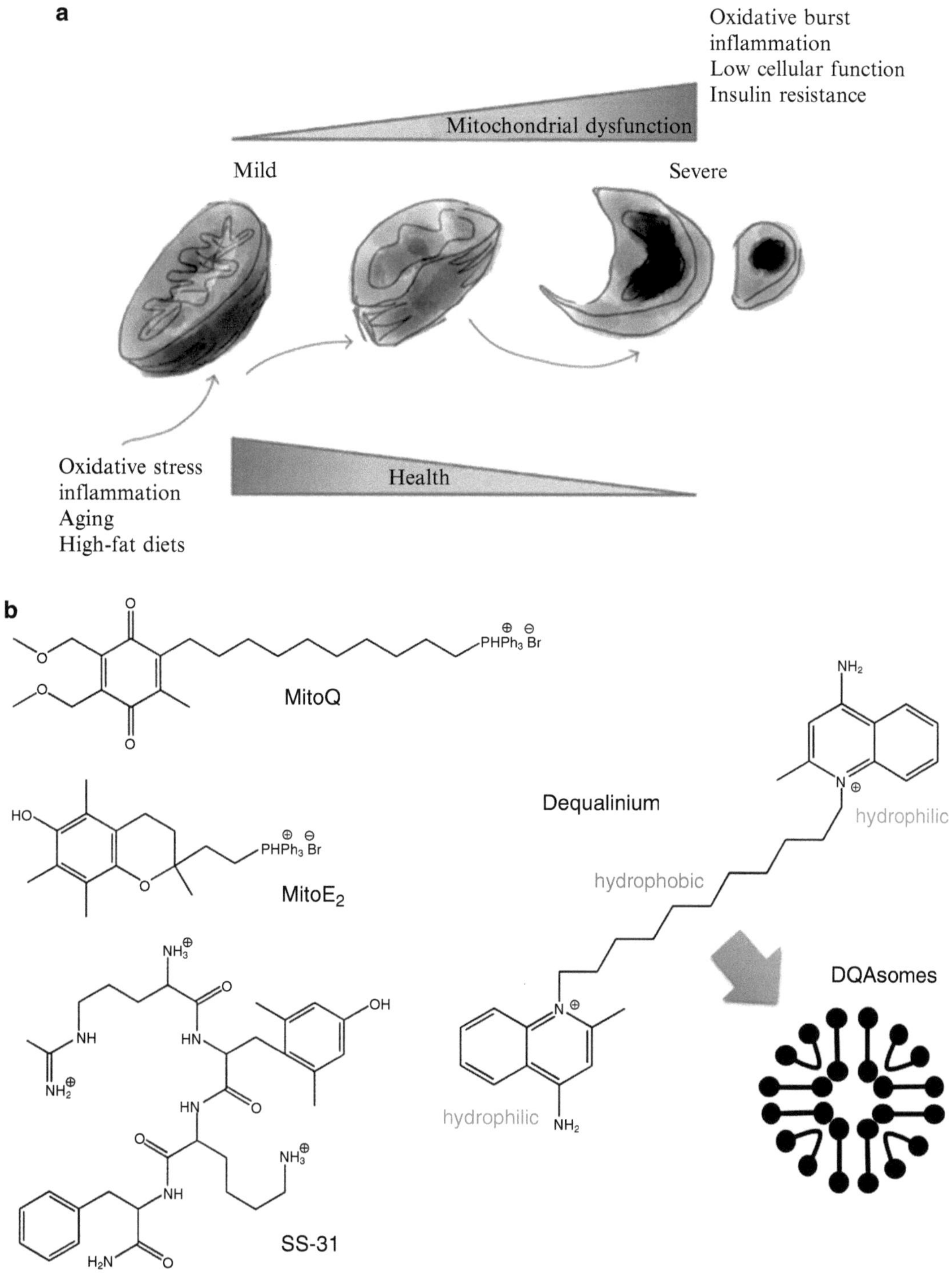

Fig. 11.4 (**a**) Mitochondrial dysfunction is a key point in the development of a number of diseases: energy excess and oxidative stress triggers the impairment of mitochondrial function that in turn increases mitochondria-derived ROS production and promotes inflammation. In this scenario, antioxidant enzymes are not capable to fight ROS, leading to mitochondrial DNA mutations and respiratory chain proteins oxidation and eventually mitochondrial failure. (**b**) A selection of synthetic compounds especially designed to target mitochondrial oxidative stress: MitoQ (a triphenylphosphonium synthetic analog of ubiquinol), MitoE2 (a triphenylphosphonium synthetic analog of α-tocopherol) and SS-peptide (a tetrapeptide containing the antioxidant 2′,6′-dimethyltyrosine residue). Dequalinium (DQA), a cationic bola-amphiphilic compound that shows self-assembly properties in liposome-like vesicles (DQAsomes) with capability to accumulate into mitochondria

chronic hepatitis C virus infection, no efficacy was observed in the progression of Parkinson's disease in human subjects. Clinical trials on SS-peptides are currently anticipated.

Another alternative for the mitochondrial targeting of compounds and macromolecules is the use of subcellular nanocarriers. Dequalinium (DQA) is an antimicrobial agent that has been used in mouthwashes, lozenges and ointments for more than 50 years (Fig. 11.4b). The cationic bola-amphiphilic structure of DQA gives it the ability to self-assemble into liposome-like vesicles containing single-chain bola lipids, structures known as DQAsomes [94]. The preferential location of DQAsomes to energised mitochondria has been shown. DQAsomes have been successfully used for the selective mitochondrial delivery of DNA fragments, which were not found in other cytosolic organelles or nuclei [95]. In this chapter, we discussed that relatively hydrophobic plant polyphenols (i.e., catechins, flavonols, flavones, flavanones, diterpenes and dicatechols) have a significant affinity for phospholipid membranes and would thus be susceptible to incorporation into DQAsomes or similar nanocarriers. Catalase and superoxide dismutase are crucial antioxidant enzymes for mitochondria viability. When these enzymes are seriously damaged by ROS, mitochondrial function is compromised. We postulate that targeting these antioxidants to intracellular sites such as the mitochondrial membrane with the use of nanoparticles opens a variety of therapeutic possibilities for the treatment of diseases triggered by intracellular ROS.

11.4 Molecular Promiscuity Enables the Synergic Effects of Polyphenols: An Evolutionary Response of xenohormetins

11.4.1 Xenohormesis, Lifespan and Human Disease

In the last decades, a considerable body of science on plant polyphenols has supported that a significant number of these compounds exert similar effects on human health. Far from being a coincidence, several authors have attempted to explain the putative connection between the presence of these compounds in nature and their multiple positive effects on human health on an evolutionary basis [96]. The Xenohormesis hypothesis [65, 97] proposes that plants synthesise signalling molecules (xenohormetins) such as polyphenols under stressful condition; when incorporated into the heterotroph diet, these compounds induce defence responses, leading to an extended lifespan [98]. Therefore, these molecules work as interspecies transference signals to prepare living beings for adversity [65]. In a manner similar to caloric restriction, some polyphenols appear to modulate the expression of certain genes, leading to increased lifespan in invertebrates and insects [99]. Within this response, PI3K/Akt down-regulation, AMPK and SIRT activation and NF-κB suppression appear to be the common signatures of the potential beneficial effects of polyphenols on lifespan as antiatherogenic or antineoplastic agents [100]. Resveratrol, quercetin, catechins and curcumin (Fig. 11.1) are just a few examples of compounds that are being thoroughly investigated.

The effects of xenohormetins appear to be under the control of the endocrine system in a coordinated manner. Thus, although most cellular studies may be useful from a mechanistic point of view, their severe limitations would require the use of animal models. Because the bioavailability of polyphenols is still an unresolved issue, the main challenge now is to prove the *in vivo* efficacy of plant products and to detect reasonable concentrations of polyphenol metabolites in plasma or tissues to account for their activity. Currently, a new human species named *Homo obesus* has been proposed, as a new phenotype of *Homo sapiens* that is deficient in metabotrophins (endogenous factors playing a role in the maintenance of energy metabolism, inflammation and wound healing) [101]. Factors most likely to contribute to this species are a sedentary lifestyle and nutrition habits. At this moment, the situation may not be evolutionarily irreversible, but it is expected to worsen. The consumption of selected polyphenols may suppose a plausible therapeutic or nutritional approach for the reversion of senescence/ageing signature, once safety issues are resolved.

11.4.2 Promiscuity and Molecular Diversity of Polyphenols: An Evolutionary Consequence

The paradigm of Western medicine in drug discovery is to find highly selective compounds against individual targets. Traditional herbal medicine has provided numerous examples of natural drugs based on this "reverse pharmacology" approach. Taxol, colchicine, digitoxin, morphine and tetrahydrocannabinol are just a few examples of natural compounds with strong specificity for their targets. Nevertheless, approximately 60 % of fairly specific drugs in their initial research fail because of toxicity or ineffectiveness in late-stage preclinical studies [102].

Plants generate a myriad of secondary metabolites that are ubiquitous in most species, which has been interpreted as positive selection of biosynthetic gene clusters [103]. This strategy of nature may be a result of cost-benefit selection. For plants, it may be more efficient to produce many small physiological mediators acting on multiple targets (biosynthetic molecular promiscuity), rather than single pharmacologically potent compounds. Therefore, those conserved and successful combinations may have been positively selected through evolution. Polyphenols are an example of this class of compounds. The polyphenol family encompasses a certain number of molecular scaffolds, but their variability increases enormously with minimal substitutions (hydroxyl or methoxy moieties) (Fig. 11.1), the ability to interact by hydrogen bonds (among them and with proteins) and the capacity to polymerise. In line with this hypothesis, it has been postulated that plants have evolved molecular promiscuity as a strategy to achieve maximal pharmacological potency, generating a limitless number of pharmacological compounds for various targets [104]. Although the cause for such variability remains unresolved, the Xenohormesis hypothesis satisfies most of the questions arisen from a theoretical point of view [65]. Surprisingly, most of these mixtures are also safe to mammals and easily metabolised and detoxified.

The current goal is to ascertain whether the polyphenolic mixtures found in plants are more effective than target-selective monosubstances. Although mixtures of compounds in botanical extracts have provided abundant evidence on their potential for the management of multifactorial diseases, only a few have shown high pharmacological potency. In the last two decades, synergy has become the only explanation for this behaviour [105], an hypothesis taking shape in recent years due to recently found evidence that will be discussed later in this chapter. Altogether, these data suggest a new direction in the strategies for drug discovery, which may be valid not only for non-communicable diseases but also for infectious diseases. Many studies focusing on the mechanism of action of botanical mixtures have failed due to the difficulty of conventional biochemical methodology to interpret the complexity of the observed effects. This does not mean that the beneficial effects of plant polyphenols are magic, but intense research must be aimed in this direction. In this scenario, the emerging "omics" technologies may give rise to a breakthrough in the understanding of plant polyphenols effects. The polypharmacology approach appears to be valid for explaining the way that some botanical mixtures target different proteins within the same signalling pathway and in several biochemical pathways at once. Network pharmacology [106], a subdiscipline of systems biology, may also provide interesting tools for the study of herbal medicine by joining complex experimental approaches and *in silico* analysis. Recent examples for these strategies have been provided to explain the biological effects of *Ruta graveolens* constituents [107], the effects of *Hibiscus sabdariffa* polyphenols on metabolic disturbances [11] and the anticancer effects of olive oil polyphenols [13].

11.4.3 Evidence for the Synergic Behaviour of Polyphenol Mixtures

Monotherapy has been the classical approach utilised to address most natural or synthetic bioactive compounds. However, there is increasing evidence that combined therapies are much more efficient than single-drug-based treatments, most likely due to the multi-causal aetiology of many diseases and/or the existence of several therapeutic targets. Multidrug therapy is now extended worldwide to treat not only infectious diseases but also AIDS, cancer, hypertension and rheumatic disorders [108]. Combined therapies may provide the advantage of synergistic effects among different drugs [105].

Several authors have highlighted the relevance of synergic effects in phytotherapy research [104, 108, 109], both for non-communicable and for infectious diseases. Recently, examples of the synergic effects of polyphenolic mixtures have been reported for several therapeutic targets, such as obesity-related pathologies [37, 110], antimicrobial susceptibility [111] and cancer [112, 113]. Synergy is especially relevant for antimicrobial therapy [114]. Natural compounds alone (e.g., terpenes, flavonoids, phenolic acids, flavanols) (Fig. 11.1) or in combination with antibiotics have been used to limit infection or to reduce microbial resistance [115–119]. The Berenbaum's isobole method is most likely the simplest method for verifying the synergy between two compounds. It supposes an affordable experimental design and provides incontrovertible graphical results (Fig. 11.5). To verify the existence of a synergistic effect of a mixture of two compounds A and B, the inhibitory concentration (IC) of mixtures at certain proportions of the compounds are determined and represented in the isobologram: when synergy is present, a concave curve towards the zero point is obtained in the isobole plot, whereas the representation becomes convex when there is antagonism between the two compounds or a straight line when no interaction is present [105, 108]. However, as the isobole method does not provide a quantitative evaluation of the synergy, calculation of the fractional inhibitory concentration index (FICI) is utilised as a complementary numerical approach [120]. The FICI value for a combination of compounds is calculated by adding $IC_{(A \text{ in the presence of } B)}/IC_{(A \text{ alone})} + IC_{(B \text{ in the presence of } A)}/IC_{(B \text{ alone})}$. A synergistic effect is considered when FICI ≤ 0.5, an additive effect when $0.5 < \text{FICI} \leq 1$, indifferent effect when $1 < \text{FICI} < 2$ and antagonism when FIC ≥ 2 [121].

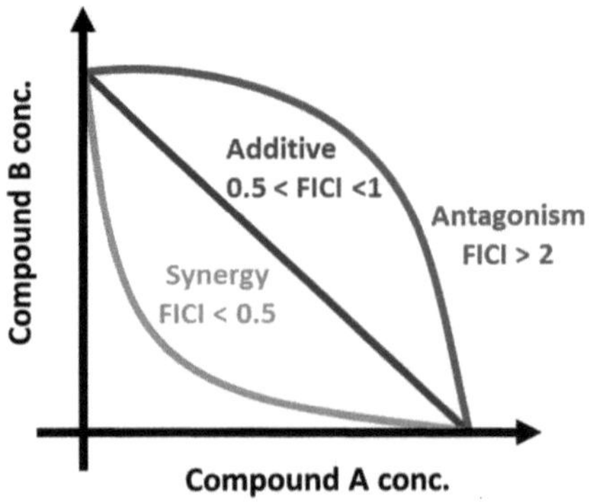

Fig. 11.5 Schematic representation of the Berenbaum's isobole method for the verification of synergy between two hypothetical compounds A and B: a concave curve indicates synergy (over-additive effect), whereas convex curve or straight line indicate antagonism or no interaction respectively. The fractional inhibitory concentration index (FICI) is a quantitative measurement to confirm synergy (synergistic effect, FICI ≤ 0.5; additive effect $0.5 < \text{FICI} \leq 1$; indifferent effect, $1 < \text{FICI} < 2$; antagonism, FIC ≥ 2)

11.5 Conclusions and Perspectives

Non-communicable diseases account for more than three out of five deaths in developed countries. Current remedies and pharmaceutical drugs for major human diseases, such as obesity, diabetes, cancer and cardiovascular diseases, are of limited efficacy. Although one-third of the current top 20 drugs on the market are plant derived, only one out of 5,000–10,000 of the new synthetic molecules in development becomes a commercial pharmaceutical drug [65]. Hence, the search for new natural therapies is an interesting research field that can improve the results of existing treatments. However, polyphenol therapy has been controversial due to questions raised about adequate polyphenolic sources, purity and dosage. In this chapter, we have provided substantial evidence to support that the modulation of signalling pathways and the epigenetic regulation of gene expression are underlying mechanisms for plant dietary polyphenols. These compounds have pleiotropic effects reaching multiple targets, such as ROS, membrane receptors, membrane phospholipid domains, nuclear receptors, signalling proteins, energy sensors, histone acetylation, miRNA expression, chemokines and inflammatory mediators.

A large body of evidence has revealed that mixtures of polyphenols in plants may be valid for the management of age-related diseases. In some cases, these mixtures work in a synergistic manner, and special combinations are exceptionally effective. Shall we then pay more attention to attempt to discover the magic polyphenolic potions that reach their multiple targets with maximum efficacy? Current limitations to the discovery of the mechanisms of action of polyphenols may due to the use of conventional molecular biology techniques, which hamper the identification of their complex pleiotropic effects. Therefore, much research remains to be performed using imaginative approaches that may lead to a significant breakthrough. To reach these targets and answer all these questions, we may need to implement polypharmacology and network pharmacology. Plant polyphenols have been used for centuries by human beings as food and medicine, and the time for their rational use in human health has most likely arrived. In the future, in our race for drug discovery, we may look back to herbal medicine for new candidates.

Acknowledgements Concepts expressed in this review have been discussed and approved by investigators from the Bioactive Food Component Platform which is currently being supported by competitive public grants from different Institutions (CD08/00283, SAF2009-11579, PI08/1381; PI08/1032, PI08/1175, PI011/130, P11-CTS-7625, GREIB.PT.2011.18, AGL2011-29857-C03-03, PROMETEO/2012/007, ACOMP/2013/093 CIBER CB12/03/30038).

Conflict of Interest Statement The authors declare that there are no conflicts of interest.

References

1. Pauling L. The discovery of the superoxide radical. Trends Biochem Sci. 1979;4:N270–1.
2. Pauling L. The relation between longevity and obesity in human beings. Proc Natl Acad Sci U S A. 1958;44:619–22.
3. Halliwell B, Rafter J, Jenner A. Health promotion by flavonoids, tocopherols, tocotrienols, and other phenols: direct or indirect effects? Antioxidant or not? Am J Clin Nutr. 2005;81:268S–76.
4. Howitz KT, Bitterman KJ, Cohen HY, Lamming DW, Lavu S, Wood JG, et al. Small molecule activators of sirtuins extend Saccharomyces cerevisiae lifespan. Nature. 2003;425:191–6.
5. Gledhill JR, Montgomery MG, Leslie AG, Walker JE. Mechanism of inhibition of bovine F1-ATPase by resveratrol and related polyphenols. Proc Natl Acad Sci U S A. 2007;104:13632–7.
6. Guo W, Kong E, Meydani M. Dietary polyphenols, inflammation, and cancer. Nutr Cancer. 2009; 61:807–10.
7. Fresco P, Borges F, Diniz C, Marques MP. New insights on the anticancer properties of dietary polyphenols. Med Res Rev. 2006;26:747–66.
8. Funes L, Laporta O, Cerdán-Calero M, Micol V. Effects of verbascoside, a phenylpropanoid glycoside from lemon verbena, on phospholipid model membranes. Chem Phys Lipids. 2010;163:190–9.
9. Alves DS, Perez-Fons L, Estepa A, Micol V. Membrane-related effects underlying the biological activity of the anthraquinones emodin and barbaloin. Biochem Pharmacol. 2004;68:549–61.
10. Caturla N, Vera-Samper E, Villalain J, Mateo CR, Micol V. The relationship between the antioxidant and the antibacterial properties of galloylated catechins and the structure of phospholipid model membranes. Free Radic Biol Med. 2003;34:648–62.
11. Joven J, Rull A, Rodriguez-Gallego E, Camps J, Riera-Borrull M, Hernandez-Aguilera A, et al. Multifunctional targets of dietary polyphenols in disease: a case for the chemokine network and energy metabolism. Food Chem Toxicol. 2013; 51:267–79.
12. Hernandez-Aguilera A, Rull A, Rodriguez-Gallego E, Riera-Borrull M, Luciano-Mateo F, Camps J, et al. Mitochondrial dysfunction: a basic mechanism in inflammation-related non-communicable diseases and therapeutic opportunities. Med Inflamm. 2013;2013:135698.
13. Menendez JA, Joven J, Aragones G, Barrajon-Catalan E, Beltran-Debon R, Borras-Linares I, et al.

Xenohormetic and anti-aging activity of secoiridoid polyphenols present in extra virgin olive oil: a new family of gerosuppressant agents. Cell Cycle. 2013;12:555–78.
14. Rull A, Garcia R, Fernandez-Sender L, Beltran-Debon R, Aragones G, Alegret JM, et al. The role of combined assessment of defense against oxidative stress and inflammation in the evaluation of peripheral arterial disease. Curr Mol Med. 2011;11:453–64.
15. Aviram M, Rosenblat M. Paraoxonases 1, 2, and 3, oxidative stress, and macrophage foam cell formation during atherosclerosis development. Free Radic Biol Med. 2004;37:1304–16.
16. Libby P, Ridker PM, Hansson GK. Progress and challenges in translating the biology of atherosclerosis. Nature. 2011;473:317–25.
17. Coll B, Alonso-Villaverde C, Joven J. Monocyte chemoattractant protein-1 and atherosclerosis: is there room for an additional biomarker? Clin Chim Acta. 2007;383:21–9.
18. Rovin BH, Lu L, Saxena R. A novel polymorphism in the MCP-1 gene regulatory region that influences MCP-1 expression. Biochem Biophys Res Commun. 1999;259:344–8.
19. Carrera-Quintanar L, Funes L, Viudes E, Tur J, Micol V, Roche E, et al. Antioxidant effect of lemon verbena extracts in lymphocytes of university students performing aerobic training program. Scand J Med Sci Sports. 2012;22:454–61.
20. Funes L, Carrera-Quintanar L, Cerdan-Calero M, Ferrer MD, Drobnic F, Pons A, et al. Effect of lemon verbena supplementation on muscular damage markers, proinflammatory cytokines release and neutrophils' oxidative stress in chronic exercise. Eur J Appl Physiol. 2011;111:695–705.
21. Quinones M, Miguel M, Aleixandre A. Beneficial effects of polyphenols on cardiovascular disease. Pharmacol Res. 2013;68:125–31.
22. Benavente-Garcia O, Castillo J. Update on uses and properties of citrus flavonoids: new findings in anticancer, cardiovascular, and anti-inflammatory activity. J Agric Food Chem. 2008;56:6185–205.
23. Siveen KS, Kuttan G. Role of macrophages in tumour progression. Immunol Lett. 2009;123:97–102.
24. Conti I, Rollins BJ. CCL2 (monocyte chemoattractant protein-1) and cancer. Semin Cancer Biol. 2004;14:149–54.
25. Solinas G, Germano G, Mantovani A, Allavena P. Tumor-associated macrophages (TAM) as major players of the cancer-related inflammation. J Leukoc Biol. 2009;86:1065–73.
26. Saji H, Koike M, Yamori T, Saji S, Seiki M, Matsushima K, et al. Significant correlation of monocyte chemoattractant protein-1 expression with neovascularization and progression of breast carcinoma. Cancer. 2001;92:1085–91.
27. Ueno T, Toi M, Saji H, Muta M, Bando H, Kuroi K, et al. Significance of macrophage chemoattractant protein-1 in macrophage recruitment, angiogenesis, and survival in human breast cancer. Clin Cancer Res. 2000;6:3282–9.
28. Qian BZ, Li J, Zhang H, Kitamura T, Zhang J, Campion LR, et al. CCL2 recruits inflammatory monocytes to facilitate breast-tumour metastasis. Nature. 2011;475:222–5.
29. Condeelis J, Pollard JW. Macrophages: obligate partners for tumor cell migration, invasion, and metastasis. Cell. 2006;124:263–6.
30. Jarde T, Perrier S, Vasson MP, Caldefie-Chezet F. Molecular mechanisms of leptin and adiponectin in breast cancer. Eur J Cancer. 2011;47:33–43.
31. Khandekar MJ, Cohen P, Spiegelman BM. Molecular mechanisms of cancer development in obesity. Nat Rev Cancer. 2011;11:886–95.
32. Hotamisligil GS, Arner P, Caro JF, Atkinson RL, Spiegelman BM. Increased adipose tissue expression of tumor necrosis factor-alpha in human obesity and insulin resistance. J Clin Invest. 1995;95:2409–15.
33. Hildenbrand R, Allgayer H, Marx A, Stroebel P. Modulators of the urokinase-type plasminogen activation system for cancer. Expert Opin Investig Drugs. 2010;19:641–52.
34. Blajecka K, Borgstrom A, Arcaro A. Phosphatidylinositol 3-kinase isoforms as novel drug targets. Curr Drug Targets. 2011;12:1056–81.
35. Engelman JA. Targeting PI3K signalling in cancer: opportunities, challenges and limitations. Nat Rev Cancer. 2009;9:550–62.
36. Joven J, Espinel E, Rull A, Aragones G, Rodriguez-Gallego E, Camps J, et al. Plant-derived polyphenols regulate expression of miRNA paralogs miR-103/107 and miR-122 and prevent diet-induced fatty liver disease in hyperlipidemic mice. Biochim Biophys Acta. 1820;2012:894–9.
37. Herranz-Lopez M, Fernandez-Arroyo S, Perez-Sanchez A, Barrajon-Catalan E, Beltran-Debon R, Menendez JA, et al. Synergism of plant-derived polyphenols in adipogenesis: perspectives and implications. Phytomedicine. 2012;19:253–61.
38. Fernandez-Arroyo S, Herranz-Lopez M, Beltran-Debon R, Borras-Linares I, Barrajon-Catalan E, Joven J, et al. Bioavailability study of a polyphenol-enriched extract from Hibiscus sabdariffa in rats and associated antioxidant status. Mol Nutr Food Res. 2012;56:1590–5.
39. Hardie DG, Ross FA, Hawley SA. AMP-activated protein kinase: a target for drugs both ancient and modern. Chem Biol. 2012;19:1222–36.
40. Hardie DG, Ross FA, Hawley SA. AMPK: a nutrient and energy sensor that maintains energy homeostasis. Nat Rev Mol Cell Biol. 2012;13: 251–62.
41. Vazquez-Martin A, Oliveras-Ferraros C, del Barco S, Martin-Castillo B, Menendez JA. The antidiabetic drug metformin: a pharmaceutical AMPK activator to overcome breast cancer resistance to HER2 inhibitors while decreasing risk of cardiomyopathy. Ann Oncol. 2009;20:592–5.

42. Fulco M, Cen Y, Zhao P, Hoffman EP, McBurney MW, Sauve AA, et al. Glucose restriction inhibits skeletal myoblast differentiation by activating SIRT1 through AMPK-mediated regulation of Nampt. Dev Cell. 2008;14:661–73.
43. Chen S, Xiao X, Feng X, Li W, Zhou N, Zheng L, et al. Resveratrol induces Sirt1-dependent apoptosis in 3T3-L1 preadipocytes by activating AMPK and suppressing AKT activity and survivin expression. J Nutr Biochem. 2012;23:1100–12.
44. Beltran-Debon R, Alonso-Villaverde C, Aragones G, Rodriguez-Medina I, Rull A, Micol V, et al. The aqueous extract of *Hibiscus sabdariffa* calices modulates the production of monocyte chemoattractant protein-1 in humans. Phytomedicine. 2010;17:186–91.
45. Caligiuri A, Bertolani C, Guerra CT, Aleffi S, Galastri S, Trappoliere M, et al. Adenosine monophosphate-activated protein kinase modulates the activated phenotype of hepatic stellate cells. Hepatology. 2008;47:668–76.
46. Koh KK, Han SH, Quon MJ. Inflammatory markers and the metabolic syndrome: insights from therapeutic interventions. J Am Coll Cardiol. 2005;46:1978–85.
47. Blagosklonny MV. Revisiting the antagonistic pleiotropy theory of aging: TOR-driven program and quasi-program. Cell Cycle. 2010;9:3151–6.
48. Menendez JA, Lupu R. Fatty acid synthase and the lipogenic phenotype in cancer pathogenesis. Nat Rev Cancer. 2007;7:763–77.
49. Sarbassov DD, Guertin DA, Ali SM, Sabatini DM. Phosphorylation and regulation of Akt/PKB by the rictor-mTOR complex. Science. 2005;307:1098–101.
50. Dunlop EA, Tee AR. Mammalian target of rapamycin complex 1: signalling inputs, substrates and feedback mechanisms. Cell Signal. 2009;21:827–35.
51. Egan DF, Shackelford DB, Mihaylova MM, Gelino S, Kohnz RA, Mair W, et al. Phosphorylation of ULK1 (hATG1) by AMP-activated protein kinase connects energy sensing to mitophagy. Science. 2011;331:456–61.
52. Long YC, Zierath JR. AMP-activated protein kinase signaling in metabolic regulation. J Clin Invest. 2006;116:1776–83.
53. Hattori Y, Nakano Y, Hattori S, Tomizawa A, Inukai K, Kasai K. High molecular weight adiponectin activates AMPK and suppresses cytokine-induced NF-kappaB activation in vascular endothelial cells. FEBS Lett. 2008;582:1719–24.
54. Iwabu M, Yamauchi T, Okada-Iwabu M, Sato K, Nakagawa T, Funata M, et al. Adiponectin and AdipoR1 regulate PGC-1alpha and mitochondria by Ca(2+) and AMPK/SIRT1. Nature. 2010;464:1313–9.
55. Eseberri I, Lasa A, Churruca I, Portillo MP. Resveratrol metabolites modify adipokine expression and secretion in 3T3-L1 pre-adipocytes and mature adipocytes. PLoS ONE. 2013;8:e63918.
56. Tian C, Ye X, Zhang R, Long J, Ren W, Ding S, et al. Green tea polyphenols reduced fat deposits in high fat-fed rats via erk1/2-PPARgamma-adiponectin pathway. PLoS ONE. 2013;8:e53796.
57. Bouskila M, Pajvani UB, Scherer PE. Adiponectin: a relevant player in PPARgamma-agonist-mediated improvements in hepatic insulin sensitivity? Int J Obes (Lond). 2005;29 Suppl 1:S17–2323.
58. Funes L, Fernández-Arroyo S, Laporta O, Pons A, Roche E, Segura-Carretero A, et al. Correlation between plasma antioxidant capacity and verbascoside levels in rats after oral administration of lemon verbena extract. Food Chem. 2009;117:589–98.
59. Quirantes-Pine R, Herranz-Lopez M, Funes L, Borras-Linares I, Micol V, Segura-Carretero A, et al. Phenylpropanoids and their metabolites are the major compounds responsible for blood-cell protection against oxidative stress after administration of Lippia citriodora in rats. Phytomedicine. 2013;20:1112–8.
60. Rubiolo JA, Vega FV. Resveratrol protects primary rat hepatocytes against necrosis induced by reactive oxygen species. Biomed Pharmacother. 2008;62:606–12.
61. Ungvari Z, Bagi Z, Feher A, Recchia FA, Sonntag WE, Pearson K, et al. Resveratrol confers endothelial protection via activation of the antioxidant transcription factor Nrf2. Am J Physiol Heart Circ Physiol. 2010;299:H18–2424.
62. Bayram B, Ozcelik B, Grimm S, Roeder T, Schrader C, Ernst IM, et al. A diet rich in olive oil phenolics reduces oxidative stress in the heart of SAMP8 mice by induction of Nrf2-dependent gene expression. Rejuvenation Res. 2012;15:71–81.
63. Izzi V, Masuelli L, Tresoldi I, Sacchetti P, Modesti A, Galvano F, et al. The effects of dietary flavonoids on the regulation of redox inflammatory networks. Front Biosci (Landmark Ed). 2012;17:2396–418.
64. Yan Y, Gao YY, Liu BQ, Niu XF, Zhuang Y, Wang HQ. Resveratrol-induced cytotoxicity in human Burkitt's lymphoma cells is coupled to the unfolded protein response. BMC Cancer. 2010;10:445.
65. Howitz KT, Sinclair DA. Xenohormesis: sensing the chemical cues of other species. Cell. 2008; 133:387–91.
66. Morselli E, Galluzzi L, Kepp O, Criollo A, Maiuri MC, Tavernarakis N, et al. Autophagy mediates pharmacological lifespan extension by spermidine and resveratrol. Aging (Albany NY). 2009;1:961–70.
67. Joven J, Micol V, Segura-Carretero A, Alonso-Villaverde C, Menendez JA. For The Bioactive Food Components P. Polyphenols and the modulation of gene expression pathways: can we eat our way out of the danger of chronic disease? Crit Rev Food Sci Nutr. 2014;54:985–1001.
68. Serganova I, Rizwan A, Ni X, Thakur SB, Vider J, Russell J, et al. Metabolic imaging: a link between lactate dehydrogenase A, lactate, and tumor phenotype. Clin Cancer Res. 2011;17:6250–61.
69. Garcia-Garcia J, Micol V, de Godos A, Gomez-Fernandez JC. The cancer chemopreventive agent resveratrol is incorporated into model membranes and inhibits protein kinase C alpha activity. Arch Biochem Biophys. 1999;372:382–8.

70. Caturla N, Perez-Fons L, Estepa A, Micol V. Differential effects of oleuropein, a biophenol from Olea europaea, on anionic and zwiterionic phospholipid model membranes. Chem Phys Lipids. 2005;137:2–17.
71. Perez-Fons L, Aranda FJ, Guillen J, Villalain J, Micol V. Rosemary (Rosmarinus officinalis) diterpenes affect lipid polymorphism and fluidity in phospholipid membranes. Arch Biochem Biophys. 2006;453:224–36.
72. Perez-Fons L, Garzon MT, Micol V. Relationship between the antioxidant capacity and effect of rosemary (*Rosmarinus officinalis* L.) polyphenols on membrane phospholipid order. J Agric Food Chem. 2010;58:161–71.
73. Laporta O, Funes L, Garzon MT, Villalain J, Micol V. Role of membranes on the antibacterial and anti-inflammatory activities of the bioactive compounds from Hypoxis rooperi corm extract. Arch Biochem Biophys. 2007;467:119–31.
74. Tarahovsky YS, Muzafarov EN, Kim YA. Rafts making and rafts braking: how plant flavonoids may control membrane heterogeneity. Mol Cell Biochem. 2008;314:65–71.
75. Tachibana H, Fujimura Y, Yamada K. Tea polyphenol epigallocatechin-3-gallate associates with plasma membrane lipid rafts: lipid rafts mediate anti-allergic action of the catechin. Biofactors. 2004;21:383–5.
76. Adachi S, Nagao T, Ingolfsson HI, Maxfield FR, Andersen OS, Kopelovich L, et al. The inhibitory effect of (−)-epigallocatechin gallate on activation of the epidermal growth factor receptor is associated with altered lipid order in HT29 colon cancer cells. Cancer Res. 2007;67:6493–501.
77. Patra SK, Rizzi F, Silva A, Rugina DO, Bettuzzi S. Molecular targets of (−)-epigallocatechin-3-gallate (EGCG): specificity and interaction with membrane lipid rafts. J Physiol Pharmacol. 2008;59 Suppl 9:217–35.
78. Duhon D, Bigelow RL, Coleman DT, Steffan JJ, Yu C, Langston W, et al. The polyphenol epigallocatechin-3-gallate affects lipid rafts to block activation of the c-Met receptor in prostate cancer cells. Mol Carcinog. 2010;49:739–49.
79. Bernal P, Lemaire S, Pinho MG, Mobashery S, Hinds J, Taylor PW. Insertion of epicatechin gallate into the cytoplasmic membrane of methicillin-resistant Staphylococcus aureus disrupts penicillin-binding protein (PBP) 2a-mediated beta-lactam resistance by delocalizing PBP2. J Biol Chem. 2010;285:24055–65.
80. Cushnie TP, Taylor PW, Nagaoka Y, Uesato S, Hara Y, Lamb AJ. Investigation of the antibacterial activity of 3-O-octanoyl-(−)-epicatechin. J Appl Microbiol. 2008;105:1461–9.
81. Stapleton PD, Shah S, Ehlert K, Hara Y, Taylor PW. The beta-lactam-resistance modifier (−)-epicatechin gallate alters the architecture of the cell wall of Staphylococcus aureus. Microbiology. 2007;153:2093–103.
82. Psahoulia FH, Drosopoulos KG, Doubravska L, Andera L, Pintzas A. Quercetin enhances TRAIL-mediated apoptosis in colon cancer cells by inducing the accumulation of death receptors in lipid rafts. Mol Cancer Ther. 2007;6:2591–9.
83. Kaneko M, Takimoto H, Sugiyama T, Seki Y, Kawaguchi K, Kumazawa Y. Suppressive effects of the flavonoids quercetin and luteolin on the accumulation of lipid rafts after signal transduction via receptors. Immunopharmacol Immunotoxicol. 2008;30:867–82.
84. Xia M, Ling W, Zhu H, Wang Q, Ma J, Hou M, et al. Anthocyanin prevents CD40-activated proinflammatory signaling in endothelial cells by regulating cholesterol distribution. Arterioscler Thromb Vasc Biol. 2007;27:519–24.
85. Catania A, Barrajon-Catalan E, Nicolosi S, Cicirata F, Micol V. Immunoliposome encapsulation increases cytotoxic activity and selectivity of curcumin and resveratrol against HER2 overexpressing human breast cancer cells. Breast Cancer Res Treat. 2013;141:55–65.
86. Micol V. Nanofood: a revolution for the nutrition market. Agro Food Industry Hi-Tech. 2008;19:4–5.
87. Edeas M, Micol V. Mitochondrial generation of reactive oxygen species (ROS) and its targeting by antioxidants: a future vision for obesity. Agro Food Industry Hi-Tech. 2007;18:16–20.
88. Mustata GT, Rosca M, Biemel KM, Reihl O, Smith MA, Viswanathan A, et al. Paradoxical effects of green tea (Camellia sinensis) and antioxidant vitamins in diabetic rats: improved retinopathy and renal mitochondrial defects but deterioration of collagen matrix glycoxidation and cross-linking. Diabetes. 2005;54:517–26.
89. Bouayed J, Bohn T. Exogenous antioxidants – Double-edged swords in cellular redox state: health beneficial effects at physiologic doses versus deleterious effects at high doses. Oxid Med Cell Longev. 2010;3:228–37.
90. Sheu SS, Nauduri D, Anders MW. Targeting antioxidants to mitochondria: a new therapeutic direction. Biochim Biophys Acta. 2006;1762:256–65.
91. Anders MW. Exploiting endobiotic metabolic pathways to target xenobiotic antioxidants to mitochondria. Mitochondrion. 2013;13:454–63.
92. Murphy MP, Smith RA. Targeting antioxidants to mitochondria by conjugation to lipophilic cations. Annu Rev Pharmacol Toxicol. 2007;47:629–56.
93. Smith RA, Porteous CM, Gane AM, Murphy MP. Delivery of bioactive molecules to mitochondria in vivo. Proc Natl Acad Sci U S A. 2003;100:5407–12.
94. Weissig V. From serendipity to mitochondria-targeted nanocarriers. Pharm Res. 2011;28:2657–68.
95. D'Souza GG, Rammohan R, Cheng SM, Torchilin VP, Weissig V. DQAsome-mediated delivery of plasmid DNA toward mitochondria in living cells. J Control Release. 2003;92:189–97.

96. Micol V. Polyphenols as xenohormetics: the future approach to prevent human chronic. Agro Food Industry Hi-Tech. 2010;21:4–5.
97. Lamming DW, Wood JG, Sinclair DA. Small molecules that regulate lifespan: evidence for xenohormesis. Mol Microbiol. 2004;53:1003–9.
98. Westphal CH, Dipp MA, Guarente L. A therapeutic role for sirtuins in diseases of aging? Trends Biochem Sci. 2007;32:555–60.
99. Zahn JM, Kim SK. Systems biology of aging in four species. Curr Opin Biotechnol. 2007;18:355–9.
100. Cherniack EP. The potential influence of plant polyphenols on the aging process. Forsch Komplementmed. 2010;17:181–7.
101. Chaldakov GN, Fiore M, Tonchev AB, Dimitrov D, Pancheva R, Rancic G, et al. Homo obesus: a metabotrophin-deficient species. Pharmacology and nutrition insight. Curr Pharm Des. 2007;13:2176–9.
102. Chen B, Wild D, Guha R. PubChem as a source of polypharmacology. J Chem Inf Model. 2009; 49:2044–55.
103. McLean S, Duncan AJ. Pharmacological perspectives on the detoxification of plant secondary metabolites: implications for ingestive behavior of herbivores. J Chem Ecol. 2006;32:1213–28.
104. Gertsch J. Botanical drugs, synergy, and network pharmacology: forth and back to intelligent mixtures. Planta Med. 2011;77:1086–98.
105. Berenbaum MC. What is synergy? Pharmacol Rev. 1989;41:93–141.
106. Hopkins AL. Network pharmacology: the next paradigm in drug discovery. Nat Chem Biol. 2008;4:682–90.
107. Rollinger JM, Schuster D, Danzl B, Schwaiger S, Markt P, Schmidtke M, et al. In silico target fishing for rationalized ligand discovery exemplified on constituents of Ruta graveolens. Planta Med. 2009;75:195–204.
108. Wagner H. Synergy research: approaching a new generation of phytopharmaceuticals. Fitoterapia. 2011;82:34–7.
109. Gilbert B, Ferreira Alves L. Synergy in plant medicines. Curr Med Chem. 2003;10:13–20.
110. Kurin E, Atanasov AG, Donath O, Heiss EH, Dirsch VM, Nagy M. Synergy study of the inhibitory potential of red wine polyphenols on vascular smooth muscle cell proliferation. Planta Med. 2012; 78:772–8.
111. Tomás-Menor L, Morales-Soto A, Barrajón-Catalán E, Roldán-Segura C, Segura-Carretero A, Micol V. Correlation between the antibacterial activity and the composition of extracts derived from various Spanish Cistus species. Food Chem Toxicol. 2013;55:313–22.
112. Darvesh AS, Bishayee A. Chemopreventive and therapeutic potential of tea polyphenols in hepatocellular cancer. Nutr Cancer. 2013;65:329–44.
113. Khandelwal AR, Hebert VY, Kleinedler JJ, Rogers LK, Ullevig SL, Asmis R, et al. Resveratrol and quercetin interact to inhibit neointimal hyperplasia in mice with a carotid injury. J Nutr. 2012;142:1487–94.
114. Qin C, Tan KL, Zhang CL, Tan CY, Chen YZ, Jiang YY. What does it take to synergistically combine sub-potent natural products into drug-level potent combinations? PLoS ONE. 2012;7:e49969.
115. Rosato A, Vitali C, De Laurentis N, Armenise D, Antonietta Milillo M. Antibacterial effect of some essential oils administered alone or in combination with Norfloxacin. Phytomedicine. 2007;14:727–32.
116. Brehm-Stecher BF, Johnson EA. Sensitization of Staphylococcus aureus and Escherichia coli to antibiotics by the sesquiterpenoids nerolidol, farnesol, bisabolol, and apritone. Antimicrob Agents Chemother. 2003;47:3357–60.
117. Betts JW, Kelly SM, Haswell SJ. Antibacterial effects of theaflavin and synergy with epicatechin against clinical isolates of Acinetobacter baumannii and Stenotrophomonas maltophilia. Int J Antimicrob Agents. 2011;38:421–5.
118. Mandalari G, Bisignano C, D'Arrigo M, Ginestra G, Arena A, Tomaino A, et al. Antimicrobial potential of polyphenols extracted from almond skins. Lett Appl Microbiol. 2010;51:83–9.
119. Taylor PW. Alternative natural sources for a new generation of antibacterial agents. Int J Antimicrob Agents. 2013;42:195–201.
120. Odds FC. Synergy, antagonism, and what the chequerboard puts between them. J Antimicrob Ther. 2003;52:1.
121. EUCAST, European Committee for Antimicrobial Susceptibility Testing (EUCAST) of the European Society of Clinical Microbiology and Infectious Diseases (ESCMID). Terminology relating to methods for the determination of susceptibility of bacteria to antimicrobial agents. EUCAST definitive document. 2000.

Postprandial Inflammation: Targeting Glucose and Lipids

12

Marijke A. de Vries, Boudewijn Klop, Hans W. Janssen, Tjin L. Njo, Elsbeth M. Westerman, and Manuel Castro Cabezas

Abstract

Many risk factors have been identified as being responsible for the process of atherogenesis. Several of these risk factors are related to inflammation, which is an obligatory feature of the atherosclerotic plaque. Increasing evidence suggests that postprandial lipoproteins and glucose may be involved in the inflammatory process preceding the development of atherosclerosis. During the postprandial situation, remnants of chylomicrons and very low-density lipoproteins bind to circulating leukocytes and endothelial cells, leading to a state of acute activation with the expression of integrins on different cells, the generation of oxidative stress, production of cytokines and complement activation. Elevated plasma glucose levels may also induce leukocyte activation in humans. In addition, advanced glycation end products, formed during hyperglycemia, cause inflammation and endothelial damage. This chain of events results in a situation of acute inflammation causing endothelial dysfunction, which may be one of the earliest defects in atherogenesis. Interestingly, while this may occur several times each day after each meal, there is only limited information on the contribution of different nutrients on the postprandial inflammatory processes. In this review, we will focus on the available evidence and we will discuss the role of lifestyle and pharmaceutical interventions in modulating postprandial inflammation.

Keywords

Adiposity • Apolipoprotein B48 • Atherosclerosis • Chylomicron • Neutrophil • Triglycerides

M.A. de Vries • B. Klop • M. Castro Cabezas (✉)
Department of Internal Medicine, Center for Diabetes and Vascular Medicine, Sint Franciscus Gasthuis, PO BOX 10900, Rotterdam 3004 BA, The Netherlands
e-mail: m.devries@sfg.nl; boudewijn.klop@gmail.com; m.castrocabezas@sfg.nl

H.W. Janssen • T.L. Njo
Department of Clinical Chemistry, Sint Franciscus Gasthuis, Rotterdam, The Netherlands
e-mail: h.janssen@sfg.nl; t.njo@sfg.nl

E.M. Westerman
Department of Clinical Pharmacy, Sint Franciscus Gasthuis, Rotterdam, The Netherlands
e-mail: e.westerman@sfg.nl

J. Camps (ed.), *Oxidative Stress and Inflammation in Non-communicable Diseases – Molecular Mechanisms and Perspectives in Therapeutics*, Advances in Experimental Medicine and Biology 824, DOI 10.1007/978-3-319-07320-0_12, © Springer International Publishing Switzerland 2014

12.1 Introduction

Atherosclerosis is the major cause of death in the Western World. The classical risk factors involved are smoking, diabetes mellitus, hyperlipidemia, hypertension and obesity. In fact, a limited set of risk factors seems to explain most part of the cardiovascular risk [1]. In the past decades, it has become evident that inflammation plays a key role in the development of atherosclerosis. Several inflammatory markers, including C-reactive protein (CRP), total leukocyte count and the third component of complement (C3) have been associated with cardiovascular disease [2–5].

Postprandial inflammation starts with the increase of remnant lipoproteins and glucose in the circulation. This leads to activation of circulating leukocytes, which interact with the endothelium [6–10]. When stimulated by cytokines, endothelial cells express adhesion molecules, such as vascular cell adhesion molecule 1 (VCAM-1), which allow activated leukocytes to bind to the vessel wall. The expression of certain molecules on activated leukocytes, such as the CD11b integrin, is obligatory in this process. Once leukocytes have adhered to the endothelium, they are stimulated by chemoattractant signals, in particular monocyte chemotactic protein-1 (MCP-1), to migrate into the arterial wall [10]. Subendothelial migrated monocytes differentiate into macrophages, which can take up modified lipoproteins, leading to foam cell formation and the development of an atherosclerotic plaque [6].

Here we will provide an overview of the processes involved in postprandial inflammation and potential lifestyle and pharmaceutical interventions, which may help to modulate postprandial inflammation.

12.2 Chylomicron and VLDL Synthesis

Lipids and fat-soluble vitamins ingested with the diet are transported from the intestine to the circulation by chylomicrons. In the intestine, dietary triglycerides are hydrolyzed into free fatty acids (FFAs) and 2-monoacylglycerols (MAG) [11]. These FFAs and MAG are then transported from the intestinal lumen to the smooth endoplasmatic reticulum (SER) of the enterocyte, where they are resynthesized into triglycerides [11, 12]. The formation of chylomicrons starts in the rough endoplasmatic reticulum (RER), where nascent apolipoprotein (apo) B48 is translocated into the lumen of the endoplasmatic reticulum (ER) [13]. The microsomal transfer protein (MTP) shuttles triglycerides from within the ER to an acceptor apo B48 molecule, giving rise to a primordial apo B-containing lipoprotein [13]. This particle is then enriched and expanded with lipids present within the ER, resulting in a prechylomicron, with a core containing triglycerides, cholesteryl esters and fat-soluble vitamins, and an outer layer with phospholipids, free cholesterol, apo B48 and apo A-IV [11, 13]. The prechylomicron leaves the ER and is transported to the Golgi apparatus in prechylomicron transport vesicles. Several proteins, including apo B48 and CD36, are involved in this transport [14]. Within the Golgi, apo A-I is attached, resulting in the formation of a mature chylomicron, which is secreted into the lymphatic system [15, 16].

In the circulation, the triglycerides present in chylomicrons are hydrolyzed into MAG and FFAs by the endothelium-bound enzyme lipoprotein lipase (LPL), giving rise to chylomicron remnant particles [17]. The dietary FFAs and chylomicron remnants can be taken up by the liver, to be used for the synthesis of very low-density lipoproteins (VLDL) [18]. The assembly of VLDL begins in the ER of hepatocytes. In humans, the liver lacks an editing enzyme complex such as in the intestine and therefore, hepatic triglyceride-rich lipoproteins (TRLs) contain solely apo B100 as structural apolipoprotein [19]. Apo B100 is synthesized in the RER and under the influence of MTP, the nascent lipoprotein becomes enriched with triglycerides which are synthesized in the SER [19]. This gives rise to pre-VLDL, which can be further processed to form triglyceride-poor $VLDL_2$. The VLDL synthesis in the liver is a continuous process, and VLDL secretion is controlled mainly by intracellular degradation, depending largely on lipidation of the nascent VLDL particles [20]. Therefore, pre-VLDL that is not transformed to $VLDL_2$ is subject to posttranslational degradation [20]. Otherwise, $VLDL_2$ is transported to the Golgi, where it is either secreted into the circulation, or converted to triglyceride-rich $VLDL_1$, which is then released into the circulation.

The large VLDL particles in the circulation are hydrolyzed by LPL, resulting in the formation of intermediate-density lipoproteins (IDL) and eventually low-density lipoproteins (LDL) [21]. However, since chylomicrons and VLDL largely share the same lipolytic pathway, competition at the level of several steps involved in the lipolysis and removal of TRLs from the circulation may occur. These include not only LPL [17], but also hepatic lipase, the low-density lipoprotein receptor (LDL-R), the LDL receptor-related protein 1 (LRP-1) and heparan sulfate proteoglycans [21, 22]. This competition may lead to accumulation of TRLs in the circulation.

12.3 Postprandial Lipids Induce Inflammation

Lipids induce a state of inflammation, with increased levels of adhesion molecules, cytokines, oxidative stress and leukocyte activation. An in vitro study demonstrated that TRLs with a high triglyceride and cholesterol content stimulate human aortic endothelial cells to express VCAM-1 [23], and during hypertriglyceridemia, the adhesion of monocytes to VCAM-1 is increased [24]. Not only TRLs itself, but also the FFAs released by hydrolysis of these particles stimulate endothelial cells to express adhesion molecules and to produce inflammatory cytokines [25]. In an in vivo study, ingestion of a high-fat meal resulted in an increase in neutrophil count, interleukin-6 (IL-6) and hydroperoxides, with a simultaneous decrease in endothelial function, reflected by flow-mediated endothelium-dependent dilatation [26]. Others showed that ingestion of a high-fat meal by healthy volunteers increased serum bacterial endotoxin, or lipopolysaccharide (LPS), which may also lead to leukocyte activation and inflammation [27]. The ingestion of fat therefore results in increased leukocyte activation, which is reflected by an increase of surface expression of CD11b, CD11c and CD14 on monocytes, and CD11b, CD66b and CD62L on neutrophils, as has been shown in vitro [28, 29] and in vivo [8, 24, 30, 31]. These data point at a pro-inflammatory effect of dietary lipids on circulating inflammatory cells with detrimental effects on the vessel wall.

Apo B-containing lipoproteins have been shown to adhere to postprandial leukocytes in the circulation [7]. By adding labeled palmitic acid to an oral fat load, it was demonstrated that leukocytes become enriched with dietary FFAs [7]. The uptake of TRLs by leukocytes is most likely mediated by different receptors, such as the LDL-R, LRP-1 and the apo B48 receptor [24, 32, 33]. Opsonization of remnants in the circulation may be directly related to leukocyte activation.

The molecular mechanism behind lipid-induced inflammation probably involves activation of the transcription factor nuclear factor kappaB (NF-κB). The remnants of TRLs migrate into the subendothelial space, where they induce the production of reactive oxygen species (ROS). This may lead to the oxidative modification of LDL, and these oxidized lipoproteins (oxLDL) are easily taken up by macrophages, smooth muscle cells and endothelial cells via scavenger receptors, such as CD36, SR-AI/II, SR-BI and the lectin-like oxidized LDL receptor 1 (LOX-1) [34, 35]. Internalized lipoproteins can induce activation of the protein kinase C (PKC) pathway, resulting in activation of the IκB kinase complex and subsequently NF-κB. Activated NF-κB will induce the transcription of genes encoding for several cytokines, including tumor necrosis factor-alpha (TNF-α), interferon-gamma (IFN-γ), interleukin-1 beta (IL-1β), IL-6 and IL-8, chemokines, such as MCP-1 and macrophage inflammatory protein-1 alpha (MIP-1α) and cellular adhesion molecules, such as VCAM-1 (Fig. 12.1).

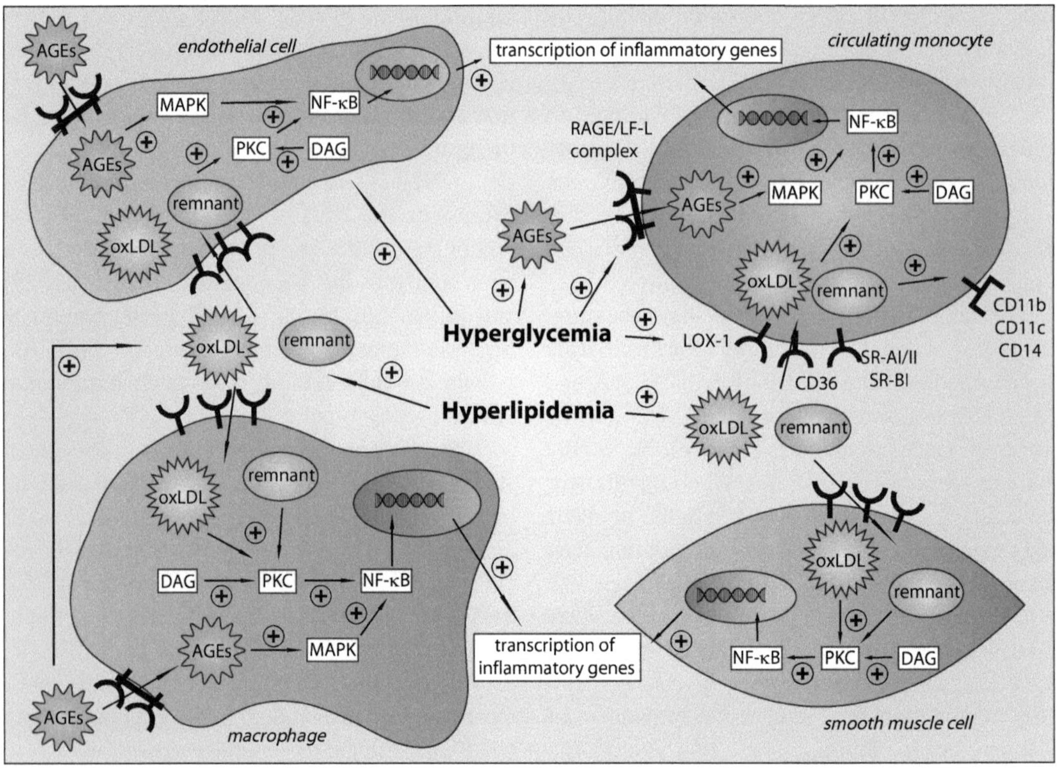

Fig. 12.1 The cellular processes involved in lipid- and glucose-induced inflammation: The remnants of triglyceride-rich lipoproteins and oxidized low-density lipoprotein (*oxLDL*) are taken up by circulating leukocytes, macrophages, endothelial cells and smooth muscle cells via several scavenger receptors, including CD36, SR-AI/II, SR-BI and lectin-like oxidized LDL receptor 1 (*LOX-1*). Intracellular, remnants and oxLDL activate the protein kinase C (*PKC*) pathway, resulting in activation of nuclear factor kappaB (*NF-κB*). NF-κB induces the transcription of several inflammatory genes, including genes encoding for cytokines, chemokines and adhesion molecules. During hyperglycemia, the intracellular synthesis of diacylglycerol (*DAG*) in endothelial cells, smooth muscle cells, monocytes an macrophages is increased, leading to further activation of the PKC pathway. Activation of the PKC pathway in monocytes also results in the release of the integrins CD11b, CD11c and CD14 stored in intracellular vesicles. In addition, during hyperglycemia, advanced glycation end products (*AGEs*) are formed. These AGEs are taken up by endothelial cells, monocytes and macrophages via the receptor for AGE (*RAGE*) and lactoferrin-like polypeptide (*LF-L*) complex, resulting in activation of mitogen-activated protein kinase (*MAPK*) and subsequently NF-κB activation. AGEs also enhance the formation of oxLDL, and during hyperglycemia the expression of LOX-1 on monocytes and macrophages increases. These processes further facilitate the uptake of oxLDL by macrophages and thus enhance the inflammatory process

12.4 Glucose and Postprandial Inflammation

Not only lipids, but also glucose may be involved in postprandial inflammation. In fact, the generation of oxidative stress due to hyperglycemia has been proposed to be a key event in the development of diabetic complications [36]. In vivo studies have demonstrated that ingestion of glucose results in increased production of TNF-α and IL-1β by peripheral blood mononucleated cells [37], reduced flow-mediated endothelium-dependent dilatation [26, 38], increased formation of ROS and oxidative stress [38] and increased leukocyte activation [9].

Hyperglycemia may induce inflammation via several mechanisms. One is the increased activation of the PKC pathway in endothelial cells, macrophages and smooth muscle cells. Activation of PKC results in increased cytokine production via activation of NF-κB [39]. This increased PKC activity is probably stimulated by an increased de novo synthesis of diacylglycerol in endothelial cells during hyperglycemia, which is an important activator of PKC [40].

Glucose can modify proteins to form advanced glycation end products (AGEs), which exert a pro-inflammatory effect in several ways [41]. AGEs can interact with cellular binding sites, such as the receptor for AGE (RAGE) and lactoferrin-like polypeptide (LF-L) on endothelial cells and macrophages. This AGEs-RAGE interaction results in uptake of AGEs by these cells and subsequently generation of ROS and activation of NF-κB via activation of mitogen-activated protein kinase (MAPK) [42–45]. The interaction between AGEs and RAGE on circulating monocytes also induces chemotaxis, leading to increased migration of these cells into the subendothelial space [44]. The presence of AGEs induces increased expression of RAGE on endothelial cells [46], which functions as a receptor for the integrin CD11b on monocytes and neutrophils, promoting leukocyte adhesion to the vessel wall [47].

Furthermore, AGEs induce the glycosylation of apo B and phospholipids on LDL particles, making these lipoproteins more susceptible to oxidative modifications, resulting in enhanced uptake by macrophages via scavenger receptors [43]. During hyperglycemia, the surface expression of LOX-1 on macrophages is increased, thereby enhancing the uptake of oxLDL by these cells [39].

Moreover, glycation of the regulatory membrane protein CD59 on endothelial cells, which inhibits the deposition of the membrane attack complex (MAC) of complement, leads to reduced activity of this protein and thus a higher susceptibility of the endothelium for the MAC-induced release of pro-inflammatory cytokines [48].

Finally, hyperglycemia results in a reduced antioxidant capacity. Hyperglycemia leads to increased reduction of glucose to sorbitol by aldose reductase, and during this process NADPH is consumed [41]. Since the cellular antioxidant capacity depends on the energy provided by NADPH to the antioxidants gluthathione and thioredoxin, reduction of NADPH will result in increased oxidative stress [41, 49]. The cellular processes involved in inflammation induced by lipids and glucose are summarized in Fig. 12.1.

12.5 Lifestyle and Pharmaceutical Interventions Modulating Postprandial Inflammation

Evidence from clinical trials is starting to emerge, demonstrating that postprandial inflammation can be reduced by lifestyle modifications. Several studies have established a positive effect of dietary antioxidants on postprandial inflammation. Foods rich in polyphenols, such as strawberries and black raspberries, reduce postprandial inflammation [50, 51]. The daily consumption of strawberry beverages during 6 weeks, reduced the postprandial oxidative modification of LDL in hyperlipidemic patients [50]. The consumption of black raspberries for 4 days attenuated the high-fat meal-induced production of IL-6 [51]. Flavonoids seem to reduce postprandial inflammation as well. An in vitro study demonstrated that the flavonoid epigallocatechin-3-gallate suppressed the hyperglycemia-induced expression of VCAM-1 on human umbilical vein endothelial cells and reduced the adhesion of monocytes to these cells, via inhibition of PKC and NF-κB activation [52]. This beneficial effect of flavonoids on postprandial inflammation has been confirmed in vivo. In a randomized trial with 30 healthy volunteers, the consumption of orange juice, which contains flavonoids, with a high-fat meal reduced the postprandial production of ROS and prevented the postprandial increase in LPS [53]. Food containing carotenoids may also decrease postprandial inflammation. The consumption of tomatoes containing this antioxidant reduced postprandial IL-6 production and prevented LDL oxidation, even though the tomatoes exaggerated postprandial lipemia [54]. This anti-inflammatory effect, despite an increase in serum lipids, has also been observed with the consumption of red wine. Although the addition of red wine to a meal resulted in a higher increase in serum lipids, red wine prevented the production of NF-κB by peripheral blood mononucleated cells [55]. In contrast, addition of vodka to the meal did not attenuate the NF-κB activation, indicating that this effect was not due to the alcohol content of the wine [55]. The anti-inflammatory effect of red wine is more likely the result of the antioxidants quercetin and α-tocopherol succinate, since treatment of human mononuclear cells with VLDL in the presence of these antioxidants inhibited the activation of NF-κB [55]. Collectively, these data suggest a beneficial effect of dietary antioxidants, such as polyphenols, flavonoids, carotenoids, quercetin and α-tocopherol succinate, on postprandial inflammation.

In addition to antioxidants, a diet rich in monounsaturated fatty acids (MUFA), as opposed to saturated fatty acids (SFA), reduces postprandial inflammation. In a small study, the consumption of a meal based on extra virgin olive oil, which is rich in MUFA, did not elicit NF-κB activation in peripheral blood mononucleated cells, in contrast to ingestion of a meal based on butter, rich in SFA, or walnuts, which contained an equal amount of MUFA, SFA and polyunsaturated fatty acids (PUFA) [56]. This beneficial effect of MUFA was confirmed in another trial with 20 elderly healthy subjects. In this study, a Mediterranean diet with a high virgin olive oil content not only reduced total cholesterol, LDL-C and apo B, but also the postprandial expression of MCP-1, matrix metalloproteinase 9 (MMP-9) and the NF-κB subunit p65 in peripheral blood mononucleated cells [57]. In another small study, the addition of avocado, rich in MUFA, to a hamburger meal attenuated the postprandial increase of IL-6, via reduced activity of the NF-κB pathway [58].

The effect of exercise on postprandial inflammation remains controversial. In a study among 20 cigarette smokers, those who reported to exercise two or more hours a week had lower postprandial levels of malondialdehyde than untrained smokers [59]. However, in a small randomized trial, exercise before a high-fat meal did not reduce postprandial serum lipid hydroperoxides [60]. In a cross-sectional study among middle-aged men, active men had lower fasting levels of IL-6 and CRP than sedentary men, but no difference in postprandial inflammation was observed [61].

Weight loss seems to have a favorable effect on postprandial inflammation. Moderate weight loss reduced the postprandial soluble intercellular adhesion molecule (s-ICAM), MCP-1 and high-sensitivity CRP (hs-CRP) increment in 11 normolipidemic moderately obese men [62]. In another study, weight loss reduced postprandial IL-6 levels in eight men with impaired glucose tolerance [63].

Several lipid-lowering drugs are effective in lowering postprandial inflammation as well. Simvastatin, atorvastatin and pitavastatin have all shown to attenuate postprandial inflammation [64–66]. In addition, fenofibrate reduces the production of TNF-α, IL-1β, IL-6, MCP-1 and MIP-1α [67, 68]. Anti-diabetic drugs may also reduce postprandial inflammation. Rosiglitazone and metformin have been shown to increase levels of the antioxidant enzyme paraoxonase (PON)-1 and decrease postprandial levels of MCP-1 [69, 70]. Treatment of patients with type 2 diabetes mellitus with nateglinide, an anti-diabetic drug belonging to the meglitinides class of drugs, significantly reduced postprandial oxidative stress and increased endothelial function [38]. Infusion of glucagon-like peptide 1 during hyperglycemia significantly reduced oxidative stress, inflammation and endothelial dysfunction in patients with type 1 diabetes mellitus [71]. To the best of our knowledge, no studies have compared the effect of lipid-lowering versus glucose-lowering therapy on postprandial inflammation in patients with diabetes mellitus. The effects of different dietary and pharmaceutical interventions are summarized in Table 12.1.

12.6 Conclusion

In the postprandial phase, lipids and glucose induce a state of systemic inflammation, with the increased production of cytokines, generation of oxidative stress, complement activation and enhanced adhesion of monocytes to the endothelium, resulting in transient endothelial dysfunction. A diet rich in MUFAs and antioxidants, such as polyphenols, flavonoids and carotenoids,

Table 12.1 Overview of interventions effective in reducing postprandial inflammation

	Intervention	Compound	Effect	Reference
Diet	Strawberries	Polyphenols	↓oxLDL	[50]
	Black raspberries	Polyphenols	↓IL-6	[51]
	Orange juice	Flavonoids	↓ROS, ↓LPS	[53]
	Tomatoes	Carotenoids	↓oxLDL, ↓IL-6	[54]
	Red wine	Quercetin	↓NF-κB	[55]
		A-tocopherol succinate		
	Extra virgin olive oil	MUFA	↓NF-κB, ↓MCP-1, ↓MMP-9	[56, 57]
	Avocado	MUFA	↓IL-6, ↓NF-κB	[58]
Weight loss			↓s-ICAM-1, ↓MCP-1, ↓hs-CRP, ↓IL-6	[62, 63]
Drugs	Statins	Simvastatin	↓C3	[64]
		Atorvastatin	↓C3	[65]
		Pitavastatin	↓urinary isoprostane	[66]
	Fibrates	Fenofibrate	↓TNF-α, ↓IL-1β, ↓IL-6, ↓MCP-1 ↓MIP-1α	[67, 68]
	Antidiabetic drugs	Rosiglitazone	↑PON-1, ↓MCP-1	[69, 70]
		Metformin	↑PON-1, ↓MCP-1	[70]
		Nateglinide	↓MDA	[38]
		GLP-1	↓sICAM-1, ↓IL-6, ↓nitrotyrosine, ↓serum isoprostane	[71]

C3 complement component 3, *hs-CRP* high-sensitivity C-reactive protein, *IL-1β* interleukin-1 beta, *IL-6* interleukin-6, *LPS* lipopolysaccharide, *MCP-1* monocyte chemotactic protein-1, *MDA* malondialdehyde, *MIP-1α* macrophage inflammatory protein-1 alpha, *MMP-9* matrix metalloproteinase 9, *NF-κB* nuclear factor kappaB, *oxLDL* oxidized low-density lipoprotein, *PON-1* paraoxonase 1, *ROS* reactive oxygen species, *s-ICAM-1* soluble intercellular adhesion molecule 1, *TNF-α* tumor necrosis factor alpha

weight loss and several drugs, such as statins, fibrates and anti-diabetic drugs, are effective in reducing postprandial inflammation.

References

1. Yusuf S, Hawken S, Ounpuu S, Dans T, Avezum A, Lanas F, et al. Effect of potentially modifiable risk factors associated with myocardial infarction in 52 countries (the INTERHEART study): case-control study. Lancet. 2004;364:937–52.
2. Danesh J, Whincup P, Walker M, Lennon L, Thomson A, Appleby P, et al. Low grade inflammation and coronary heart disease: prospective study and updated meta-analyses. BMJ. 2000;321:199–204.
3. Friedman GD, Klatsky AL, Siegelaub AB. The leukocyte count as a predictor of myocardial infarction. N Engl J Med. 1974;290:1275–8.
4. Muscari A, Massarelli G, Bastagli L, Poggiopollini G, Tomassetti V, Drago G, et al. Relationship of serum C3 to fasting insulin, risk factors and previous ischaemic events in middle-aged men. Eur Heart J. 2000;21:1081–90.
5. Muscari A, Bozzoli C, Puddu GM, Sangiorgi Z, Dormi A, Rovinetti C, et al. Association of serum C3 levels with the risk of myocardial infarction. Am J Med. 1995;98:357–64.
6. Ross R. Atherosclerosis–an inflammatory disease. N Engl J Med. 1999;340:115–26.
7. Alipour A, van Oostrom AJHHM, Izraeljan A, Verseyden C, Collins JM, Frayn KN, et al. Leukocyte activation by triglyceride-rich lipoproteins. Arterioscler Thromb Vasc Biol. 2008;28:792–7.
8. Van Oostrom AJHHM, Rabelink TJ, Verseyden C, Sijmonsma TP, Plokker HWM, De Jaegere PPT, et al. Activation of leukocytes by postprandial lipemia in healthy volunteers. Atherosclerosis. 2004;177:175–82.
9. Sampson MJ, Davies IR, Brown JC, Ivory K, Hughes DA. Monocyte and neutrophil adhesion molecule expression during acute hyperglycemia and after antioxidant treatment in type 2 diabetes and control patients. Arterioscler Thromb Vasc Biol. 2002;22:1187–93.
10. Libby P. Inflammation in atherosclerosis. Arterioscler Thromb Vasc Biol. 2012;32:2045–51.
11. Iqbal J, Hussain MM. Intestinal lipid absorption. Am J Physiol Endocrinol Metab. 2009;296:E1183–94.
12. Wang TY, Liu M, Portincasa P, Wang DQ-H. New insights into the molecular mechanism of intestinal

fatty acid absorption. Eur J Clin Investig. 2013;43:1203–23.
13. Abumrad NA, Davidson NO. Role of the gut in lipid homeostasis. Physiol Rev. 2012;92:1061–85.
14. Siddiqi S, Saleem U, Abumrad NA, Davidson NO, Storch J, Siddiqi SA, et al. A novel multiprotein complex is required to generate the prechylomicron transport vesicle from intestinal ER. J Lipid Res. 2010;51:1918–28.
15. Mansbach CM, Gorelick F. Development and physiological regulation of intestinal lipid absorption. II. Dietary lipid absorption, complex lipid synthesis, and the intracellular packaging and secretion of chylomicrons. Am J Physiol Gastrointest Liver Physiol. 2007;293:G645–50.
16. Mansbach CM, Siddiqi SA. The biogenesis of chylomicrons. Annu Rev Physiol. 2010;72:315–33.
17. Brunzell JD, Hazzard WR, Porte D, Bierman EL. Evidence for a common, saturable, triglyceride removal mechanism for chylomicrons and very low density lipoproteins in man. J Clin Invest. 1973;52:1578–85.
18. Nakajima K, Nakano T, Tokita Y, Nagamine T, Inazu A, Kobayashi J, et al. Postprandial lipoprotein metabolism: VLDL vs chylomicrons. Clin Chim Acta. 2011;412:1306–18.
19. Rustaeus S, Lindberg K, Stillemark P, Claesson C, Asp L, Larsson T, et al. Assembly of very low density lipoprotein: a two-step process of apolipoprotein B core lipidation. J Nutr. 1999;129:463S–6.
20. Olofsson S-O, Borén J. Apolipoprotein B secretory regulation by degradation. Arterioscler Thromb Vasc Biol. 2012;32:1334–8.
21. Bjorkegren J, Packard CJ, Hamsten A, Bedford D, Caslake M, Foster L, et al. Accumulation of large very low density lipoprotein in plasma during intravenous infusion of a chylomicron-like triglyceride emulsion reflects competition for a common lipolytic pathway. J Lipid Res. 1996;37:76–86.
22. Dallinga-Thie GM, Franssen R, Mooij HL, Visser ME, Hassing HC, Peelman F, et al. The metabolism of triglyceride-rich lipoproteins revisited: new players, new insight. Atherosclerosis. 2010;211:1–8.
23. Wang YI, Bettaieb A, Sun C, Deverse JS, Radecke CE, Mathew S, et al. Triglyceride-rich lipoprotein modulates endothelial vascular cell adhesion molecule (VCAM)-1 expression via differential regulation of endoplasmic reticulum stress. PLoS One. 2013;8:e78322.
24. Gower RM, Wu H, Foster GA, Devaraj S, Jialal I, Ballantyne CM, et al. CD11c/CD18 expression is upregulated on blood monocytes during hypertriglyceridemia and enhances adhesion to vascular cell adhesion molecule-1. Arterioscler Thromb Vasc Biol. 2011;31:160–6.
25. Higgins LJ, Rutledge JC. Inflammation associated with the postprandial lipolysis of triglyceride-rich lipoproteins by lipoprotein lipase. Curr Atheroscler Rep. 2009;11:199–205.
26. Van Oostrom AJHHM, Sijmonsma TP, Verseyden C, Jansen EHJM, de Koning EJP, Rabelink TJ, et al. Postprandial recruitment of neutrophils may contribute to endothelial dysfunction. J Lipid Res. 2003;44:576–83.
27. Erridge C, Attina T, Spickett CM, Webb DJ. A high-fat meal induces low-grade endotoxemia: evidence of a novel mechanism of postprandial inflammation. Am J Clin Nutr. 2007;86:1286–92.
28. Wanten G, van Emst-De Vries S, Naber T, Willems P. Nutritional lipid emulsions modulate cellular signaling and activation of human neutrophils. J Lipid Res. 2001;42:428–36.
29. Bentley C, Hathaway N, Widdows J, Bejta F, De Pascale C, Avella M, et al. Influence of chylomicron remnants on human monocyte activation in vitro. Nutr Metab Cardiovasc Dis. 2011;21:871–8.
30. Hiramatsu K, Arimori S. Increased superoxide production by mononuclear cells of patients with hypertriglyceridemia and diabetes. Diabetes. 1988;37:832–7.
31. Van Oostrom AJHHM, Plokker HWM, van Asbeck BS, Rabelink TJ, van Kessel KPM, Jansen EHJM, et al. Effects of rosuvastatin on postprandial leukocytes in mildly hyperlipidemic patients with premature coronary sclerosis. Atherosclerosis. 2006;185:331–9.
32. Koo C, Wernette-Hammond ME, Garcia Z, Malloy MJ, Uauy R, East C, et al. Uptake of cholesterol-rich remnant lipoproteins by human monocyte-derived macrophages is mediated by low density lipoprotein receptors. J Clin Invest. 1988;81:1332–40.
33. Bermudez B, Lopez S, Varela LM, Ortega A, Pacheco YM, Moreda W, et al. Triglyceride-rich lipoprotein regulates APOB48 receptor gene expression in human THP-1 monocytes and macrophages. J Nutr. 2012;142:227–32.
34. Pirillo A, Norata GD, Catapano AL. LOX-1, OxLDL, and atherosclerosis. Mediat Inflamm. 2013;2013:152786.
35. Shi Y, Cosentino F, Camici GG, Akhmedov A, Vanhoutte PM, Tanner FC, et al. Oxidized low-density lipoprotein activates p66Shc via lectin-like oxidized low-density lipoprotein receptor-1, protein kinase C-beta, and c-Jun N-terminal kinase kinase in human endothelial cells. Arter Thromb Vasc Biol. 2011;31:2090–7.
36. De M Bandeira S, da Fonseca LJS, da S Guedes G, Rabelo LA, Goulart MOF, Vasconcelos SML. Oxidative stress as an underlying contributor in the development of chronic complications in diabetes mellitus. Int J Mol Sci. 2013;14:3265–84.
37. Deopurkar R, Ghanim H, Friedman J, Abuaysheh S, Sia CL, Mohanty P, et al. Differential effects of cream, glucose, and orange juice on inflammation, endotoxin, and the expression of Toll-like receptor-4 and suppressor of cytokine signaling-3. Diabetes Care. 2010;33:991–7.
38. Wang L, Guo L, Zhang L, Zhou Y, He Q, Zhang Z, et al. Effects of glucose load and nateglinide intervention

on endothelial function and oxidative stress. J Diabetes Res. 2013;2013:849295.
39. Geraldes P, King GL. Activation of protein kinase C isoforms and its impact on diabetic complications. Circ Res. 2010;106:1319–31.
40. Inoguchi T, Xia P, Kunisaki M, Higashi S, Feener EP, King GL. Insulin's effect on protein kinase C and diacylglycerol induced by diabetes and glucose in vascular tissues. Am J Physiol. 1994;267:E369–79.
41. Brownlee M. Biochemistry and molecular cell biology of diabetic complications. Nature. 2001;414:813–20.
42. Yan SD, Schmidt AM, Anderson GM, Zhang J, Brett J, Zou YS, et al. Enhanced cellular oxidant stress by the interaction of advanced glycation end products with their receptors/binding proteins. J Biol Chem. 1994;269:9889–97.
43. Aronson D, Rayfield EJ. How hyperglycemia promotes atherosclerosis: molecular mechanisms. Cardiovasc Diabetol. 2002;1:1.
44. Schmidt AM, Yan SD, Brett J, Mora R, Nowygrod R, Stern D. Regulation of human mononuclear phagocyte migration by cell surface-binding proteins for advanced glycation end products. J Clin Invest. 1993;91:2155–68.
45. Fukami K, Yamagishi S-I, Okuda S. Role of AGEs-RAGE system in cardiovascular disease. Curr Pharm Des. 2014;20:2395–402.
46. Tanaka N, Yonekura H, Yamagishi S, Fujimori H, Yamamoto Y, Yamamoto H. The receptor for advanced glycation end products is induced by the glycation products themselves and tumor necrosis factor-alpha through nuclear factor-kappa B, and by 17beta-estradiol through Sp-1 in human vascular endothelial cells. J Biol Chem. 2000;275:25781–90.
47. Chavakis T, Bierhaus A, Al-Fakhri N, Schneider D, Witte S, Linn T, et al. The pattern recognition receptor (RAGE) is a counterreceptor for leukocyte integrins: a novel pathway for inflammatory cell recruitment. J Exp Med. 2003;198:1507–15.
48. Acosta J, Hettinga J, Flückiger R, Krumrei N, Goldfine A, Angarita L, et al. Molecular basis for a link between complement and the vascular complications of diabetes. Proc Natl Acad Sci U S A. 2000;97:5450–5.
49. Oka S-I, Hsu C-P, Sadoshima J. Regulation of cell survival and death by pyridine nucleotides. Circ Res. 2012;111:611–27.
50. Burton-Freeman B, Linares A, Hyson D, Kappagoda T. Strawberry modulates LDL oxidation and postprandial lipemia in response to high-fat meal in overweight hyperlipidemic men and women. J Am Coll Nutr. 2010;29:46–54.
51. Sardo CL, Kitzmiller JP, Apseloff G, Harris RB, Roe DJ, Stoner GD, et al. An open-label randomized crossover trial of lyophilized black raspberries on postprandial inflammation in older overweight males: a pilot study. Am J Ther. 2013. Epub ahead of print.
52. Yang J, Han Y, Chen C, Sun H, He D, Guo J, et al. EGCG attenuates high glucose-induced endothelial cell inflammation by suppression of PKC and NF-κB signaling in human umbilical vein endothelial cells. Life Sci. 2013;92:589–97.
53. Ghanim H, Sia CL, Upadhyay M, Upadhyay M, Korzeniewski K, Viswanathan P, et al. Orange juice neutralizes the proinflammatory effect of a high-fat, high-carbohydrate meal and prevents endotoxin increase and Toll-like receptor expression. Am J Clin Nutr. 2010;91:940–9.
54. Burton-Freeman B, Talbot J, Park E, Krishnankutty S, Edirisinghe I. Protective activity of processed tomato products on postprandial oxidation and inflammation: a clinical trial in healthy weight men and women. Mol Nutr Food Res. 2012;56:622–31.
55. Blanco-Colio LM, Valderrama M, Alvarez-Sala LA, Bustos C, Ortego M, Hernández-Presa MA, et al. Red wine intake prevents nuclear factor-kappaB activation in peripheral blood mononuclear cells of healthy volunteers during postprandial lipemia. Circulation. 2000;102:1020–6.
56. Bellido C, López-Miranda J, Blanco-Colio LM, Pérez-Martínez P, Muriana FJ, Martín-Ventura JL, et al. Butter and walnuts, but not olive oil, elicit postprandial activation of nuclear transcription factor kappaB in peripheral blood mononuclear cells from healthy men. Am J Clin Nutr. 2004;80:1487–91.
57. Camargo A, Delgado-Lista J, Garcia-Rios A, Cruz-Teno C, Yubero-Serrano EM, Perez-Martinez P, et al. Expression of proinflammatory, proatherogenic genes is reduced by the Mediterranean diet in elderly people. Br J Nutr. 2012;108:500–8.
58. Li Z, Wong A, Henning SM, Zhang Y, Jones A, Zerlin A, et al. Hass avocado modulates postprandial vascular reactivity and postprandial inflammatory responses to a hamburger meal in healthy volunteers. Food Funct. 2013;4:384–91.
59. Bloomer RJ, Fisher-Wellman KH. Postprandial oxidative stress in exercise trained and sedentary cigarette smokers. Int J Environ Res Public Health. 2009;6:579–91.
60. Clegg M, McClean C, Davison WG, Murphy HM, Trinick T, Duly E, et al. Exercise and postprandial lipaemia: effects on peripheral vascular function, oxidative stress and gastrointestinal transit. Lipids Heal Dis. 2007;6:30.
61. Dixon NC, Hurst TL, Talbot DCS, Tyrrell RM, Thompson D. Active middle-aged men have lower fasting inflammatory markers but the postprandial inflammatory response is minimal and unaffected by physical activity status. J Appl Physiol. 2009;107:63–8.
62. Plat J, Jellema A, Ramakers J, Mensink RP. Weight loss, but not fish oil consumption, improves fasting and postprandial serum lipids, markers of endothelial function, and inflammatory signatures in moderately obese men. J Nutr. 2007;137:2635–40.
63. Corpeleijn E, Saris WHM, Jansen EHJM, Roekaerts PMHJ, Feskens EJM, Blaak EE. Postprandial interleukin-6 release from skeletal muscle in men with impaired glucose tolerance can be reduced by weight loss. J Clin Endocrinol Metab. 2005;90:5819–24.

64. Halkes CJ, van Dijk H, de Jaegere PP, Plokker HW, van Der Helm Y, Erkelens DW, et al. Postprandial increase of complement component 3 in normolipidemic patients with coronary artery disease: effects of expanded-dose simvastatin. Arterioscler Thromb Vasc Biol. 2001;21:1526–30.
65. Verseyden C, Meijssen S, van Dijk H, Jansen H, Castro Cabezas M. Effects of atorvastatin on fasting and postprandial complement component 3 response in familial combined hyperlipidemia. J Lipid Res. 2003;44:2100–8.
66. Kakuda H, Kobayashi J, Nakato M, Takekoshi N. Short-term effect of pitavastatin treatment on glucose and lipid metabolism and oxidative stress in fasting and postprandial state using a test meal in Japanese men. Cholesterol. 2013;2013:314170.
67. Rosenson RS, Huskin AL, Wolff DA, Helenowski IB, Rademaker AW. Fenofibrate reduces fasting and postprandial inflammatory responses among hypertriglyceridemia patients with the metabolic syndrome. Atherosclerosis. 2008;198:381–8.
68. Okopień B, Krysiak R, Herman ZS. Effects of short-term fenofibrate treatment on circulating markers of inflammation and hemostasis in patients with impaired glucose tolerance. J Clin Endocrinol Metab. 2006;91:1770–8.
69. Van Wijk J, Coll B, Castro Cabezas M, Koning E, Camps J, Mackness B, et al. Rosiglitazone modulates fasting and post-prandial paraoxonase 1 activity in type 2 diabetic patients. Clin Exp Pharmacol Physiol. 2006;33:1134–7.
70. Coll B, van Wijk JPH, Parra S, Castro Cabezas M, Hoepelman IM, Alonso-Villaverde C, et al. Effects of rosiglitazone and metformin on postprandial paraoxonase-1 and monocyte chemoattractant protein-1 in human immunodeficiency virus-infected patients with lipodystrophy. Eur J Pharmacol. 2006;544:104–10.
71. Ceriello A, Novials A, Ortega E, Canivell S, La Sala L, Pujadas G, et al. Glucagon-like peptide 1 reduces endothelial dysfunction, inflammation, and oxidative stress induced by both hyperglycemia and hypoglycemia in type 1 diabetes. Diabetes Care. 2013;36:2346–50.

Dynamic Interplay Between Metabolic Syndrome and Immunity

13

György Paragh, Ildikó Seres, Mariann Harangi, and Péter Fülöp

Abstract

Obesity and its co-morbidities as metabolic syndrome, type 2 diabetes mellitus and cardiovascular diseases are major health problems worldwide. Several reports indicated that nutrient excess and metabolic syndrome are linked with altered immune response. Indeed, metabolic syndrome is characterized by insulin resistance and chronic low-grade inflammation, which conditions are the consequences of the complex interaction between adipocytes and immune cells. Enlarged white adipose tissue is infiltrated by immune cells and secretes various bioactive substances, like adipokines, cytokines and other inflammatory mediators. Due to its special architecture in which metabolic and immune cells are in intimate proximity, metabolic and immunologic pathways are closely integrated in adipose tissue. With the contribution of altered gut microbiota, adipokines and cytokines modulate insulin signaling and immune response leading to adipose tissue inflammation and systemic insulin resistance. In this chapter, we focus on the cellular and molecular mechanisms that lead to impaired insulin sensitivity and chronic low-grade inflammation in obesity. We also detail the potential role of adipokines and immune cells in this deleterious process, and the concerns of vaccination in metabolic syndrome. Finally, we address the links between obesity and gut microbiota as an emerging new field of interest, and scratch the surface of potential therapeutic opportunities.

G. Paragh (✉) • I. Seres • M. Harangi • P. Fülöp
Division of Metabolism, Department of Internal Medicine, Faculty of Medicine,
University of Debrecen, Nagyerdei krt. 98,
Debrecen H-4032, Hungary
e-mail: paragh@belklinika.com;
seres@belklinika.com; mharangi@hotmail.com;
pfulop@belklinika.com

J. Camps (ed.), *Oxidative Stress and Inflammation in Non-communicable Diseases – Molecular Mechanisms and Perspectives in Therapeutics*, Advances in Experimental Medicine and Biology 824,
DOI 10.1007/978-3-319-07320-0_13,

Keywords
Adipokines • Gut microbiota • Immunity • Metabolic syndrome • Obesity • Obesity therapy • Vaccination

13.1 Introduction

Affecting about the one third of the population in the Western countries, metabolic syndrome (MetS) is a cluster of conditions that, when being present together, increase the risk of type 2 diabetes mellitus (T2DM) and cardiovascular diseases (CVDs) [1]. The concept of metabolic syndrome was created in 1988 [2]. Refined several times since then, recent diagnostic criteria of MetS include abdominal (central) obesity, increased triglyceride and decreased high-density lipoprotein-cholesterol levels, increased blood pressure and elevated fasting glucose levels [3]. Although no obligatory component is articulated and the presence of three findings out of five qualifies a person for metabolic syndrome, the existence of central obesity is agreed to be useful in preliminary screening for MetS.

Being the hallmark of MetS, obesity represents a major health burden worldwide with a high prevalence in the Western countries [4]. The enlarged white adipose tissue (WAT) secretes several biological substances termed adipo(cyto) kines with various metabolic and immunologic effects. Generated upon the complex interplay of immune and fat cells (adipocytes), altered adipokine secretion and pro-inflammatory stimuli make the bed for subsequent insulin resistance, which is the central etio-pathological feature of MetS. Insulin resistance is characterized by (1) the failure of insulin to stimulate glucose uptake into the skeletal muscle, (2) impaired inhibition of hepatic glucose output and (3) the failure of insulin to inhibit lipolysis in the adipocytes.

The progression of insulin resistance shares several molecular pathways with immunologic processes and obesity is characterized by a chronic low-grade inflammation with elevated C-reactive protein (CRP) levels [5, 6]. Several human studies indicated the ties between inflammation and metabolic alterations leading to the concept of "metabolic inflammation". In fact, regulation of metabolic and inflammatory pathways, or nutrient- and pathogen-sensing are highly conserved throughout various species, while metabolic and immune systems are fundamental for the organisms to survive [7]. Also, insulin-sensitive organs (especially WAT and liver) are structured in a special tissue architecture in which metabolic and immune cells are in close proximity with excellent blood supply. In this setup, the tight integration of metabolic and immune processes fosters the development of a special milieu in which the dysfunction of one system has a pronounced impact on the other. Indeed, acute inflammatory response is considered to be a major part of the defensive mechanisms; however, long-term inflammation, as it is present in obesity, has deleterious effects on metabolic functions. On the other hand, excess nutrient intake and chronic metabolic surplus trigger an inflammatory response by activating immune cells that are typically involved in classical defensive mechanisms against pathogens. Therefore, it is not surprising that obesity and metabolic syndrome portray a "metabolic inflammation" with abnormal secretion of adipokines, cytokines and other inflammatory mediators together with an increased infiltration of pro-inflammatory immune cells into the adipose tissue.

13.2 Clinical Studies

Early observations revealed the possible involvement of inflammation in insulin resistance, when high-dose sodium salicylate administration dramatically diminished glycosuria in diabetic patients [8]. This effect was further corroborated in the middle of the twentieth century, when acetyl salicylate treatment was successful to discontinue the insulin treatment in diabetic subjects [9]; however, the impact of salicylates on insulin sensitivity was not suspected at that time. Several latter epidemiological studies also confirmed the relationship between inflammation and insulin resistance, since increased levels of inflammatory markers were found to be significant predictors of T2DM [10, 11]. The role of chronic inflammation was also confirmed in the development of atherosclerosis, which is a major consequence of insulin-resistant conditions leading to increased cardiovascular morbidity and mortality [12]. Alternatively, lifestyle modification or drug treatment of patients with impaired glucose tolerance resulted in a significant improvement in CRP levels in a 1-year period [13].

White adipose tissue is no longer considered to be a mere energy depository tissue but rather a biologically active endocrine organ that secretes various substances, like adipokines, cytokines and chemokines. Therefore, WAT plays fundamental roles in the regulation of appetite and satiety, glucose and lipid metabolism, energy expenditure, insulin sensitivity, immune response, coagulation, blood pressure and endothelial function [14]. Indeed, altered secretion of various adipokines were observed in several insulin-resistant conditions including metabolic syndrome, type 2 diabetes mellitus and cardiovascular diseases [15, 16]. Also, altered adipokine profile is already present in obese non-diabetic patients and abnormal secretion of adipokines were found in subjects with increased fasting glucose (IFG) and impaired glucose tolerance (IGT), indicating the importance of excess weight in the early stages of insulin resistance [17, 18].

13.3 Molecular Mechanisms of Insulin Signaling and Resistance

To maintain normoglycemia, insulin is secreted by the β cells of islets of Langerhans that are located in the pancreas. It acts on the above mentioned insulin sensitive tissues and promotes glucose uptake, glycogen storage and lipogenesis. Insulin also inhibits glycogenolysis and gluconeogenesis and reduces lipolysis. Several mechanisms, including inflammation, excess nutrient availability, hypoxia, oxidative stress and endoplasmatic reticulum (ER) stress are linked to the decreased sensitivity of the insulin-targeted tissues. When the subsequent increase in insulin secretion can no longer compensate for the impaired sensitivity, hyperglycemia and manifest T2DM develop.

When insulin sensitivity is not impaired, binding of insulin to the α-subunit of its receptor results in the autophosphorylation of the β-subunit and induces the recruitment and binding of a substrate adaptor insulin receptor substrate-1 (IRS-1) [19]. Downstream tyrosine phosphorylation of the IRS-1 activates phosphatidyl-inositol 3-kinase (PI3K)-Akt/protein kinase B (PKB) pathway to induce several downstream signaling mechanisms including glucose transporter-4 (GLUT4) translocation to the cell membrane and promoting glucose uptake into the cell [20]. PI3K-PKB/Akt activation is also responsible for the other metabolic effects of insulin in (1) suppressing glycogen synthase kinase-3 (GSK3) and glycogen synthesis [21] and (2) forkhead box protein O1 (FOXO)-1/phosphoenolpyruvate carboxykinase (PEPCK)-mediated gluconeogenesis [22] together with (3) modulating mammalian target of rapamycin (mTOR) activity and protein synthesis [23]. Insulin also activates the Ras-mitogen-activated protein kinase (MAPK)/extracellular signal regulated kinase (ERK) pathway and regulates cell growth [24].

Alternative serine phosphorylation of IRS-1 proteins is a key step in the negative regulation of insulin signaling that causes insulin resistance. Ser312, Ser636, or Ser 1101 residues in human IRS-1 may be phosphorylated by several kinases including inhibitor of κB (IκB) kinase (IKK, an upstream activator of nuclear factor κB [NFκB]) or c-Jun N-terminal kinase (JNK) that are key components of major pro-inflammatory pathways [25]. Inhibition of insulin signaling by these pro-inflammatory kinases provides an important link between inflammation and insulin resistance, which was first revealed by *Hotamisligil et al.* clearly indicating the role of tumor necrosis factor-α (TNF-α) on impairing insulin action [26]. In their milestone manuscript, they found increased TNF-α mRNA and protein expression locally and systemically in obese rats, while neutralization of TNF-α resulted in a significant improvement in insulin sensitivity.

TNF-α is a potent inductor of IKK and JNK signaling and several studies have indicated the crucial role of various hallmarks of MetS including fatty acid and ceramide accumulation, ER stress, hypoxia, reactive oxygen species (ROS) and certain cytokines (e.g., interleukin- [IL]-6 or IL-1β) in the activation of these pathways in models of insulin resistance [27, 28]. Conversely, genetic or chemical inhibition of IKK and JNK signal transduction abolishes insulin resistance and improves insulin sensitivity [29]. Besides serine phosphorylation of IRS-1, IKK and JNK activation enhances systemic inflammation by increasing the expression of pro-inflammatory genes through NFκB and activator protein-1 (AP-1) in the target tissues, therefore closing the vitious cycle [30]. Target genes of NFκB include various chemokines and cytokines (e.g., TNF-α, interferon-γ [IFN-γ], IL-1β, IL-6, monocyte chemoattractant protein-1 [MCP-1]), receptors of these molecules, toll-like receptors (TLRs), selectins, adhesion molecules and CRP, cyclooxygenase-2 (COX2) or inducible nitric oxide synthase (iNOS) [29]. Induction of these molecules leads to the recruitment of monocytes and their differentiation to macrophages in insulin sensitive tissues, especially in the white adipose tissue (*see later*).

The importance of IKK and JNK pathway in obesity-related insulin resistance and inflammation was demonstrated by elegant investigations [31, 32]. Briefly, myeloid cell IKKβ- and JNK-null mice were generated by different techniques to disrupt macrophage inflammatory pathways in cells derived from the bone marrow, while the expression of IKKβ- and JNK remained intact in the insulin-sensitive tissues. The authors demonstrated that the knockout animals were protected from high-fat diet (HFD)-induced insulin resistance, although they developed the same degree of obesity as their wild-type littermates. These results demonstrated the key role of macrophages in systemic insulin resistance, since in the absence of the myeloid-driven inflammatory cell activation, obesity *per se* did not lead to insulin resistance as characterized by hyperinsulinemia and glucose intolerance.

In addition, pattern recognition receptors (PRRs) like TLRs and the receptor for advanced glycation end products (RAGE) are also able to activate IKK and JNK signaling. Ligands of TLRs include saturated fatty acids (SFAs), gut microbiome-derived lipopolysaccharide (LPS) and lipopeptides, while RAGE may also bind microbial products and advanced glycation end products (AGEs), thus providing another mechanistic link between obesity, insulin resistance and inflammation [28, 29, 33].

Inflammation is also closely related with atherosclerosis, which is a major contributor to CVDs that develop in patients with metabolic syndrome. Even early lipid deposition in the arterial wall is paralleled by the upregulation of P- and E-selectin, together with the increased expression of intercellular adhesion molecule-1 (ICAM-1) and vascular cell adhesion molecule-1 (VCAM-1) to adhere immune cells [34]. Endothelial and smooth muscle cells, lymphocytes and macrophages secrete TNF-α and increase the local production of IL-6 in the atheroma [35]. Increased expression of cluster of differentiation (CD)40 and its counterpart CD40L and subsequent matrix metalloproteinase (MMP) activation lead to increased vulnerability of the atherosclerotic plaque and may result in thrombosis in the vessel wall [36].

In addition to the IKK- and JNK-mediated insulin resistance, WAT-related adipokines such as TNF-α, IL-6 and IL-1β may block insulin signaling by inducing suppressors of cytokine signaling (SOCS)1 and SOCS3, thus causing subsequent ubiquitin-mediated degradation of IRS-1 and IRS-2 [37, 38]. Overexpression of SOCS1 and SOCS3 in mouse liver leads to insulin resistance and increased expression of sterol regulatory element-binding protein-1c (SREBP-1c), which is the major regulator of fatty acid synthesis. In turn, inhibition of SOCS1 and SOCS3 in obese diabetic mice restored insulin sensitivity, normalized the expression of SREBP-1c, while ameliorated hepatic steatosis and hypertriglyceridemia [39].

13.4 The Role of Adipokines in Insulin Resistance

As mentioned above, the discovery of TNF-α in promoting insulin resistance was a breakthrough in understanding the association between obesity

Table 13.1 The potential role of selected adipokines in insulin resistance and inflammation

Adipokine	Source	Role in insulin resistance and inflammation
Leptin	Adipocytes	Induces JAK/STAT signaling
		Enhances NFκB- and JNK-mediated gene transcription
		Stimulates TNF-α and IL-6 production
		Activates monocytes and macrophages
		Enhances pro-inflammatory Th1 and suppresses anti-inflammatory Th2 cytokines
Adiponectin	Adipocytes	Inhibits NFκB activation and TNF-α production
		Promotes the production of anti-inflammatory cytokines
		Decreases macrophage IFN-γ secretion
		Promotes anti-inflammatory M2 phenotypic of macrophages
		Inhibits oxLDL-induced endothelial cell proliferation
IL-6	Adipocytes	Induces JAK/STAT signaling
	Macrophages	Promotes SOCS-3 induction
	T cells	Increases energy expenditure when acting centrally
TNF-α	Adipocytes	Induces NFκB- and JNK-mediated gene transcription
	Macrophages	Increases hepatic lipogenesis
	T and B cells	
Resistin	Macrophages	Enhances IL-6 and TNF-α secretion
	Adipocytes (?)	Activates JNK, p38 MAPK and SOCS-3 signaling when acting centrally
RBP4	Adipocytes	Interferes with GLUT4 expression and IRS-1 phosphorylation
	Macrophages	
Visfatin	Monocytes	Induces pro-inflammatory cytokine production
	Macrophages	
	Dendritic cells	
	Adipocytes (?)	
Chemerin	Adipocytes	Regulates adipocyte maturation
	β cells	Interferes with IRS-1 phosphorylation
	Neutrophils	May act towards or against inflammation
Adipsin	Adipocytes	Alternative complement activation
	Monocytes	
	Macrophages	
Omentin-1	Stromal vascular cells	Exerts insulin-mimetic effects
		Attenuates NFκB activation

and inflammation. WAT, as an active endocrine organ and recently also considered as part of the innate immune system, secretes a variety of adipokines and cytokines, thus providing a potential interplay between metabolic alterations and immune response (Table 13.1). Adipose tissue enlargement (both hyperplasia and hypertrophy of the adipocytes) and subsequent microhypoxia are accompanied by immune cell infiltration and activation [40]. Adipocytes and immune cells share several metabolic and inflammatory functions including pathogen and nutrient sensing, complement activation, phagocytosis or secretion of inflammatory mediators [40, 41]. Besides pre-adipocytes and mature adipocytes, WAT contains macrophages, leukocytes, T and B lymphocytes, fibroblasts and endothelial cells. While adipocytes seem to be the exclusive source of leptin and adiponectin (as considered to be true adipokines), the other mediators are primarily expressed by the activated macrophages and/or other cells [29]. Once immune cells are recruited and activated in the WAT, their communication with the adipocytes perpetuates the metabolic-inflammatory vitious cycle leading to self-sustained cell recruitment and activation, together with enhanced secretion of pro-inflammatory mediators and abnormal adipokine secretion.

13.4.1 Leptin

Leptin is a pro-inflammatory cytokine-like product of the *ob* gene that regulates energy expenditure and food intake [42, 43]. Serum leptin concentration was shown to increase along with the amount of body fat [44], while leptin production of adipose tissue is elevated significantly by TNF-α, IL-1 or LPS in acute infections [45]. Leptin deficiency was demonstrated to cause monogenic obesity both in humans and rodents with severe obesity and insulin resistance that can be restored by exogenous administration of leptin [46, 47]. In contrast, when obesity develops first as a consequence of sedentary lifestyle and excess nutrient intake, it results in increased serum leptin concentrations suggesting the existence of leptin resistance in obese subjects [43]. Leptin resistance can be mediated by various mechanisms including hyperleptinemia-induced SOCS3 activation and ER stress [48, 49].

The structure of leptin is similar to pro-inflammatory cytokines like IL-2 or IL-6; and leptin exerts its biological effects by binding to its receptor ObR, which is a member of the class I cytokine receptor family [50]. Indeed, leptin induces several pro-inflammatory responses by activating Janus kinase 2 (JAK2)/signal transducer and activator of transcription 3 (STAT3), PI3K/Akt and MAPK pathways and initiates JNK- and NFκB-mediated inflammatory gene transcription [51, 52]. Leptin is known to stimulate the production of TNF-α and IL-6 by activating monocytes and macrophages [53] and increases the hepatic stellate cell-derived production of CCL2 and hypoxia-inducible factor-1α (HIF-1α) [54]. Leptin was reported to enhance pro-inflammatory T helper (Th)1 and suppress anti-inflammatory Th2 cytokines; and leptin administration reversed the immunosuppressive effects of acute starvation [55]. Also, leptin-induced nutrient sensor Akt-mTOR pathway was demonstrated to set the responsiveness of regulatory T (Treg) lymphocytes, since the anergic state of Treg cells depended on the increased activity of the mTOR pathway [56]. These data also indicate the impact of metabolic/energy status on immune tolerance and autoimmunity.

13.4.2 Adiponectin

Circulating in various forms in the plasma, adiponectin shares homologies with collagens, complement factors and TNF-α, and its expression is reduced in the WAT of obese humans and mice [57]. Plasma concentration of adiponectin was reported to be decreased in T2DM patients [58] and it shows an inverse relationship with the risk of myocardial infarction [59], while it correlates positively with antioxidant markers such as paraoxonase-1 (PON1), too [60]. Adiponectin enhances glucose uptake and fatty acid β-oxidation while decreasing gluconeogenesis in the skeletal muscle and liver, by activating adenosine monophosphate-activated protein kinase (AMPK) and the peroxisome proliferator-activated receptor-γ coactivator 1-α (PGC-1α), therefore it promotes insulin sensitivity and controls energy metabolism [61].

Adiponectin negatively regulates TNF-α production in macrophages by inhibiting NFκB activation [62] and *vice versa*, TNF-α and other pro-inflammatory cytokines such as IL-6 reduce adiponectin secretion in the adipocytes [63]. Besides immune mediators, weight loss has been shown to increase adiponectin levels in obese subjects [64].

Adiponectin promotes the production of anti-inflammatory cytokines including IL-10, or IL-1 receptor agonist (IL-1RA) in monocytes, macrophages and dendritic cells (DCs), while decreasing macrophage IFN-γ secretion [65]. Additionally, adiponectin was shown to promote the anti-inflammatory M2 phenotypic switch of macrophages and apoptotic cell clearance [66] and it might serve as a negative regulator of natural killer (NK) cell function through interfering with NFκB activation and subsequent death receptor expression [67]. Providing an additional link between adipose tissue-derived mediators and atherosclerosis, adiponectin inhibits oxidized low-density lipoprotein-(oxLDL)-induced proliferation of endothelial cells [68].

Contrary to the above mentioned, recent studies have questioned the beneficial role of adiponectin in atherosclerosis or inflammation. Circulating adiponectin levels were found to be increased in patients with rheumatoid arthritis or end-stage kidney failure [69–71]. Indeed, globular adiponectin activated NFκB and induced pro-inflammatory gene expression in vascular endothelial cells [72], indicating a rather harmful role of this adipokine in inflammation. This paradox might be explained by the presence of various forms of adiponectin in the circulation that may exert opposing effect on certain inflammatory mechanisms.

13.4.3 IL-6

IL-6 is a pro-inflammatory cytokine that, similarly to leptin, activates the JAK/STAT signaling pathway in various cells and regulates energy homeostasis and inflammatory response [73]. IL-6 is secreted by adipocytes and immune cells, including macrophages and T cells [74]. About 30 % of circulating IL-6 originates from the adipose tissue in humans and IL-6 levels show positive correlations with the amount of body fat and insulin resistance [75, 76]. IL-6 administration inhibited insulin receptor autophosphorylation, IRS-1 tyrosine phosphorylation, and subsequent PI3K/Akt activation in HepG2 cells via IL-6-mediated SOCS3 induction [77]. These data indicate that SOCS3 induction in the liver may serve as an important mechanism of IL-6-mediated insulin resistance.

However, the role of IL-6 in obesity-related insulin resistance is still controversial, possibly due to its tissue-specific action. IL-6 knockout mice were reported to develop mature-onset obesity, which was partly reversible by IL-6 replacement. Also, central (intracerebroventricular) IL-6 treatment increased energy expenditure and reduced obesity, suggesting that centrally acting IL-6 exerts anti-obesity effects [78]. The complex impact of IL-6 on obesity and hepatic insulin resistance was also demonstrated in recent investigations indicating the role of IL-6 deficiency in HFD-induced hepatic insulin resistance and inflammation, due to the defects in mitochondrial metabolism including oxidative phosphorylation, electron transport chain and tricarboxylic acid cycle [79].

13.4.4 TNF-α

Providing the first evidence of the interplay between obesity and inflammation, expression of TNF-α mRNA was found to be increased in the adipose tissue in various rodent models of obesity and diabetes [26]. The authors also demonstrated that neutralization of TNF-α in obese, leptin receptor-deficient *fa/fa* rats enhanced insulin secretion, indicating the role of TNF-α in insulin resistance that accompanies obesity. Indeed, TNF-α levels were elevated in obese people and showed strong positive correlations with waist-to-hip ratios (WHRs) and insulin levels [80].

Accumulation of free fatty acids (FFAs) in the liver leads to steatosis and non-alcoholic fatty liver disease (NAFLD), which is part of the metabolic syndrome and confers a vulnerable condition to exogenous *noxas* with increased risk of insulin resistance and CVDs [81, 82]. Hepatic FFA accumulation leads to dramatic lipotoxicity and lysosomal destabilization with releasing cysteine protease cathepsin B and inducing NFκB-mediated TNF-α generation [83]. Cathepsin B and TNF-α may perpetuate the vitious cycle by inducing hepatic lipogenesis and decreasing adiponectin secretion, therefore aggravating steatosis and inflammation [84, 85]. TNF-α triggers insulin resistance by inducing IKKβ/NFκB-mediated gene transcription and by activating JNK and Ser-307 phosphorylation of IRS1 [86, 87]. Although TNF-α blockade seems to be a promising perspective to improve insulin sensitivity, the clinical outcomes of TNF-α neutralization in humans are still controversial. Indeed, administration of TNF-α-neutralizing antibody to T2DM patients resulted in unaltered insulin sensitivity and glycemic control while suppressing inflammation [88].

13.4.5 Resistin

Resistin circulates in blood as a high-molecular-weight hexamer and a low-molecular-weight trimer. The hexamer form is more abundant although less bioactive compared to the trimer form [89]. Contrary to previous findings localizing resistin expression predominantly into adipocytes, recent evidence indicate that macrophages are the major source of this adipokine which is under the negative control of peroxisome proliferator-activated receptor (PPAR)γ activators such as rosiglitazone [90]. Early studies demonstrated that circulating resistin levels are elevated in obesity and anti-resistin antibody administration improved insulin sensitivity in mice with diet-induced obesity [91]. However, human data are controversial showing variable levels of resistin in obese subjects with inconsistent findings of its association with insulin sensitivity and cardiovascular morbidity [92]. *In vitro* data suggest that resistin mRNA expression is increased by IL-1, IL-6, TNF-α and LPS in peripheral blood mononuclear cells [93], while resistin was shown to upregulate IL-6 and TNF-α secretion that could be mitigated by inhibiting NFκB signaling [94]. In addition, resistin binds to hypothalamic TLR4, leading to systemic insulin resistance through activating JNK, p38 MAPK and SOCS3 signaling [95].

13.4.6 Retinol Binding Protein-4

Expressed in adipose tissue, liver and macrophages, serum RBP4 concentrations were shown to be increased in insulin resistant mice and humans with obesity and T2DM [96]. Expressions of RBP4 and GLUT4 were also found to correlate inversely in adipocytes, whereas RBP4 interferes with insulin-stimulated phosphorylation of IRS-1 at Ser307 residue in human fat cells [97]. Human data also indicated the correlation of serum RBP4 levels with the magnitude of insulin resistance in subjects with obesity, IGT or T2DM; and serum RBP4 levels were associated with components of MetS, including increased body mass index (BMI), WHR, serum triglyceride levels, systolic blood pressure and decreased high-density lipoprotein cholesterol levels [98]. However, others questioned the role of RBP4 in insulin resistance [99], therefore further investigations are of interest to clarify the association of RBP4 with metabolic consequences.

13.4.7 Visfatin

The role of visfatin (known also as pre B cell colony-enhancing factor-1 [PBEF1]) in insulin signaling is still a matter of debate, also. Its plasma levels were measured to be increased in T2DM patients compared to non-diabetic subjects [100]; and visfatin was shown to induce MCP-1 in human adipocytes [101]. Its pro-inflammatory properties were also demonstrated on $CD14^+$ monocytes by enhancing the production of IL-1β, TNF-α and especially IL-6, together with an enhanced surface expression of costimulatory molecules CD54, CD40, and CD80 [102]. In contrast, central visfatin administration was recently reported to improve glycemic control by increasing both insulin secretion and sensitivity and β cell mass in type 2 diabetic rats [103].

13.4.8 Chemerin

Chemerin and its receptor (chemokine-like receptor 1, CMKLR1 or ChemR23) are expressed at high levels in WAT [104]. Chemerin levels are increased even in obese non-diabetic individuals and in subjects with prediabetes [17, 18]. Recent data indicate that chemerin impairs glucose uptake and promotes insulin resistance by inducing serine phosphorylation of IRS-1 and interfering with downstream insulin signaling [105]. The exact effect of chemerin on inflammation has yet to be elucidated since it might serve as a pro- and anti-inflammatory mediator, too [106]. In fact, chemerin levels were found to correlate with CRP, IL-6 and TNF-α concentrations together with the components of metabolic syndrome in patients with stable chest pain [107]. Chemerin also serves as a chemoattractant for macrophages and DCs [108] besides regulating adipogenesis and adipocyte metabolism [109]. In contrast, chemerin was able to inhibit macrophage activation, neutrophil and monocyte recruitment in zymosan-induced peritonitis [110]. A similar anti-inflammatory property of chemerin was demonstrated in a mouse model of LPS-induced acute lung inflammation [111]. These data indicate that the various isoforms of chemerin may play distinct roles in the stages of inflammation.

13.4.9 Adipsin

Derived from adipocytes and monocytes/macrophages, adipsin (complement factor D46) is a key enzyme in the alternative complement pathway [112]. It also promotes triglyceride deposition and inhibits lipolysis in adipocytes [113]. Serum levels of adipsin are found to be variable in obese humans being either unchanged or increased [114].

13.4.10 Omentin-1

Omentin-1 levels have been reported to be decreased and inversely correlated with BMI and the degree of insulin resistance [115]. Omentin-1 concentrations also show inverse association with the presence and angiographic severity of coronary artery disease in patients with MetS [116]. This novel adipokine originates predominantly from the stromal vascular cells of the omental (visceral) adipose tissue and is barely detectable in subcutaneous fat. Also, omentin-1 enhances insulin-stimulated glucose uptake in both subcutaneous and omental adipocytes and increases Akt phosphorylation [117]. Also, omentin-1 levels correlated negatively with CRP levels while attenuating NFκB activation in women with polycystic ovary syndrome (PCOS) [118].

13.5 Obesity and Innate Immunity

Adipose tissue inflammation results as a consequence of complex interplay between adipocytes and immune cells (Fig. 13.1). A decade ago, a significant interest was generated by a couple of articles demonstrating the key role of macrophage infiltration into the enlarged adipose tissue, leading to increased expression of inflammatory markers such as TNF-α and IL-6 [74, 119]. Adipose tissue macrophages (ATMs) may constitute up 40 % of the WAT cell population and their role in insulin resistance has been confirmed since then [31, 32]. As we discussed above, disrupting macrophage inflammatory pathways in mice caused a relative protection from HFD-induced insulin resistance. In line with increasing body weight, ATMs accumulate in WAT and their content correlates positively with insulin resistance [119, 120]. Adipose tissue macrophages show phenotypic heterogeneity with two distinct activation/polarization states: M1 macrophages are classified as "classically activated" pro-inflammatory cells, while M2 macrophages are defined as "alternatively activated" anti-inflammatory cells [121]. M1 macrophages are activated by pro-inflammatory mediators and show enhanced inflammatory gene expression with increased reactivity to IFN-γ and LPS/fatty acids through their TLRs; while M2 macrophages secrete anti-inflammatory substances and have low reactivity to LPS and fatty acids [121]. Also, insulin sensitizing (i.e. thiazolidinedione, TZD) drug treatment promotes M2 polarization state by activating PPARγ-mediated transcription [122]. M2 cells predominate WAT in lean state and weight gain leads to a pro-inflammatory environment in WAT by inducing M1 macrophage accumulation with enhanced expression of TNF-α, IL-1, IL-6, MCP-1 and iNOS [123]. In conclusion, both recruitment and pro-inflammatory M1 phenotypic switch are required for the development of obesity-induced insulin resistance.

Despite the expanding literature data, the exact mechanism of macrophage infiltration to WAT has yet to be elucidated. However, several mechanisms

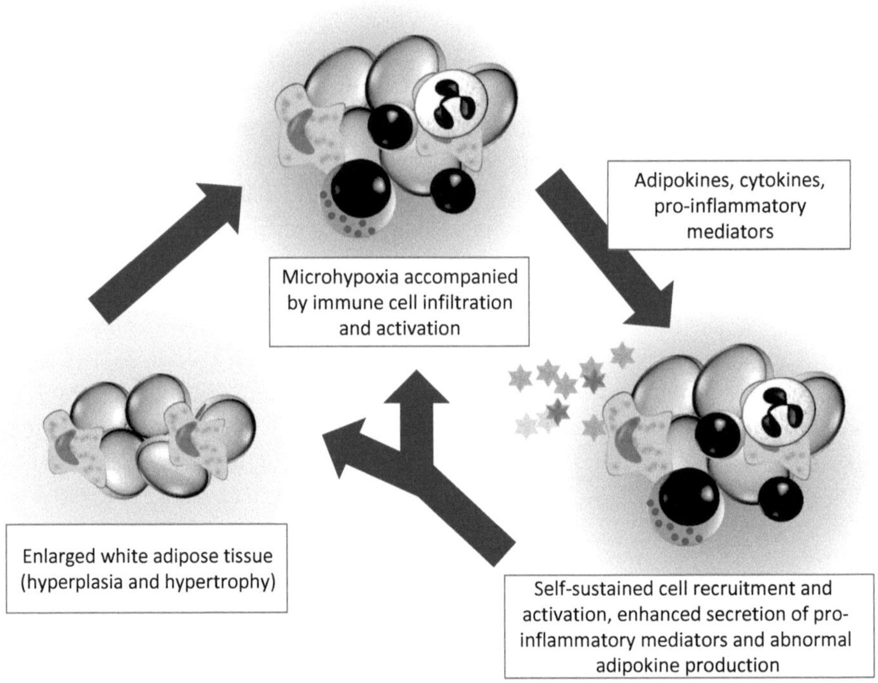

Fig. 13.1 Interaction between adipocytes and immune cells. Enlarged white adipose tissue secretes various adipokines and inflammatory mediators. Lipid-laden adipocytes become distant from the vasculature resulting in adipose tissue microhypoxia and cell death. Adipose tissue macrophages undergo a pro-inflammatory M1 phenotypic switch and scavenge cellular debris. Additionally, other types of immune cells including neutrophils, B and T lymphocytes infiltrate the adipose tissue and become activated due to their interaction with each other and as a consequence of abnormal adipokine secretion, thus resulting in a self-sustained cell recruitment and pro-inflammatory activation while aggravating altered adipokine secretion

including altered adipokine, chemokine and cytokine secretion (*see above*), microhypoxia and adipocyte death are presumed to participate in this process. Adipose tissue microhypoxia may develop when enlarged, lipid-laden adipocytes become distant from the vasculature, leading to decreased oxygen pressure and subsequent cell death and inflammatory response [124]. Indeed, HIF-1α, the major regulator of oxygen homeostasis and a known immune mediator for immune cells was found to be increased in the WAT of obese subjects [120]. It was also demonstrated that more than 90 % of ATMs reside around dead adipocytes forming crown-like structures (CLSs) and scavenging residual lipid droplets [125].

Besides the monocyte/macrophage system, other cells of the innate immunity also participate in obesity-induced WAT inflammation. Neutrophil granulocytes are one of the first cells that respond to inflammation and may exacerbate the chronic inflammation by recruiting macrophages and by interacting with antigen-presenting cells. In fact, treatment of hepatocytes with neutrophil elastase was demonstrated to result in cellular insulin resistance, and deletion of neutrophil elastase in obese mice led to a less extended tissue inflammation together with improved glucose tolerance and increased insulin sensitivity [126].

The role of mast cells and eosinophils in obesity-related inflammation is less detailed. Mast cells were found to be activated in human adipose tissue and localized preferentially in fibrosis depots. Also, mast cell number correlated with WAT fibrosis, macrophage accumulation and endothelial cell inflammation [127]. Eosinophils were reported to be the main source of IL-4 in mouse WAT and were responsible for the maintenance of M2 phenotypic state, thus indicating the protective role of eosinophils against metabolic syndrome [128]. Additionally, the number of invariant natural killer T (iNKT) cells was shown to be reduced in expanded WAT, correlating with pro-inflammatory macrophage infiltration, while iNKT cell numbers were restored after weight loss both in mice and humans. Also, iNKT cell-depleted mice displayed enhanced weight gain, steatosis and insulin resistance on high-fat diet [129].

13.6 Obesity and Adaptive Immunity

The role of adaptive immune cells in obesity-induced inflammation is less detailed; however, recent investigations indicate the significance of adaptive immunity in adipose tissue inflammation. Indeed, infiltration of adaptive immune cells is thought to precede the accumulation of macrophages in WAT [130] and adipocytes are able to activate $CD4^+$ cells independently from macrophages [131]. Depletion of $CD8^+$ lymphocytes lowered macrophage infiltration and adipose tissue inflammation and ameliorated systemic insulin resistance in obese mice on HFD, while adoptive transfer of $CD8^+$ T cells to CD8-null mice aggravated adipose tissue inflammation [130].

$CD4^+$ cells include pro-inflammatory Th1 and Th17 lymphocytes and anti-inflammatory Th2 and Treg cells. Compared to lean littermates, a higher $CD8^+$:$CD4^+$ T cell ratio was found in mice with HFD-induced obesity, whereas a dramatic increase in the number of Th1-polarized cells was detected together with a significant decrease in the Th2-polarized fraction and $Foxp3^+$ $CD4^+$ Treg cell number [132–134]. Additionally, as we discussed above, leptin enhances Th1-mediated IFN-γ cytokine secretion and suppresses Th2 cytokine profile (IL-4, IL-10, IL-13) and Treg proliferation [55, 135].

B lymphocytes were shown to accumulate in adipose tissue before T cells, shortly after the initiation of a high-fat diet [136]. B cells secrete pathogenic immunoglobulin G (IgG) and promote pro-inflammatory T cell activation, thus inducing M1 macrophage polarization and development of insulin resistance [137]. Also, impaired TLR function, increased pro-inflammatory IL-8 and lack of anti-inflammatory/protective IL-10 production were reported in the B cells of diabetic patients [138].

In conclusion, obesity appears to promote the accumulation of B cells and predisposes to a Th1-Th17 switch and $CD8^+$ T cell recruitment with a reduced Treg cell compartment in adipose tissue, therefore leading to a pro-inflammatory microenvironment and increased recruitment of macrophages that is accompanied by an abnormal adipokine secretion of the adipocytes.

13.7 The Vaccination Theory: A Potential Link with Metabolic Syndrome?

The incidence of life-threatening infectious diseases has been decreasing in the developed countries due to the wide use of vaccines with obvious benefits for both the individual and the society. The levels of inflammatory markers increase after vaccination due to the immune response that is necessary for the development of immunity. Besides stimulating the immune system and increasing cytokine secretion, vaccines also increase cortisol production [139, 140]. Influenza and pneumococcal vaccine are known to increase CRP levels [141], while diphteria-tetanus-polio-typhim and influenza vaccination boost the production of IL-6 [142, 143].

These data suggest that vaccinations might modulate the metabolic alterations that characterize MetS; and *vice versa*, metabolic changes may also have an impact on the degree and the extent of inflammation triggered by vaccinations. Activation of the immune system with vaccinations may cause a hypothalamic reprogramming leading to increased cortisol response. Indeed, metabolic syndrome shares several similarities with mild Cushing syndrome [140] and intrauterine programming of the hypothalamic-pituitary-adrenal axis was proposed to be a potential mechanism underlying the association between low birth weight and insulin resistance in the adulthood [144]. Noteworthy, the incidence of the adolescent MetS has also been increased [145] concomitantly with increasing the rate and dose of vaccinations; however, there are no large epidemiological studies confirming this association. These data raise the potential link between vaccination and MetS, although do not provide frank evidence about possible unfavorable metabolic effects of vaccinations and do not question the benefits of modern disease control. Further studies are highly needed to clarify this interesting question.

13.8 Gut Microbiota and Obesity

The human intestine contains a vast and complex microbial community named as gut microflora or microbiota [146]. Gut microbiota is estimated to consist of at least 10^{14} bacteria and archaea with approximately 1,100 species, with an estimated 160 such species per individual [147]. Gut microbiota harbors four predominant phyla: the Gram-positive *Actinobacteria* and *Firmicutes* and the Gram-negative *Bacteroidetes* and *Proteobacteria*. Leptin-deficient *ob/ob* mice were shown to have a significant reduction in the abundance of *Bacteroidetes* with a proportional increase in *Firmicutes* [148]. Consistent with these data, a similar increase in the ratio of *Firmicutes/Bacteroidetes* was found in obese humans [149], however, latter studies have questioned these alterations [150]. Overall, it is presumed that "obese microbiota" interacts with the host epithelial cells indirectly controlling energy expenditure and storage to harvest more energy from the diet [151, 152].

Diets rich in fat affects epithelial integrity of the intestine leading to increased permeability and absorption of gut-derived peptidoglycans (PGs), LPS and antigens associated with chylomicrons [147], as it is depicted on Fig. 13.2. Nucleotide-binding oligomerization domain-containing protein (NOD)-like receptors (NLRs) and TLRs are members of pattern recognition receptors that play major roles in mediating inflammation and immune mechanisms in adipocytes. PRRs are expressed by adipocytes and cells of the innate immune system recognizing bacterial carbohydrates like LPS, nucleic acids or peptidoglycans. Increased bacterial permeability leads to inflammatory cell recruitment and activation in adipose tissue and may induce systemic inflammation via PRR ligands. Also, NOD1 activation has recently been reported to modulate adipocyte differentiation while inducing NFκB activation, pro-inflammatory gene expression and leading to alternative serine phosphorylation of IRS-1, thereby causing insulin resistance [153, 154].

The importance of the toll-like receptors linking gut microbiota with metabolic syndrome has also been demonstrated. TLR5-deficient mice developed hyperphagia and MetS characterized by hyperlipidemia, hypertension, insulin resistance, and increased adiposity [155]. These metabolic alterations correlated with the changes in the composition of gut microbiota, and transfer of gut microbiota from TLR5-deficient mice to wild-type germ-free mice conferred several features of MetS to the recipients. In addition, mRNA expressions of PAI-1, IL-1 and TNF-α were increased in the visceral adipose tissue of mice with HFD-induced obesity and this pro-inflammatory induction was abolished by antibiotic treatment (ampicillin and neomycin) [156]. Also, prebiotic carbohydrates and antibiotics were effective in reducing systemic endotoxin concentration and hepatic inflammatory cytokine expression [156, 157]. Notably, widespread use of antibiotics in young children are supposed to contribute to the obesity epidemic by causing alterations in the composition of their intestinal microbiota [158].

13.9 Therapeutic Options

The beneficial effect of salicylates on diabetic complications dates back to the nineteenth century, however, the molecular target has just recently been identified to be the IKKβ/NFκB axis [159–161]. Aspirin treatment seems to be a promising option by decreasing metabolic and immunologic complications, although its use is

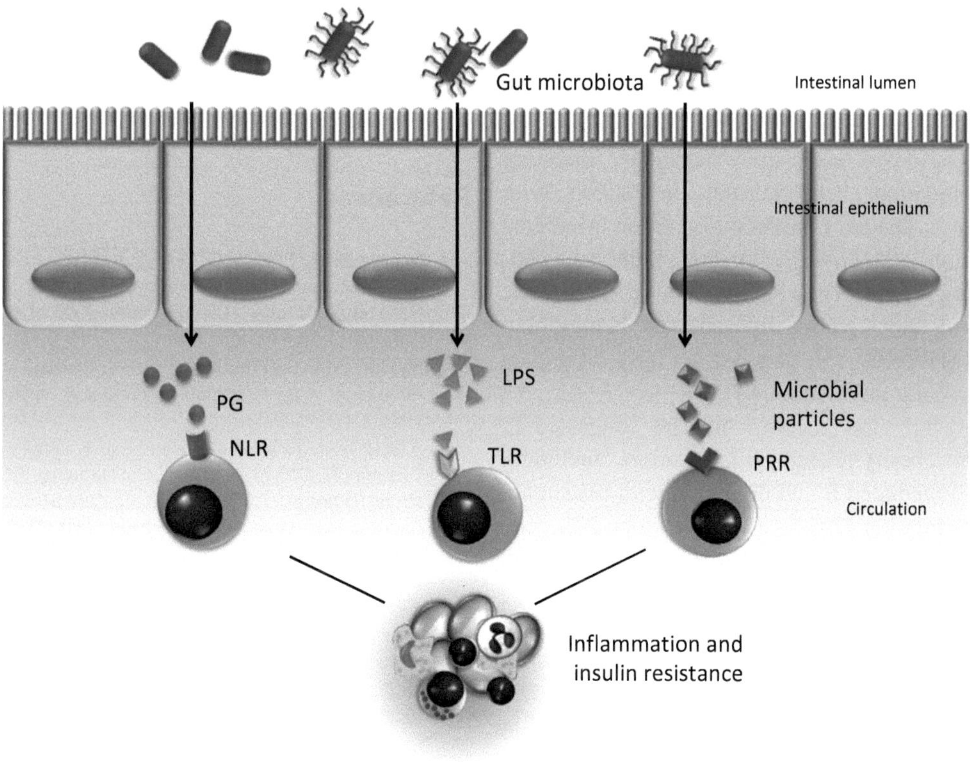

Fig. 13.2 The role of gut microbiota in obesity. Damaged epithelial barrier leads to increased intestinal permeability and absorption of peptidoglycans, lipopolysaccharide and microbial particles. Originating from gut microbiota, these molecules are ligands of various pattern recognition receptors including NLRs and TLRs that are located on adipocytes and immune cells. Subsequent activation of these cells results in a local and systemic inflammation and insulin resistance. *Abbreviations*: *LPS* lipopolysaccharide, *NLR* NOD-like receptor, *PG* peptidoglycan, *PRR* pattern recognition receptor, *TLR* toll-like receptor

limited due its various side effects. Investigations on TNF-α blockade have also been inconclusive [162]. Therefore, only with a mastery of the above mentioned molecular and cellular interplay would make us able to develop specific and individually tailored immunologic treatments against inflammation and insulin resistance without severe complications.

In fact, several drugs with established anti-inflammatory effects are already in clinical use. Thiazolidinediones are administered to type 2 diabetics; however, their use have been questioned due to unexpected side effects [163]. TZDs are potent PPARγ-activators with pronounced insulin sensitizing effects [164]. PPARγ is known to be expressed not only in adipocytes, but also in hepatocytes or endothelial cells, and it is required for the infiltration of alternatively activated (M2) macrophages into the adipose tissue [165]. The anti-inflammatory action of TZDs are thought to be mediated via the trans-repression of NFκB causing decreased expression of the target genes; a mechanism that is possibly carried out by small ubiquitin-like modifier (SUMO) proteins and SUMOylation of the nuclear PPARγ [166]. PPARγ-independent nuclear translocation of glucocorticoid receptor has also been proposed to be, at least partly, responsible for the anti-inflammatory properties of TZDs [167].

Statin are widely used lipid lowering drugs that inhibit 3-hydroxy-3-methyl-glutaryl coenzyme A (HMGCoA) reductase, which is the rate-limiting enzyme of cholesterol biosynthesis. Statins are reported to be effective in decreasing CRP levels and downregulating the activities of NFκB, AP-1 and HIF-1α [168]. Concerns have recently been arisen regarding statin-induced diabetes mellitus; however, benefits of lipid lowering therapy seem to outweigh potential diabetic consequences [169].

Together with acetyl salicylate (aspirin), non-acetylated salicylates (sodium salicylate, salsalate) also inhibit NFκB action [159], although they do not prolong bleeding time. Despite the expectations, a recent study named as "Targeting Inflammation Using Salsalate in Patients With Type 2 Diabetes: Effects on Flow-Mediated Dilation" (TINSAL-FMD) has resulted in conflicting results. While being effective in glycemic control and significantly reducing hemoglobin A1C (HbA1C), salsalate did not have an impact on vascular inflammation [170].

Several other mechanisms are suggested to target obesity-related inflammation, including CD20-mediated B cell depletion with rituximab, or modulation of transcription factors, NK cells or interleukins; however, to date, there are no convincing data about the effectiveness and safety of these methods. The safest and most effective of all is still the combination of exercise and gradual weight loss. Indeed, such lifestyle intervention was reported to reduce inflammation and macrophage infiltration in obese subjects and improved insulin sensitivity [171].

13.10 Concluding Remarks

Obesity and its connotations as metabolic syndrome, type 2 diabetes mellitus and cardiovascular diseases share a special metabolic-immunologic milieu that is built upon the complex interplay of adipocytes and immune cells. These cells and their secretory products mutually target each other resulting in insulin resistance and chronic low-grade inflammation. The vast majority of the known adipokines impair insulin signaling and interfere with inflammatory pathways. Immune cells infiltrate adipose tissue in a chronological sequence and enhance the pro-inflammatory microenvironment together with gut-derived signals. Vaccination has recently been suggested to alter metabolic pathways and may lead to MetS; however, further studies are needed to clarify this issue. In this regard, a deeper knowledge of these complex regulatory mechanisms would provide novel therapeutic approaches to control obesity-related diseases and metabolic syndrome.

Acknowledgments This work was supported by the TÁMOP-4.2.2.A-11/1/KONV-2012-0031 project. The TÁMOP project is co-financed by the European Union and the European Social Fund.

References

1. Mozumdar A, Liguori G. Persistent increase of prevalence of metabolic syndrome among U.S. adults: NHANES III to NHANES 1999–2006. Diabetes Care. 2011;34:216–9.
2. Reaven GM. Banting lecture 1988. Role of insulin resistance in human disease. Diabetes. 1988;37:1595–607.
3. Alberti KG, Eckel RH, Grundy SM, Zimmet PZ, Cleeman JI, Donato KA, et al. Harmonizing the metabolic syndrome: a joint interim statement of the International Diabetes Federation Task Force on Epidemiology and Prevention; National Heart, Lung, and Blood Institute; American Heart Association; World Heart Federation; International Atherosclerosis Society; and International Association for the Study of Obesity. Circulation. 2009;120:1640–5.
4. Flegal KM, Carroll MD, Kit BK, Ogden CL. Prevalence of obesity and trends in the distribution of body mass index among US adults, 1999–2010. JAMA. 2012;307:491–7.
5. Hak AE, Stehouwer CD, Bots ML, Polderman KH, Schalkwijk CG, Westendorp IC, et al. Associations of C-reactive protein with measures of obesity, insulin resistance, and subclinical atherosclerosis in healthy, middle-aged women. Arterioscler Thromb Vasc Biol. 1999;19:1986–91.
6. Visser M, Bouter LM, McQuillan GM, Wener MH, Harris TB. Elevated C-reactive protein levels in overweight and obese adults. JAMA. 1999;282:2131–5.
7. Hotamisligil GS. Inflammation and metabolic disorders. Nature. 2006;444:860–7.
8. Ebstein W. Zur therapie des Diabetes mellitus, insbesondere über die Anwendung des salicylsauren Natron bei demselben. Berliner Klinische Wochenschrift. 1876;13:337–40.
9. Reid J, Macdougall AI, Andrews MM. On the efficacy of salicylate in treating diabetes mellitus. Br Med J. 1957;2:1071–4.
10. Spranger J, Kroke A, Möhlig M, Hoffmann K, Bergmann MM, Ristow M, et al. Inflammatory cytokines and the risk to develop type 2 diabetes: results of the prospective population-based European Prospective Investigation into Cancer and Nutrition (EPIC)-Potsdam Study. Diabetes. 2003;52:812–7.
11. Freeman DJ, Norrie J, Caslake MJ, Gaw A, Ford I, Lowe GD, et al. C-reactive protein is an independent predictor of risk for the development of diabetes in the West of Scotland Coronary Prevention Study. Diabetes. 2002;51:1596–600.

12. Wilson PW. Evidence of systemic inflammation and estimation of coronary artery disease risk: a population perspective. Am J Med. 2008;121 Suppl 10:S15–20.
13. Haffner S, Temprosa M, Crandall J, Fowler S, Goldberg R, Horton E, et al. Intensive lifestyle intervention or metformin on inflammation and coagulation in participants with impaired glucose tolerance. Diabetes. 2005;54:1566–72.
14. Hajer GR, van Haeften TW, Visseren FL. Adipose tissue dysfunction in obesity, diabetes, and vascular diseases. Eur Heart J. 2008;29:2959–71.
15. Eckel RH, Grundy SM, Zimmet PZ. The metabolic syndrome. Lancet. 2005;365:1415–28.
16. Blüher M. Clinical relevance of adipokines. Diabetes Metab J. 2012;36:317–27.
17. Lőrincz H, Katkó M, Harangi M, Somodi S, Gaál K, Fülöp P, et al. Strong correlations between circulating chemerin levels and lipoprotein subfractions in nondiabetic obese and nonobese subjects. Clin Endocrinol (Oxf). 2013. doi:10.1111/cen.12363.
18. Tönjes A, Fasshauer M, Kratzsch J, Stumvoll M, Blüher M. Adipokine pattern in subjects with impaired fasting glucose and impaired glucose tolerance in comparison to normal glucose tolerance and diabetes. PLoS One. 2010. doi:10.1371/journal.pone.0013911.
19. Sun XJ, Rothenberg P, Kahn CR, Backer JM, Araki E, Wilden PA, et al. Structure of the insulin receptor substrate IRS-1 defines a unique signal transduction protein. Nature. 1991;352:73–7.
20. Taniguchi CM, Emanuelli B, Kahn CR. Critical nodes in signalling pathways: insights into insulin action. Nat Rev Mol Cell Biol. 2006;7:85–96.
21. Cross DA, Alessi DR, Cohen P, Andjelkovich M, Hemmings BA. Inhibition of glycogen synthase kinase-3 by insulin mediated by protein kinase B. Nature. 1995;378:785–9.
22. Nakae J, Kitamura T, Silver DL, Accili D. The forkhead transcription factor Foxo1 (Fkhr) confers insulin sensitivity onto glucose-6-phosphatase expression. J Clin Invest. 2001;108:1359–67.
23. Harrington LS, Findlay GM, Gray A, Tolkacheva T, Wigfield S, Rebholz H, et al. The TSC1-2 tumor suppressor controls insulin-PI3K signaling via regulation of IRS proteins. J Cell Biol. 2004;166:213–23.
24. Boura-Halfon S, Zick Y. Phosphorylation of IRS proteins, insulin action, and insulin resistance. Am J Physiol Endocrinol Metab. 2009;296:E581–91.
25. Gual P, Le Marchand-Brustel Y, Tanti JF. Positive and negative regulation of insulin signaling through IRS-1 phosphorylation. Biochimie. 2005;87:99–109.
26. Hotamisligil GS, Shargill NS, Spiegelman BM. Adipose expression of tumor necrosis factor-alpha: direct role in obesity-linked insulin resistance. Science. 1993;259:87–91.
27. Schenk S, Saberi M, Olefsky JM. Insulin sensitivity: modulation by nutrients and inflammation. J Clin Invest. 2008;118:2992–3002. doi:10.1172/JCI34260.
28. Kwon H, Pessin JE. Adipokines mediate inflammation and insulin resistance. Front Endocrinol (Lausanne). 2013;4:71. doi:10.3389/fendo.2013.00071.
29. Shoelson SE, Lee J, Goldfine AB. Inflammation and insulin resistance. J Clin Invest. 2006;116:1793–801.
30. Samuel VT, Shulman GI. Mechanisms for insulin resistance: common threads and missing links. Cell. 2012;148:852–71.
31. Arkan MC, Hevener AL, Greten FR, Maeda S, Li ZW, Long JM, et al. IKK-beta links inflammation to obesity-induced insulin resistance. Nat Med. 2005;11:191–8.
32. Solinas G, Vilcu C, Neels JG, Bandyopadhyay GK, Luo JL, Naugler W, et al. JNK1 in hematopoietically derived cells contributes to diet-induced inflammation and insulin resistance without affecting obesity. Cell Metab. 2007;6:386–97.
33. Akira S, Uematsu S, Takeuchi O. Pathogen recognition and innate immunity. Cell. 2006;124:783–801.
34. Adams DH, Shaw S. Leucocyte-endothelial interactions and regulation of leucocyte migration. Lancet. 1994;343:831–6.
35. Seino Y, Ikeda U, Ikeda M, Yamamoto K, Misawa Y, Hasegawa T, et al. Interleukin 6 gene transcripts are expressed in human atherosclerotic lesions. Cytokine. 1994;6:87–91.
36. Schönbeck U, Libby P. CD40 signaling and plaque instability. Circ Res. 2001;89:1092–103.
37. Emanuelli B, Peraldi P, Filloux C, Chavey C, Freidinger K, Hilton DJ, et al. SOCS-3 inhibits insulin signaling and is up-regulated in response to tumor necrosis factor-alpha in the adipose tissue of obese mice. J Biol Chem. 2001;276:47944–9.
38. Rui L, Yuan M, Frantz D, Shoelson S, White MF. SOCS-1 and SOCS-3 block insulin signaling by ubiquitin-mediated degradation of IRS1 and IRS2. J Biol Chem. 2002;277:42394–8.
39. Ueki K, Kondo T, Tseng YH, Kahn CR. Central role of suppressors of cytokine signaling proteins in hepatic steatosis, insulin resistance, and the metabolic syndrome in the mouse. Proc Natl Acad Sci U S A. 2004;101:10422–7.
40. Sell H, Habich C, Eckel J. Adaptive immunity in obesity and insulin resistance. Nat Rev Endocrinol. 2012;8:709–16. doi:10.1038/nrendo.2012.114.
41. Cousin B, Munoz O, Andre M, Fontanilles AM, Dani C, Cousin JL, et al. A role for preadipocytes as macrophage-like cells. FASEB J. 1999;13:305–12.
42. Zhang Y, Proenca R, Maffei M, Barone M, Leopold L, Friedman JM. Positional cloning of the mouse obese gene and its human homologue. Nature. 1994;372:425–32.
43. Friedman JM, Halaas JL. Leptin and the regulation of body weight in mammals. Nature. 1998;395:763–70.
44. Considine RV, Sinha MK, Heiman ML, Kriauciunas A, Stephens TW, Nyce MR, et al. Serum immunoreactive-leptin concentrations in normal-weight and obese humans. N Engl J Med. 1996;334:292–5.

45. Grunfeld C, Zhao C, Fuller J, Pollack A, Moser A, Friedman J, Feingold KR. Endotoxin and cytokines induce expression of leptin, the ob gene product, in hamsters. J Clin Invest. 1996;97:2152–7.
46. Farooqi IS, Jebb SA, Langmack G, Lawrence E, Cheetham CH, Prentice AM, et al. Effects of recombinant leptin therapy in a child with congenital leptin deficiency. N Engl J Med. 1999;341:879–84.
47. Halaas JL, Gajiwala KS, Maffei M, Cohen SL, Chait BT, Rabinowitz D, et al. Weight-reducing effects of the plasma protein encoded by the obese gene. Science. 1995;269:543–6.
48. Bjørbaek C, El-Haschimi K, Frantz JD, Flier JS. The role of SOCS-3 in leptin signaling and leptin resistance. J Biol Chem. 1999;274:30059–65.
49. Ozcan L, Ergin AS, Lu A, Chung J, Sarkar S, Nie D, et al. Endoplasmic reticulum stress plays a central role in development of leptin resistance. Cell Metab. 2009;9:35–51.
50. Procaccini C, De Rosa V, Galgani M, Carbone F, La Rocca C, Formisano L, Matarese G. Role of adipokines signaling in the modulation of T cells function. Front Immunol. 2013;4:332. doi:10.3389/fimmu.2013.00332.
51. Kloek C, Haq AK, Dunn SL, Lavery HJ, Banks AS, Myers Jr MG. Regulation of Jak kinases by intracellular leptin receptor sequences. J Biol Chem. 2002;277:41547–55.
52. Sweeney G. Leptin signalling. Cell Signal. 2002;14:655–63.
53. Gainsford T, Willson TA, Metcalf D, Handman E, McFarlane C, Ng A, et al. Leptin can induce proliferation, differentiation, and functional activation of hemopoietic cells. Proc Natl Acad Sci U S A. 1996;93:14564–8.
54. Aleffi S, Petrai I, Bertolani C, Parola M, Colombatto S, Novo E, et al. Upregulation of proinflammatory and proangiogenic cytokines by leptin in human hepatic stellate cells. Hepatology. 2005;42:1339–48.
55. Lord GM, Matarese G, Howard JK, Baker RJ, Bloom SR, Lechler RI. Leptin modulates the T-cell immune response and reverses starvation-induced immunosuppression. Nature. 1998;394:897–901.
56. Procaccini C, De Rosa V, Galgani M, Abanni L, Calì G, Porcellini A, et al. An oscillatory switch in mTOR kinase activity sets regulatory T cell responsiveness. Immunity. 2010;33:929–41. doi:10.1016/j.immuni.2010.11.024.
57. Hu E, Liang P, Spiegelman BM. AdipoQ is a novel adipose-specific gene dysregulated in obesity. J Biol Chem. 1996;271:10697–703.
58. Hotta K, Funahashi T, Arita Y, Takahashi M, Matsuda M, Okamoto Y, et al. Plasma concentrations of a novel, adipose-specific protein, adiponectin, in type 2 diabetic patients. Arterioscler Thromb Vasc Biol. 2000;20:1595–9.
59. Pischon T, Girman CJ, Hotamisligil GS, Rifai N, Hu FB, Rimm EB. Plasma adiponectin levels and risk of myocardial infarction in men. JAMA. 2004;291:1730–7.
60. Bajnok L, Csongradi E, Seres I, Varga Z, Jeges S, Peti A, et al. Relationship of adiponectin to serum paraoxonase 1. Atherosclerosis. 2008;197:363–7.
61. Díez JJ, Iglesias P. The role of the novel adipocyte-derived hormone adiponectin in human disease. Eur J Endocrinol. 2003;48:293–300.
62. Yokota T, Oritani K, Takahashi I, Ishikawa J, Matsuyama A, Ouchi N, et al. Adiponectin, a new member of the family of soluble defense collagens, negatively regulates the growth of myelomonocytic progenitors and the functions of macrophages. Blood. 2000;96:1723–32.
63. Maeda N, Shimomura I, Kishida K, Nishizawa H, Matsuda M, Nagaretani H, et al. Diet-induced insulin resistance in mice lacking adiponectin/ACRP30. Nat Med. 2002;8:731–7.
64. Bruun JM, Lihn AS, Verdich C, Pedersen SB, Toubro S, Astrup A, Richelsen B. Regulation of adiponectin by adipose tissue-derived cytokines: in vivo and in vitro investigations in humans. Am J Physiol Endocrinol Metab. 2003;285:E527–33.
65. Wolf AM, Wolf D, Rumpold H, Enrich B, Tilg H. Adiponectin induces the anti-inflammatory cytokines IL-10 and IL-1RA in human leukocytes. Biochem Biophys Res Commun. 2004;323:630–5.
66. Takemura Y, Ouchi N, Shibata R, Aprahamian T, Kirber MT, Summer RS, et al. Adiponectin modulates inflammatory reactions via calreticulin receptor-dependent clearance of early apoptotic bodies. J Clin Invest. 2007;117:375–86.
67. Kim KY, Kim JK, Han SH, Lim JS, Kim KI, Cho DH, et al. Adiponectin is a negative regulator of NK cell cytotoxicity. J Immunol. 2006;176:5958–64.
68. Motoshima H, Wu X, Mahadev K, Goldstein BJ. Adiponectin suppresses proliferation and superoxide generation and enhances eNOS activity in endothelial cells treated with oxidized LDL. Biochem Biophys Res Commun. 2004;315:264–71.
69. Otero M, Lago R, Gomez R, Lago F, Dieguez C, Gómez-Reino JJ, Gualillo O. Changes in plasma levels of fat-derived hormones adiponectin, leptin, resistin and visfatin in patients with rheumatoid arthritis. Ann Rheum Dis. 2006;65:1198–201.
70. Shoji T, Shinohara K, Hatsuda S, Kimoto E, Fukumoto S, Emoto M, et al. Altered relationship between body fat and plasma adiponectin in end-stage renal disease. Metabolism. 2005;54:330–4.
71. Sztanek F, Seres I, Harangi M, Lőcsey L, Koncsos P, Paragh G. Effect of nutritional status on human paraoxonase-1 activity in patients with chronic kidney disease. Kidney Blood Press Res. 2012;36:310–9.
72. Hattori Y, Hattori S, Kasai K. Globular adiponectin activates nuclear factor-kappaB in vascular endothelial cells, which in turn induces expression of proinflammatory and adhesion molecule genes. Diabetes Care. 2006;29:139–41.
73. Horn F, Henze C, Heidrich K. Interleukin-6 signal transduction and lymphocyte function. Immunobiology. 2000;202:151–67.

74. Weisberg SP, McCann D, Desai M, Rosenbaum M, Leibel RL, Ferrante Jr AW. Obesity is associated with macrophage accumulation in adipose tissue. J Clin Invest. 2003;112:1796–808.
75. Mohamed-Ali V, Goodrick S, Rawesh A, Katz DR, Miles JM, Yudkin JS, et al. Subcutaneous adipose tissue releases interleukin-6, but not tumor necrosis factor-alpha, in vivo. J Clin Endocrinol Metab. 1997;82:4196–200.
76. Bastard JP, Jardel C, Bruckert E, Blondy P, Capeau J, Laville M, et al. Elevated levels of interleukin 6 are reduced in serum and subcutaneous adipose tissue of obese women after weight loss. J Clin Endocrinol Metab. 2000;85:3338–42.
77. Senn JJ, Klover PJ, Nowak IA, Zimmers TA, Koniaris LG, Furlanetto RW, Mooney RA. Suppressor of cytokine signaling-3 (SOCS-3), a potential mediator of interleukin-6-dependent insulin resistance in hepatocytes. J Biol Chem. 2003;278:13740–6.
78. Wallenius V, Wallenius K, Ahrén B, Rudling M, Carlsten H, Dickson SL, et al. Interleukin-6-deficient mice develop mature-onset obesity. Nat Med. 2002;8:75–9.
79. Matthews VB, Allen TL, Risis S, Chan MH, Henstridge DC, Watson N, et al. Interleukin-6-deficient mice develop hepatic inflammation and systemic insulin resistance. Diabetologia. 2010;53:2431–41.
80. Tsigos C, Kyrou I, Chala E, Tsapogas P, Stavridis JC, Raptis SA, Katsilambros N. Circulating tumor necrosis factor alpha concentrations are higher in abdominal versus peripheral obesity. Metabolism. 1999;48:1332–5.
81. Fülöp P, Derdák Z, Sheets A, Sabo E, Berthiaume EP, Resnick MB, et al. Lack of UCP2 reduces Fas-mediated liver injury in ob/ob mice and reveals importance of cell-specific UCP2 expression. Hepatology. 2006;44:592–601.
82. Bugianesi E, Moscatiello S, Ciaravella MF, Marchesini G. Insulin resistance in nonalcoholic fatty liver disease. Curr Pharm Des. 2010;16:1941–51.
83. Feldstein AE, Werneburg NW, Canbay A, Guicciardi ME, Bronk SF, Rydzewski R, et al. Free fatty acids promote hepatic lipotoxicity by stimulating TNF-alpha expression via a lysosomal pathway. Hepatology. 2004;40:185–94.
84. Grunfeld C, Feingold KR. The metabolic effects of tumor necrosis factor and other cytokines. Biotherapy. 1991;3:143–58.
85. Kern PA, Di Gregorio GB, Lu T, Rassouli N, Ranganathan G. Adiponectin expression from human adipose tissue: relation to obesity, insulin resistance, and tumor necrosis factor-alpha expression. Diabetes. 2003;52:1779–85.
86. Hotamisligil GS, Peraldi P, Budavari A, Ellis R, White MF, Spiegelman BM. IRS-1-mediated inhibition of insulin receptor tyrosine kinase activity in TNF-alpha- and obesity-induced insulin resistance. Science. 1996;271:665–8.
87. Wellen KE, Hotamisligil GS. Inflammation, stress, and diabetes. J Clin Invest. 2005;115:1111–9.
88. Ofei F, Hurel S, Newkirk J, Sopwith M, Taylor R. Effects of an engineered human anti-TNF-alpha antibody (CDP571) on insulin sensitivity and glycemic control in patients with NIDDM. Diabetes. 1996;45:881–5.
89. Patel SD, Rajala MW, Rossetti L, Scherer PE, Shapiro L. Disulfide-dependent multimeric assembly of resistin family hormones. Science. 2004;304:1154–8.
90. Patel L, Buckels AC, Kinghorn IJ, Murdock PR, Holbrook JD, Plumpton C, et al. Resistin is expressed in human macrophages and directly regulated by PPAR gamma activators. Biochem Biophys Res Commun. 2003;300:472–6.
91. Steppan CM, Bailey ST, Bhat S, Brown EJ, Banerjee RR, Wright CM, et al. The hormone resistin links obesity to diabetes. Nature. 2001;409:307–12.
92. Choi SH, Hong ES, Lim S. Clinical implications of adipocytokines and newly emerging metabolic factors with relation to insulin resistance and cardiovascular health. Front Endocrinol (Lausanne). 2013;4:97. doi:10.3389/fendo.2013.00097.
93. Kaser S, Kaser A, Sandhofer A, Ebenbichler CF, Tilg H, Patsch JR. Resistin messenger-RNA expression is increased by proinflammatory cytokines in vitro. Biochem Biophys Res Commun. 2003;309:286–90.
94. Bokarewa M, Nagaev I, Dahlberg L, Smith U, Tarkowski A. Resistin, an adipokine with potent proinflammatory properties. J Immunol. 2005;174:5789–95.
95. Benomar Y, Gertler A, De Lacy P, Crépin D, Ould Hamouda H, Riffault L, Taouis M. Central resistin overexposure induces insulin resistance through Toll-like receptor 4. Diabetes. 2013;62:102–14.
96. Yang Q, Graham TE, Mody N, Preitner F, Peroni OD, Zabolotny JM, et al. Serum retinol binding protein 4 contributes to insulin resistance in obesity and type 2 diabetes. Nature. 2005;436:356–62.
97. Ost A, Danielsson A, Lidén M, Eriksson U, Nystrom FH, Strålfors P. Retinol-binding protein-4 attenuates insulin-induced phosphorylation of IRS1 and ERK1/2 in primary human adipocytes. FASEB J. 2007;21:3696–704.
98. Graham TE, Yang Q, Blüher M, Hammarstedt A, Ciaraldi TP, Henry RR, et al. Retinol-binding protein 4 and insulin resistance in lean, obese, and diabetic subjects. N Engl J Med. 2006;354:2552–63.
99. Promintzer M, Krebs M, Todoric J, Luger A, Bischof MG, Nowotny P, et al. Insulin resistance is unrelated to circulating retinol binding protein and protein C inhibitor. J Clin Endocrinol Metab. 2007;92:4306–12.
100. Chen MP, Chung FM, Chang DM, Tsai JC, Huang HF, Shin SJ, Lee YJ. Elevated plasma level of

visfatin/pre-B cell colony-enhancing factor in patients with type 2 diabetes mellitus. J Clin Endocrinol Metab. 2006;91:295–9.
101. Sommer G, Kralisch S, Kloting N, Kamprad M, Schrock K, Kratzsch J, et al. Visfatin is a positive regulator of MCP-1 in human adipocytes in vitro and in mice in vivo. Obesity (Silver Spring). 2010;18:1486–92. doi:10.1038/oby.2009.462.
102. Moschen AR, Kaser A, Enrich B, Mosheimer B, Theurl M, Niederegger H, Tilg H. Visfatin, an adipocytokine with proinflammatory and immunomodulating properties. J Immunol. 2007;178:1748–58.
103. da Kim S, Kang S, Moon NR, Park S. Central visfatin potentiates glucose-stimulated insulin secretion and β-cell mass without increasing serum visfatin levels in diabetic rats. Cytokine. 2014;65:159–66. doi:10.1016/j.cyto.2013.11.008.
104. Bozaoglu K, Bolton K, McMillan J, Zimmet P, Jowett J, Collier G, et al. Chemerin is a novel adipokine associated with obesity and metabolic syndrome. Endocrinology. 2007;148:4687–94.
105. Sell H, Laurencikiene J, Taube A, Eckardt K, Cramer A, Horrighs A, et al. Chemerin is a novel adipocyte-derived factor inducing insulin resistance in primary human skeletal muscle cells. Diabetes. 2009;58:2731–40.
106. Roman AA, Parlee SD, Sinal CJ. Chemerin: a potential endocrine link between obesity and type 2 diabetes. Endocrine. 2012;42:243–51.
107. Lehrke M, Becker A, Greif M, Stark R, Laubender RP, von Ziegler F, et al. Chemerin is associated with markers of inflammation and components of the metabolic syndrome but does not predict coronary atherosclerosis. Eur J Endocrinol. 2009;161:339–44. doi:10.1530/EJE-09-0380.
108. Wittamer V, Franssen JD, Vulcano M, Mirjolet JF, Le Poul E, Migeotte I, et al. Specific recruitment of antigen-presenting cells by chemerin, a novel processed ligand from human inflammatory fluids. J Exp Med. 2003;198:977–85.
109. Goralski KB, McCarthy TC, Hanniman EA, Zabel BA, Butcher EC, Parlee SD, et al. Chemerin, a novel adipokine that regulates adipogenesis and adipocyte metabolism. J Biol Chem. 2007;282:28175–88.
110. Cash JL, Hart R, Russ A, Dixon JP, Colledge WH, Doran J, et al. Synthetic chemerin-derived peptides suppress inflammation through ChemR23. J Exp Med. 2008;205:767–75.
111. Luangsay S, Wittamer V, Bondue B, De Henau O, Rouger L, Brait M, et al. Mouse ChemR23 is expressed in dendritic cell subsets and macrophages, and mediates an anti-inflammatory activity of chemerin in a lung disease model. J Immunol. 2009;183:6489–99.
112. White RT, Damm D, Hancock N, Rosen BS, Lowell BB, Usher P, et al. Human adipsin is identical to complement factor D and is expressed at high levels in adipose tissue. J Biol Chem. 1992;267:9210–3.
113. Ronti T, Lupattelli G, Mannarino E. The endocrine function of adipose tissue: an update. Clin Endocrinol (Oxf). 2006;64:355–65.
114. Cianflone K, Xia Z, Chen LY. Critical review of acylation-stimulating protein physiology in humans and rodents. Biochim Biophys Acta. 2003; 1609:127–43.
115. de Souza Batista CM, Yang RZ, Lee MJ, Glynn NM, Yu DZ, Pray J, et al. Omentin plasma levels and gene expression are decreased in obesity. Diabetes. 2007;56:1655–61.
116. Shang FJ, Wang JP, Liu XT, Zheng QS, Xue YS, Wang B, Zhao LY. Serum omentin-1 levels are inversely associated with the presence and severity of coronary artery disease in patients with metabolic syndrome. Biomarkers. 2011;16:657–62.
117. Yang RZ, Lee MJ, Hu H, Pray J, Wu HB, Hansen BC, et al. Identification of omentin as a novel depot-specific adipokine in human adipose tissue: possible role in modulating insulin action. Am J Physiol Endocrinol Metab. 2006;290:E1253–61.
118. Tan BK, Adya R, Farhatullah S, Chen J, Lehnert H, Randeva HS. Metformin treatment may increase omentin-1 levels in women with polycystic ovary syndrome. Diabetes. 2010;59:3023–31. doi:10.2337/db10-0124.
119. Xu H, Barnes GT, Yang Q, Tan G, Yang D, Chou CJ, et al. Chronic inflammation in fat plays a crucial role in the development of obesity-related insulin resistance. J Clin Invest. 2003;112:1821–30.
120. Cancello R, Henegar C, Viguerie N, Taleb S, Poitou C, Rouault C, et al. Reduction of macrophage infiltration and chemoattractant gene expression changes in white adipose tissue of morbidly obese subjects after surgery-induced weight loss. Diabetes. 2005;54:2277–86.
121. Mantovani A, Sica A, Sozzani S, Allavena P, Vecchi A, Locati M. The chemokine system in diverse forms of macrophage activation and polarization. Trends Immunol. 2004;25:677–86.
122. Hevener AL, Olefsky JM, Reichart D, Nguyen MT, Bandyopadyhay G, Leung HY, et al. Macrophage PPAR gamma is required for normal skeletal muscle and hepatic insulin sensitivity and full antidiabetic effects of thiazolidinediones. J Clin Invest. 2007;117:1658–69.
123. Lumeng CN, Bodzin JL, Saltiel AR. Obesity induces a phenotypic switch in adipose tissue macrophage polarization. J Clin Invest. 2007;117:175–84.
124. Trayhurn P, Wood IS. Adipokines: inflammation and the pleiotropic role of white adipose tissue. Br J Nutr. 2004;92:347–55.
125. Cinti S, Mitchell G, Barbatelli G, Murano I, Ceresi E, Faloia E, et al. Adipocyte death defines macrophage localization and function in adipose tissue of obese mice and humans. J Lipid Res. 2005;46:2347–55.
126. Talukdar S, da Oh Y, Bandyopadhyay G, Li D, Xu J, McNelis J, et al. Neutrophils mediate insulin resistance in mice fed a high-fat diet through secreted elastase. Nat Med. 2012;18:1407–12.
127. Divoux A, Moutel S, Poitou C, Lacasa D, Veyrie N, Aissat A, et al. Mast cells in human adipose tissue: link with morbid obesity, inflammatory status, and

diabetes. J Clin Endocrinol Metab. 2012;97:E1677–85. doi:10.1210/jc.2012-1532.
128. Wu D, Molofsky AB, Liang HE, Ricardo-Gonzalez RR, Jouihan HA, Bando JK, et al. Eosinophils sustain adipose alternatively activated macrophages associated with glucose homeostasis. Science. 2011;332:243–7.
129. Lynch L, Nowak M, Varghese B, Clark J, Hogan AE, Toxavidis V, et al. Adipose tissue invariant NKT cells protect against diet-induced obesity and metabolic disorder through regulatory cytokine production. Immunity. 2012;37:574–87. doi:10.1016/j.immuni.2012.06.016.
130. Nishimura S, Manabe I, Nagasaki M, Eto K, Yamashita H, Ohsugi M, et al. CD8+ effector T cells contribute to macrophage recruitment and adipose tissue inflammation in obesity. Nat Med. 2009;15:914–20.
131. Meijer K, de Vries M, Al-Lahham S, Bruinenberg M, Weening D, Dijkstra M, et al. Human primary adipocytes exhibit immune cell function: adipocytes prime inflammation independent of macrophages. PLoS One. 2011. doi:10.1371/journal.pone.0017154.
132. Feuerer M, Herrero L, Cipolletta D, Naaz A, Wong J, Nayer A, et al. Lean, but not obese, fat is enriched for a unique population of regulatory T cells that affect metabolic parameters. Nat Med. 2009;15:930–9.
133. Winer S, Chan Y, Paltser G, Truong D, Tsui H, Bahrami J, et al. Normalization of obesity-associated insulin resistance through immunotherapy. Nat Med. 2009;15:921–9.
134. Sell H, Eckel J. Adipose tissue inflammation: novel insight into the role of macrophages and lymphocytes. Curr Opin Clin Nutr Metab Care. 2010;13:366–70.
135. De Rosa V, Procaccini C, Calì G, Pirozzi G, Fontana S, Zappacosta S, et al. A key role of leptin in the control of regulatory T cell proliferation. Immunity. 2007;26:241–55.
136. Duffaut C, Galitzky J, Lafontan M, Bouloumié A. Unexpected trafficking of immune cells within the adipose tissue during the onset of obesity. Biochem Biophys Res Commun. 2009;384:482–5. doi:10.1016/j.bbrc.2009.05.002.
137. Winer DA, Winer S, Shen L, Wadia PP, Yantha J, Paltser G, et al. B cells promote insulin resistance through modulation of T cells and production of pathogenic IgG antibodies. Nat Med. 2011;17:610–7.
138. Jagannathan M, McDonnell M, Liang Y, Hasturk H, Hetzel J, Rubin D, et al. Toll-like receptors regulate B cell cytokine production in patients with diabetes. Diabetologia. 2010;53:1461–71.
139. Rowe J, Yerkovich ST, Richmond P, Suriyaarachchi D, Fisher E, Feddema L, et al. Th2-associated local reactions to the acellular diphtheria-tetanus-pertussis vaccine in 4- to 6-year-old children. Infect Immun. 2005;73:8130–5.
140. Classen JB. Review of evidence that epidemics of type 1 diabetes and type 2 diabetes/metabolic syndrome are polar opposite responses to iatrogenic inflammation. Curr Diabetes Rev. 2012;8:413–8.
141. Posthouwer D, Voorbij HA, Grobbee DE, Numans ME, van der Bom JG. Influenza and pneumococcal vaccination as a model to assess C-reactive protein response to mild inflammation. Vaccine. 2004;23:362–5.
142. El Yousfi M, Mercier S, Breuillé D, Denis P, Papet I, Mirand PP, Obled C. The inflammatory response to vaccination is altered in the elderly. Mech Ageing Dev. 2005;126:874–81.
143. Bernstein ED, Gardner EM, Abrutyn E, Gross P, Murasko DM. Cytokine production after influenza vaccination in a healthy elderly population. Vaccine. 1998;16:1722–31.
144. Phillips DI, Barker DJ, Fall CH, Seckl JR, Whorwood CB, Wood PJ, Walker BR. Elevated plasma cortisol concentrations: a link between low birth weight and the insulin resistance syndrome? J Clin Endocrinol Metab. 1998;83:757–60.
145. Duncan GE, Li SM, Zhou XH. Prevalence and trends of a metabolic syndrome phenotype among U.S. adolescents, 1999–2000. Diabetes Care. 2004;27:2438–43.
146. Hooper LV, Gordon JI. Commensal host-bacterial relationships in the gut. Science. 2001;292:1115–8.
147. Tilg H, Kaser A. Gut microbiome, obesity, and metabolic dysfunction. J Clin Invest. 2011;121:2126–32.
148. Ley RE, Bäckhed F, Turnbaugh P, Lozupone CA, Knight RD, Gordon JI. Obesity alters gut microbial ecology. Proc Natl Acad Sci U S A. 2005;102:11070–5.
149. Ley RE, Turnbaugh PJ, Klein S, Gordon JI. Microbial ecology: human gut microbes associated with obesity. Nature. 2006;444:1022–3.
150. Schwiertz A, Taras D, Schäfer K, Beijer S, Bos NA, Donus C, Hardt PD. Microbiota and SCFA in lean and overweight healthy subjects. Obesity (Silver Spring). 2010;18:190–5.
151. Bäckhed F, Ding H, Wang T, Hooper LV, Koh GY, Nagy A, et al. The gut microbiota as an environmental factor that regulates fat storage. Proc Natl Acad Sci U S A. 2004;101:15718–23.
152. Turnbaugh PJ, Ley RE, Mahowald MA, Magrini V, Mardis ER, Gordon JI. An obesity-associated gut microbiome with increased capacity for energy harvest. Nature. 2006;444:1027–31.
153. Zhou YJ, Zhou H, Li Y, Song YL. NOD1 activation induces innate immune responses and insulin resistance in human adipocytes. Diabetes Metab. 2012;38:538–43.
154. Purohit J, Hu P, Burke SJ, Collier JJ, Chen J, Zhao L. The effects of NOD activation on adipocyte differentiation. Obesity (Silver Spring). 2013;21:737–47.
155. Vijay-Kumar M, Aitken JD, Carvalho FA, Cullender TC, Mwangi S, Srinivasan S, et al. Metabolic syndrome and altered gut microbiota in mice lacking Toll-like receptor 5. Science. 2010;328:228–31.
156. Cani PD, Bibiloni R, Knauf C, Waget A, Neyrinck AM, Delzenne NM, Burcelin R. Changes in gut microbiota control metabolic endotoxemia-induced

inflammation in high-fat diet-induced obesity and diabetes in mice. Diabetes. 2008;57:1470–81.
157. Frazier TH, DiBaise JK, McClain CJ. Gut microbiota, intestinal permeability, obesity-induced inflammation, and liver injury. JPEN J Parenter Enteral Nutr. 2011;35 Suppl 5:14S–20.
158. Blaser MJ, Falkow S. What are the consequences of the disappearing human microbiota? Nat Rev Microbiol. 2009;7:887–94.
159. Kopp E, Ghosh S. Inhibition of NF-kappa B by sodium salicylate and aspirin. Science. 1994;265:956–9.
160. Yuan M, Konstantopoulos N, Lee J, Hansen L, Li ZW, Karin M, Shoelson SE. Reversal of obesity- and diet-induced insulin resistance with salicylates or targeted disruption of Ikkbeta. Science. 2001;293:1673–7.
161. Hundal RS, Petersen KF, Mayerson AB, Randhawa PS, Inzucchi S, Shoelson SE, Shulman GI. Mechanism by which high-dose aspirin improves glucose metabolism in type 2 diabetes. J Clin Invest. 2002;109:1321–6.
162. Dominguez H, Storgaard H, Rask-Madsen C, Steffen Hermann T, Ihlemann N, Baunbjerg Nielsen D, et al. Metabolic and vascular effects of tumor necrosis factor-alpha blockade with etanercept in obese patients with type 2 diabetes. J Vasc Res. 2005;42:517–25.
163. Nissen SE, Wolski K. Effect of rosiglitazone on the risk of myocardial infarction and death from cardiovascular causes. N Engl J Med. 2007;356:2457–71.
164. Yki-Järvinen H. Thiazolidinediones. N Engl J Med. 2004;351:1106–18.
165. Stienstra R, Duval C, Keshtkar S, van der Laak J, Kersten S, Müller M. Peroxisome proliferator-activated receptor gamma activation promotes infiltration of alternatively activated macrophages into adipose tissue. J Biol Chem. 2008;283:22620–7.
166. Pascual G, Fong AL, Ogawa S, Gamliel A, Li AC, Perissi V, et al. A SUMOylation-dependent pathway mediates transrepression of inflammatory response genes by PPAR-gamma. Nature. 2005;437:759–63.
167. Ialenti A, Grassia G, Di Meglio P, Maffia P, Di Rosa M, Ianaro A. Mechanism of the anti-inflammatory effect of thiazolidinediones: relationship with the glucocorticoid pathway. Mol Pharmacol. 2005;67:1620–8.
168. Castrillo A, Tontonoz P. Nuclear receptors in macrophage biology: at the crossroads of lipid metabolism and inflammation. Annu Rev Cell Dev Biol. 2004;20:455–80.
169. Wang KL, Liu CJ, Chao TF, Chen SJ, Wu CH, Huang CM, Chang CC, Wang KF, Chen TJ, Lin SJ, Chiang CE. Risk of new-onset diabetes mellitus versus reduction in cardiovascular events with statin therapy. Am J Cardiol. 2013. doi:10.1016/j.amjcard.2013.10.043.
170. Goldfine AB, Buck JS, Desouza C, Fonseca V, Chen YD, Shoelson SE, et al. Targeting inflammation using salsalate in patients with type 2 diabetes: effects on flow-mediated dilation (TINSAL-FMD). Diabetes Care. 2013;36:4132–9. doi:10.2337/dc13-0859.
171. Bruun JM, Helge JW, Richelsen B, Stallknecht B. Diet and exercise reduce low-grade inflammation and macrophage infiltration in adipose tissue but not in skeletal muscle in severely obese subjects. Am J Physiol Endocrinol Metab. 2006;290:E961–7.

The Axis AGE-RAGE-Soluble RAGE and Oxidative Stress in Chronic Kidney Disease

14

Alejandro Gugliucci and Teresita Menini

Abstract

Chronic kidney disease (CKD) has been shown to be associated with high oxidative stress and cardiovascular disease. In this chapter our focus will be on the role of advanced glycation end products (AGE) and their receptor, RAGE in CKD progression and their role on cardiovascular complications. We provide a succinct, yet comprehensive summary of the current knowledge, the challenges and the future therapeutic avenues that are stemming out from novel recent findings. We first briefly review glycation and AGE formation and the role of the kidney in their metabolism. Next, we focus on the RAGE, its signaling and role in oxidative stress. We address the possible role of soluble RAGEs as decoys and the controversy regarding this issue. We then provide the latest information on the specific role of both AGE and RAGE in inflammation and perpetuation of kidney damage in diabetes and in CKD without diabetes, which is the main purpose of the review. Finally, we offer an update on new avenues to target the AGE-RAGE axis in CKD.

Keywords

Advanced glycation • AGE receptor • Atherosclerosis • Cardiovascular disease • Hemodialysis • Nephropathy • Oxidative stress • Renal failure

14.1 Introduction

Chronic kidney disease (CKD) can be described in terms of etiology and in terms of the loss of the kidney homeostatic function [1, 2]. The main etiologies are diabetes, hypertension, pyelonephritis and a variety of immune nephritis. Regarding renal function loss, CKD is currently classified into five stages considering the level of glomerular filtration rate (estimated GFR). From these

A. Gugliucci (✉) • T. Menini
Glycation, Oxidation and Disease Laboratory, Department of Research, College of Osteopathic Medicine, Touro University-California, 1310 Club Drive, 94592 Vallejo, CA, USA
e-mail: alejandro.gugliucci@tu.edu; teresita.menini@tu.edu

J. Camps (ed.), *Oxidative Stress and Inflammation in Non-communicable Diseases – Molecular Mechanisms and Perspectives in Therapeutics*, Advances in Experimental Medicine and Biology 824, DOI 10.1007/978-3-319-07320-0_14, five stages mild CKD (CKD 1-2) is probably the most common in the elderly while end-stage renal disease (ESRD) or CKD 5 is less common and is predominantly associated with atherosclerosis, diabetes and glomerular diseases [1, 2].

Cardiovascular disease is very prominent in CKD mostly in stages 4 and 5 with patients being treated with hemodialysis (HD) having a shorter life expectancy than same age healthy subjects. More than one million patients with ESRD across the globe are surviving with the support of renal replacement therapy. Over 80 % patients receive hemodialysis (HD), the most common modality. Survival on HD has increased gradually, even though vascular complications, such as hemorrhagic stroke and ischemic heart disease continue to be key problems. All patients with chronic renal failure (CRF) have increased risk for death from cardiovascular disease. They have multiple metabolic abnormalities that may accelerate atherosclerosis, such as hypertension, insulin resistance, and dys-lipoproteinemia, along with other ESRD-related risk factors. In fact cardiovascular disease (CVD) is the leading cause of death in this patient population. It is important then to focus on the prevention of the causes of renal failure but also on the prevention of the associated complications [2–6].

Advanced glycation-end-products (AGE) accumulate in ESRD patients as part of a group of molecules called uremic-toxins [7–11]. The accumulation of AGEs in these patients occurs mostly due to two mechanisms, increased production and impaired excretion. AGE formation refers to the non-enzymatic reaction between proteins and glucose in the Maillard reaction, forming reversible early glycation products that with time became AGEs by a slow and complex rearrangement [9, 10, 12–14]. Other mechanisms by which AGEs are formed include lipoxidation and oxidative stress, a process that involves carbonyl compounds like methylglyoxal (MG). There is a specific pathway, the glyoxalase system that acts as defense mechanism against MG [15–17].

The importance of exogenous AGEs from food and smoke needs to be considered in CKD patients [18–21]. In healthy individuals, circulating AGEs, mostly derived from degradation of AGE-attached proteins, which are referred to as glycation free adducts and glycation adduct residues of proteins, are excreted via the kidney as we and others demonstrated in the nineties [9, 10, 22]. In the case of renal failure patients, this excretion mechanism is defective or nonexistent. AGEs will accumulate in the tissues where they may cross-link with other proteins perpetuating the damage. They are also able to interact with the receptor for AGE (RAGE) or other receptors producing the release of cytokines and further inflammation [3].

Several excellent reviews can be found in the literature on specific aspects of this complex issue [4, 23–25]. In this chapter we aimed at providing a succinct, yet comprehensive summary of the current knowledge, the challenges and the future therapeutic avenues that are stemming out from novel recent findings. Our main focus is on the role of AGE and RAGE in CKD independent of the presence of diabetes. We will first briefly review the glycation pathway and AGE production and the role of the kidney in their metabolism. Next, we will focus on the RAGE, its signaling mechanisms and the key role in oxidative stress. We will address the role of soluble RAGEs as putative decoy molecules and the controversy regarding this issue. We will then provide the latest information on the specific role of both AGE and RAGE in inflammation and perpetuation of kidney damage in diabetes and in CKD without diabetes, which is the main purpose of the review. Finally we will provide an update on new avenues to target the AGE-RAGE axis in CKD.

14.2 AGE Production and Renal Metabolism

Advanced glycation end products (AGE) are produced in the classical Maillard reaction, discovered in 1913 [11–14, 26, 27]. In humans AGE have two sources, endogenous and exogenous.

14.2.1 Endogenous AGE

Protein glycation is a complex series of sequential reactions collectively called the Maillard reaction,

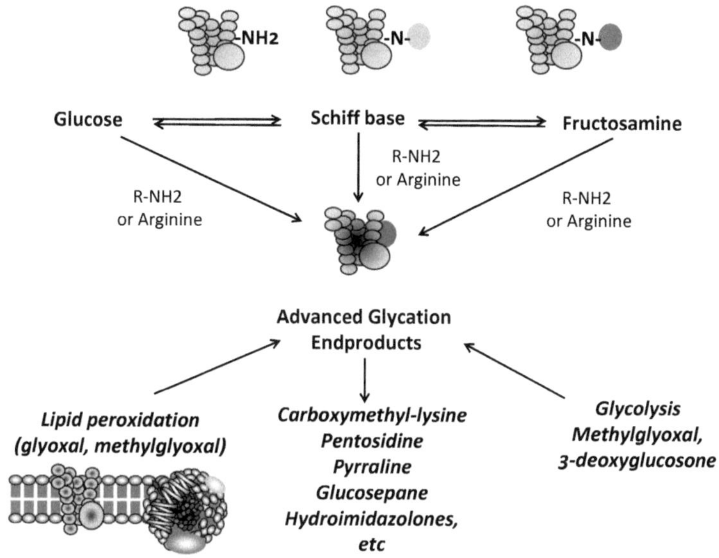

Fig. 14.1 The glycation reaction and the production of advanced glycation endproducts (AGE). Protein glycation is a complex series of sequential reactions collectively called the Maillard reaction. In early stages the glycation adduct fructosyl-lysine (FL) is formed, the reaction then proceeds to form advanced glycation end products (AGE). Degradation of glycated proteins, glycolytic intermediates, and lipid peroxidation give rise to glyoxal, methylglyoxal, and 3-deoxyglucosone which are also potent glycating agents. Other important sources of AGE products are hydroimidazolones derived from arginine residues modified by glyoxal, MG and 3-DG. Other key AGEs compounds are Nε-carboxymethyl-lysine, Nε-carboxyethyl-lysine, pentosidine, pyrraline and glucosepane

present in all tissues and fluids where enough concentration of glucose reacts with proteins [11–14, 26–30]. In early stages the glycation adduct fructosyl-lysine (FL) is formed, the reaction then proceeds to form advanced glycation end products (AGE) as depicted in Fig. 14.1. Degradation of glycated proteins, glycolytic intermediates, and lipid peroxidation give rise to glyoxal (G), methylglyoxal (MG), and 3-deoxyglucosone (3-DG) which are also potent glycating agents [31]. These compounds are able to react with proteins to form more AGEs directly. Other important sources of AGE products are hydroimidazolones derived from arginine residues modified by glyoxal, MG and 3-DG. Other key AGEs compounds are Nε-carboxymethyl-lysine (CML), Nε-carboxyethyl-lysine (CEL), pentosidine, pyrraline and glucosepane [32–34].

14.2.2 Exogenous AGE

AGEs are also present in ingested food. The content importantly depends on the nutrient composition and on the way food is processed, roasting, smoking and baking producing the highest levels of AGE [18–21]. This exogenous source of AGEs seems to be important since modulation of the intake has been proved to modify the circulating levels of AGE. In CKD patients there are increased concentrations of many α-oxoaldehydes, particularly glyoxal, MG, and 3-DG [15, 16, 31, 35–39]. Even short-term modifications of dietary AGE intake can significantly alter circulating AGE levels in renal failure patients on maintenance peritoneal dialysis [20]. These findings support the hypothesis that exogenous glycotoxins derived from common diets, including those recommended for dialysis patients, play a significant role in maintaining high circulating AGE levels in uremia. Recently, it has been shown that circulating AGE levels decreased after dietary AGE restriction in diabetic patients with normal renal function as well as in non-diabetic subjects with renal failure [19, 20]. In a study on CKD, the patients consumed self-selected diets prepared by them at home indicating that these dietary modifications are feasible as an

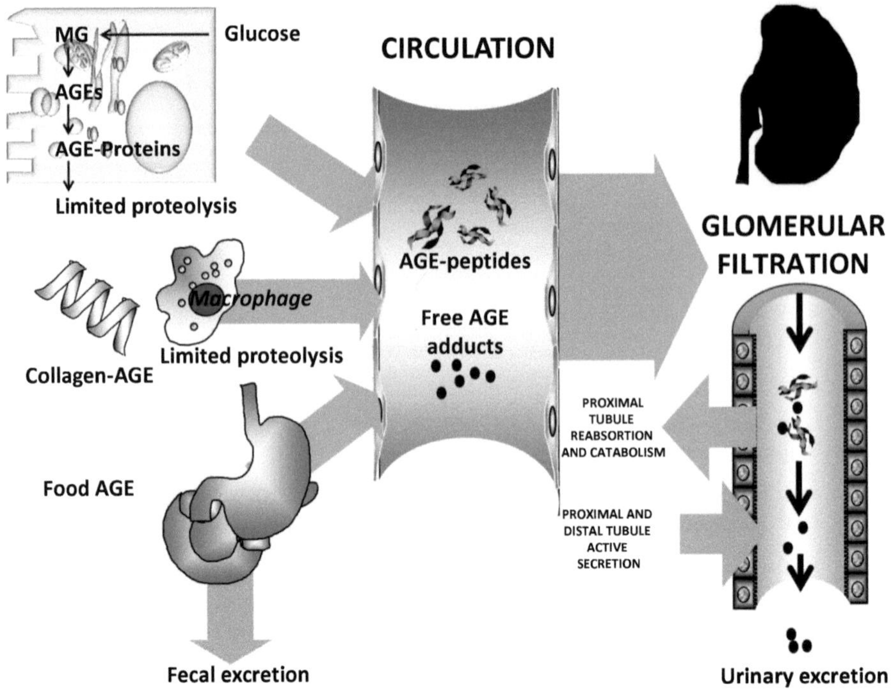

Fig. 14.2 The role of the kidney in overall metabolism of AGEs. In the circulation and tissues, AGEs have several sources, intracellular as shown in Fig. 14.1, plasma or extracellular matrix and finally AGEs that come from food. Human serum contains partially hydrolyzed AGE peptides and free AGE adducts, stemming from incomplete cellular or macrophage catabolism. Endogenous AGE peptides are filtered and then reabsorbed in the proximal tubules followed by excretion of free AGE. These processes, largely dependent on GFR and tubular function, are, by definition, severely impaired in CKD and ESRD. Therefore, as can be seen CKD patients are prone to and certainly have the highest level of AGEs, AGE peptides and free adducts, due to increased production via oxidative stress and impaired clearance

intervention in this population [21, 40]. Dietary restriction of AGE may prove a reasonable method to reduce the excessive burden of toxic AGE in vital tissues and presumably the morbidity and mortality associated with their accumulation.

14.2.3 Iatrogenic Sources of AGEs in CKD

Specifically in CKD patients under replacement therapy, there is a third source of AGE to be considered. This is the α-oxoaldehydes present in fluids used for peritoneal dialysis (PD) [23, 41]. These compounds formed by thermal sterilization are introduced into the peritoneal cavity and greatly contribute to the glycation reaction and AGE formation in these patients [1, 4]. This is aggravated by the fact that the renal clearance of AGE free adducts is impaired which triggers an increase in the concentrations of free AGE in plasma. Contributing factors to aggravate oxidative stress and related endothelial dysfunction could also include defective dialysis membranes.

14.2.4 Systemic Metabolism and Handling of AGEs, the Key Role of the Kidney

Two decades ago it became apparent that human serum contains partially hydrolyzed AGE peptides and free AGE adducts [10, 29, 42]. They are increased in diabetes due to excess production and much more so in ESRD even in the absence of diabetes. We have shown that endogenous AGE peptides are filtered and then reabsorbed in the proximal tubules followed by excretion of free AGE adducts [10], as depicted in Fig. 14.2. These processes, largely dependent on GFR and tubular function, are, by definition, severely impaired in CKD and ESRD [15, 16, 35, 36]. Therefore, CKD patients are prone to and certainly have the highest plasma level of AGEs, AGE peptides and free adducts, due to increased production via oxidative stress and impaired clearance. In that sense AGEs qualify as one of the classical "middle toxins" typical of uremia [42, 43]. Precisely, we have shown that HD increases clearance of AGE peptides and that is associated with an increase in the antioxidant enzyme PON1, suggesting that ROS increase by AGEs further compromise the patient's antioxidant defenses, adding one more element to the vicious cycle we have just described [44, 45]. Patients on dialysis regardless if diabetic or not have high plasma pentosidine and CML levels and HD techniques are unable to completely clear the AGE from plasma [36, 37, 40]. AGE and RAGE, as we will see on the next section may contribute to amplification and perpetuation of inflammatory processes in non-diabetic patients with renal failure.

14.2.5 Pathogenic Consequences of AGE Accumulation

AGEs accumulate predominantly in tissues with slow turnover and as a consequence these tissues will show more damage. AGE crosslinking of collagen and other ECM proteins leads to matrix changes, hardening of arteries and complex sequences of signaling via integrins that aggravate the damage [12, 27, 46]. Increased production and retention of AGEs directly leads to vascular damage in ESRD, which aggravates the condition, and this is compounded by the effects mediated by RAGE, as we will discuss below.

14.2.6 Circulating AGEs May Underestimate Tissue AGE Accumulation: Alternative Practical Approaches

Because in plasma the AGEs have a relatively high turnover rate, several authors have suggested that plasma levels of AGEs may not accurately reflect tissue damage. Although the levels of plasma AGE are very high in ESRD patients, skin autofluorescence (SAF), which correlates with AGE damage to dermal collagen, may be a better marker of tissue damage in these patients [1, 4]. In a recent report it was shown that in ESRD patients both SAF and serum pentosidine correlated with carotid intima-media thickness, and SAF also inversely correlated with endothelial progenitor cells, but there was no correlation with serum pentosidine [1, 4].

14.2.7 A Special Case for Diabetics: The "Glycemic Memory"

The Diabetes Control and Complications Trial-Epidemiology of Diabetes Interventions and Complications (DCCTEDIC) study evaluated the effectiveness of intense therapy. This study has revealed that in type 1 diabetic patients the risk of developing progressive nephropathy persist for several years after the end of aggressive treatment [47]. There was also a 50 % reduced risk of cardiovascular events up to 11 years after the end of the trial. When AGE levels where studied in skin biopsies as part of a sub-study of DCCT, it was found that higher levels of AGE were independent predictors of worse renal and cardiovascular outcomes. It is apparent that vascular abnormalities are not easily reversed, even with a better control in the blood glucose levels [47]. This fact has been suggested by these clinical trials and is referred to as "glycemic memory". It is likely that glycemic memory is related to the vascular complications derived from the glycation of proteins, AGE formation and accumulation [2]. AGEs in long-lived proteins may be a link to this puzzling finding. These adducts remain on these long-lived proteins even when hyperglycemia has been corrected and may perpetuate the damage. Indeed, studies that examined the relevance of skin autofluorescence (SAF) as a surrogate marker of tissue damage on cardiovascular morbidity and mortality provide strong evidence of the importance of the AGE accumulation in ESRD patients. SAF is then a strong and independent predictor of cardiovascular mortality in ESRD patients [1, 4]. Even in non-diabetic CKD patients, initial accumulation of AGEs in long-lived proteins would perpetuate the damage even under optimal therapy.

In renal patients with ESRD the mechanism of increased presence of AGE products is mostly related to oxidative stress compared to the diabetic patients in whom it is related to both glycation and oxidative stress [5, 23, 48]. In ESRD patients there is increased lipid peroxidation with a decrease ratio of oxidized to reduced glutathione and an increase in malondialdehyde-lysine [49, 50]. This increased oxidative stress is probably related to impaired renal function or to replacement therapy.

14.3 The RAGE Signaling Pathway and Oxidative Stress

AGEs are recognized by a wide array of cellular receptors, some pro- and some anti-inflammatory. In this review we focus on the RAGE. The receptor for advanced glycation end-products (receptor for AGEs, RAGE) is a multi-ligand protein first isolated from bovine lung [6, 51–56]. RAGE integrates the immunoglobulin superfamily of receptors. As depicted in Fig. 14.3 it has an extracellular

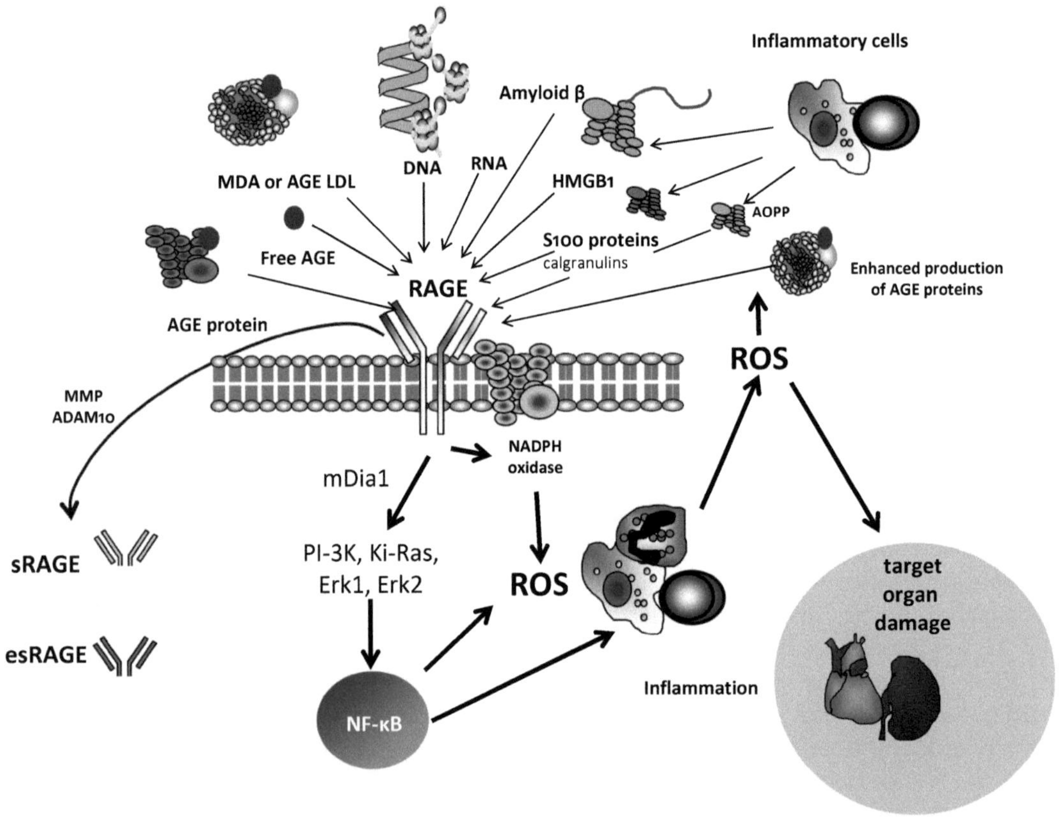

Fig. 14.3 The RAGE, its ligands and main signaling pathways enhance oxidative stress. The receptor for advanced glycation end-products (receptor for AGEs, RAGE) is a multi-ligand protein that integrates the immunoglobulin superfamily of receptors. RAGE recognizes a variety of ligands including high mobility group box 1 protein (HMGB1), the leukocyte integrin Mac-1, S100/calgranulins, modified LDL, DNA, RNA and amyloid fibrils. When activated, RAGE leads to a sequence of signaling with activation of NF-kB, oxidative stress and inflammation. RAGE signals via phosphatidylinositol-3 kinase (PI-3 K), Ki-Ras and the MAPKs, Erk1 and Erk2 In a coordinated fashion these pathways promote and sustain the translocation of nuclear factor-κB (NF-κB) from the cytoplasm to the nucleus. Activation promotes inflammation and tissue injury sustained by a RAGE-dependent expression of pro-inflammatory mediators such as monocyte chemoattractant protein-1 (MCP-1) and vascular cell adhesion molecule-1 (VCAM-1). RAGE can also trigger spurs a surge of reactive oxygen species (ROS)via reduced nicotinamide adenine dinucleotide phosphate, NAD(P)H oxidase

region containing one 'V'-type immunoglobulin domain and two 'C'-type immunoglobulin domains. The receptor has an hydrophobic trans-membrane segment and a greatly charged, cytoplasmic domain. The latter orchestrates intracellular RAGE signaling [3, 25, 57–59]. RAGE recognizes a variety of ligands including high mobility group box 1 protein (HMGB1), the leukocyte integrin Mac-1, S100/calgranulins, modified LDL, DNA, RNA and amyloid fibrils. These ligands share structural features: multiple β-sheets, and RAGE recognizes its ligands through them [25, 60, 61].

14.3.1 What Are the Consequences of RAGE Occupancy?

When activated, RAGE leads to a sequence of signaling with activation of NF-kB, oxidative stress and inflammation as we show in Fig. 14.3. RAGE signals via phosphatidylinositol-3 kinase (PI-3K), Ki-Ras and the MAPKs, Erk1 and Erk2 [24, 25, 57, 60, 62, 63]. In a coordinated fashion these pathways promote and sustain the translocation of nuclear factor-κB (NF-κB) from the cytoplasm to the nucleus. This occurs in a variety of cell types: monocytes, endothelial cells, microglia, podocytes, etc. Activation promotes inflammation and tissue injury sustained by a RAGE-dependent expression of pro-inflammatory mediators such as monocyte chemoattractant protein-1 (MCP-1) and vascular cell adhesion molecule-1 (VCAM-1). In this way, RAGE activity is associated with diabetic microvascular complications including nephropathy, retinopathy and neuropathy [24, 25, 57, 60, 62, 63]. In the special case of CKD, given the fact that AGE levels are even higher than in diabetes, those events are magnified and compounded by an increase in oxidative stress from multiple sources associated with the impaired renal homeostatic function and of an iatrogenic nature (HD and PD). In humans, RAGE gene polymorphisms are indeed associated with the incidence of diabetic nephropathy [57].

14.3.2 Activation of NADPH Oxidase and ROS Generation

In parallel with the aforementioned cascades leading to inflammation, in endothelial (glomerular) and mesangial cells, RAGE triggers a surge of reactive oxygen species (ROS). As we depict in Fig. 14.3, this pathway involves reduced nicotinamideadenine dinucleotide phosphate, NAD(P)H oxidase, which, akin to other situations (e.g.: neutrophils) is a key player in oxidative stress and dysfunction [61, 64]. A cardinal feature of progressive CKD is an exaggerated chronic inflammation associated with persistent activation of monocytes/macrophages as well as mesangial cells: accumulation of AGEs as described in Sect. 14.2 surely plays a significant function via engagement of RAGE [23, 49, 50, 57, 65]. Indeed, in monocytes from subjects with CKD, RAGE expression is intimately associated with its severity or stage.

Beyond the local renal level, whether RAGE is augmented in atherosclerotic lesions in CKD remains unclear for humans. CKD prominently hastens atherogenesis in apo E-deficient mice. Blocking RAGE decreases the pro-atherogenic effects of CKD, conceivably mediated by a reduction in oxidative stress.

14.3.3 Soluble RAGE May Act as Decoy, Protective Molecules in Some Cases

The C-terminal truncated form of RAGE mRNA lacks the sequences encoding the transmembrane and intracytoplasmic domains [66–68]. The extracellular domain of RAGE thereby produced, is released from cells, found in the circulation in humans. It has been named endogenous secretory RAGE (esRAGE) and may play a role in cardiovascular disease (Fig. 14.3). This esRAGE cancels the effects of AGEs on cells in culture [24]. Overexpression of esRAGE in mice reverses diabetic vascular dysfunction. EsRAGE may then exert a decoy function: a feedback mechanism has been proposed by which esRAGE prevents RAGE signaling [24, 66, 69–72]. It has also been suggested that some sRAGE isoforms that could act as decoy receptors may be cleaved proteolytically from the native RAGE expressed on the cell surface, suggesting heterogeneity of the origin and nature of sRAGE [57, 61, 64, 70]. This proteolytic generation of sRAGE was initially described as occurring in mice. Recent findings by screening chemical inhibitors and genetically modified mouse embryonic fibroblasts suggest that a disintegrin and metalloprotease 10 (ADAM10) and metalloprotease9 (MMP 9) are the membrane proteases responsible for RAGE cleavage (Fig. 14.3). ADAM is known to shed several inflammatory receptors and can be involved in regulation of RAGE⁄sRAGE balance [57, 61, 64, 70]. Thus, the molecular heterogeneity of the diverse types of sRAGE in human plasma could exert significant protective effects against RAGE-mediated toxicity. However, the endogenous action of sRAGE may not be confined to a decoy function against RAGE-signaling. In an HMGB1-induced arthritis model, for example, sRAGE is found to interact with Mac-1 and act as an important proinflammatory and chemotactic molecule [57]. Further analyses are warranted to understand more about the endogenous activity of sRAGE. ELISA assays are available to assess the levels of these circulating forms of RAGE and multiple studies have been conducted and others are under way.

14.3.4 Clinical Significance of Serum sRAGE Measurement

As sRAGE and esRAGE may be involved in feedback regulation of the toxic effects of RAGE-mediated signaling, recent clinical studies have focused on the potential significance of circulating sRAGE and esRAGE in a variety of pathophysiological conditions [66, 69–71, 73–79]. Total sRAGE levels were shown to be significantly lower in non-diabetic patients with angiographically proven coronary artery disease than in age-matched healthy controls [78]. In contrast to non-diabetic population, evidence so far with regard to a mechanistic association between sRAGE and vascular disease in diabetes is contradictory.

A study shows that serum sRAGE is positively associated with coronary artery disease in type 1 diabetic patients [79]. However, a recent longitudinal study in small numbers of type 1 diabetic subjects showed that annual increase in carotid IMT was inversely associated with arithmetic average of plasma sRAGE, although this report failed to show a significant relation between basal sRAGE and IMT. Similarly, the findings regarding plasma levels of the sRAGE in diabetes are quite confusing; many reports showed increased levels, whereas substantial contradictory findings also exist. These discrepancies might be the results of co-existing renal insufficiency, which markedly influences plasma sRAGE levels [70].

14.3.5 Can sRAGE/RAGE Be Modulated by Current Therapies?

It is important to determine whether currently available pharmacological agents can regulate plasma sRAGE or esRAGE. Inhibition of angiotensin-converting enzyme (ACE) in rats increased renal expression of sRAGE, and this was associated with the decreases in expression of renal full-length RAGE protein [68, 80]. Plasma sRAGE levels were significantly increased by inhibition of ACE in both diabetic rats and in human subjects with type 1 diabetes. Thus, one attractive scenario is that the protective effect of ACE inhibition against progression of renal dysfunction is mediated through regulation of RAGE versus soluble RAGE production [68, 80]. Other potential agents that may affect circulating soluble RAGE include the thiazolidinediones and statins both of which are known to modulate the AGEs-RAGE system in culture [81–83]. An open-label, parallel group study was performed with 64 participants randomized to receive add-on therapy with either rosiglitazone or sulfonylurea to observe the effect on plasma soluble RAGE [84–86]. At 6 months, both rosiglitazone and sulfonylurea resulted in a significant reduction in HbA1c, fasting glucose and AGE. However, significant increases in total sRAGE and esRAGE were only seen in the rosiglitazone group. Thus, thiazolidinedione could be one promising candidate drug to increase circulating levels of esRAGE and sRAGE [84–86]. Changes in serum levels of sRAGE and esRAGE were assessed on a clinical trial that explored the cardiovascular effects of atorvastatin in hypercholesterolemic Chinese type 2 diabetic patients. Atorvastatin indeed increased circulating esRAGE levels in these groups of patients [85]. A randomized clinical trial is underway comparing the effect of pioglitazone with glimepiride on plasma sRAGE and esRAGE, expression of RAGE on peripheral mononuclear cells and RAGE shedase gene expression in type 2 diabetic patients [85].

14.4 Pathogenic Roles of RAGE Activation in CKD

By virtue of its multi-ligand nature RAGE participates in diverse facets of the pathogenesis and progression of renal disease [2, 5, 23, 40, 57, 60, 61, 63, 64, 82, 87]. In diabetes and in experimental high-fat feeding, AGE, advanced oxidation protein products (AOPP) and AGE- oxidized LDL form swiftly and contribute to initial cell stress and signaling, all of which are elements theoretically linked to origination of renal disease [4, 5, 25, 61]. Once renal disease is initiated, the subsequent attraction of inflammatory cells to the nephron and their activation, results in release of pro-inflammatory RAGE ligands such as S100/calgranulins and HMGB1 [5, 64, 88, 89]. The unimpeded feed-forward cycle of oxidative stress and inflammatory signaling generates more AGE, leads to crosslinking of local ECM and to the formation of amyloid fibrils, further perturbing kidney function [63, 64]. Intriguingly, post-hoc glycation of pre-existing kidney amyloid might itself augment the burden of ligands triggering RAGE signaling [5, 64, 88, 89].

14.4.1 Local Feed Forward Mechanisms in the Glomerulus

The kidney glomerulus indeed offers a very special microenvironment, whereby RAGE-dependent inflammation and oxidative stress, fuels the accretion of numerous ligand families that in turn activate RAGE.

The podocyte is the principal RAGE-expressing cell in the glomerulus, followed by the endothelium [63, 64]. Studies in vitro suggest that RAGE-dependent signal transduction in podocytes produces apoptosis, production of monocyte chemoattractant peptide-1 and inflammatory mediators at least in part via NF-κB [63, 64]. These pathways enhance the progress of glomerular sclerosis [8, 58, 63]. Multiple evidence points to RAGE-dependent Rac-1 as a crucial downstream target of RAGE, given its fundamental roles in oxidative stress and cytoskeleton/structural integrity. RAGE modulates podocyte vascular endothelial growth factor (VEGF) expression and the upregulation of VEGF receptors in the glomerular endothelial cell can damage glomerular basement membrane integrity leading to albuminuria [8, 58, 63]. There is actual physiologic evidence for "back-flow" across the glomerular basement membrane (GBM) at the extremes of pulse pressure at least in animal models [8, 58, 63]. If this occurs in humans, cytokines and growth factors produced by the podocyte might engage receptors on endothelial and mesangial cells [63]. Accordingly, VEGF and MCP-1 upregulation by the podocyte could be pivotal steps in inducing leakiness of the GBM, influx of inflammatory cell and development of established renal disease. Curtailing the cycle of RAGE-dependent stress in the glomerulus may thus provide a means to treat renal disease of diverse etiologies.

AGE–RAGE is up-regulated in the kidney of diabetic and lupus patients, where key roles for RAGE in T lymphocyte priming and early differentiation were revealed [53, 64, 90]. In murine models of inflammation as well as in the diabetic kidney, non-AGE ligands of RAGE accumulate, such as the S100/calgranulins, HMGB1, and amyloid-β peptide and β-sheet fibrils. HMGB1 (which interacts with RAGE and toll-like receptors) is a key component of DNA-containing immune complexes that stimulate cytokine production, a process that involves RAGE as well. AGEs and RAGE were identified in the glomeruli and tubules and correlated with amyloid deposits [53, 64, 90].

14.4.2 Soluble RAGEs and CVD in CKD

Association of circulating sRAGE or esRAGE with vascular diseases in CKD or ESRD subjects is an important topic to be elucidated. So far, only limited reports are available. In a cohort study with 206 ESRD patients including 35 diabetics the clinical significance of circulating esRAGE on cardiovascular outcomes was assessed. In patients with ESRD, plasma esRAGE levels at base line were inversely associated with body mass index as was shown for healthy and diabetic populations [91]. Plasma esRAGE was also significantly associated with plasma adiponectin in these populations. Serum esRAGE was significantly and inversely associated with carotid IMT. In another study, sRAGE was inversely associated with IMT or plaque numbers in the carotid artery of 142 CKD patients [67, 70, 72]. Prominently, the subjects in the lowest tertile of plasma esRAGE levels displayed significantly higher cardiovascular mortality, but not non-cardiovascular mortality even though the plasma esRAGE levels at baseline were higher in ESRD patients than in those devoid of kidney disease. In the subpopulation of non-diabetic subjects alone, low circulating esRAGE level was also a predictor of cardiovascular mortality, independent of the other classical risk factors. It is not known at present how esRAGE is involved in cardiovascular mortality, they may simply be valuable surrogate marker. In these studies the AGEs pentosidine or carboxymethyl-lysine did not predict cardiovascular mortality. Moreover, the inverse correlation between low circulating esRAGE level and cardiovascular mortality was not dependent of plasma AGEs levels.

14.4.3 Non-AGE Ligands Are Equally Important in CKD

Thus, the protective effect of esRAGE against cardiovascular mortality may not be entirely dependent on neutralization of toxic AGEs. As we depict in Fig. 14.3, other endogenous ligands for RAGE, for instance calgranulins such as S100A12, may also be involved in the function of esRAGE. The plasma level of S100A12 has been shown to be increased in diabetes and inversely correlated with serum sRAGE level [66, 74]. Low sRAGE levels are also shown to be associated with a 2–3 times higher risk for mortality especially after correction for creatinine clearance in a large cohort after renal transplantation. Thus, *low* circulating esRAGE level appears to be a predictor for atherosclerosis and cardiovascular events in patients with ESRD.

14.4.4 Podocytes Are Activated by RAGE

AGEs are associated with podocyte apoptosis via the FOXO4 transcription factor. Incubation of cultured podocytes with AGE or CML-AGE, or AGE retrieved from the serum of subjects with chronic kidney disease resulted in their apoptosis [58, 59, 63]. The cytoplasmic domain of RAGE is devoid of endogenous tyrosine kinase activity, suggesting that different molecules in the intracellular space are involved in signal transduction. Extracellular signal-regulated kinases (ERK) and diaphanous-1 ormDia-1, are the two suggested candidates that demonstrate strong links to RAGE biology [58, 59, 63]. In models of glomerular sclerosis RAGE plays key roles in cellular migration, and ligand–RAGE interaction stimulates activation of cdc42 and Rac-1 in podocytes which promotes cytoskeletal reorganization resulting in foot process effacement [63]. Given the multi-ligand nature of RAGE, it is reasonable to test the premise that RAGE is linked to the pathogenesis and/or acceleration of non-diabetic kidney disease, driven by immune/inflammatory stimuli and/or oxidative stress.

Diabetes is the main cause of end-stage renal failure. Studies testing sRAGE showed chief roles for RAGE in the pathogenesis of diabetes-associated nephropathy [77, 79, 86, 90, 92]. Neutralizing anti-RAGE antibodies beneficially impacted the course of renal disease in mouse models of type 1 and type 2 diabetes [6]. However, results in mice models are to be taken with caution. Humans are not rodents and a well-known drawback of mouse models of diabetes, is the failure of progression to advanced stages of

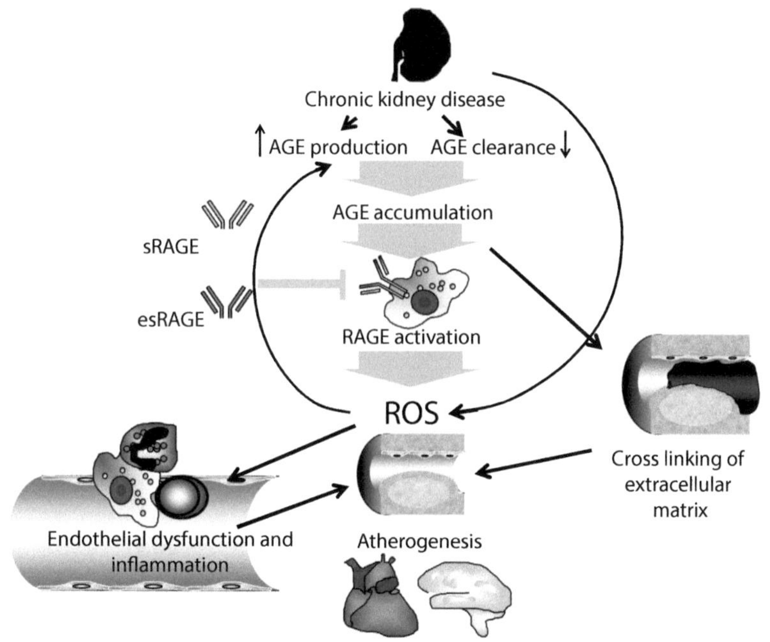

Fig. 14.4 Pathogenic roles of the AGE-RAGE-soluble RAGE axis in CKD. By virtue of its multi-ligand nature RAGE participates in diverse facets of the pathogenesis and progression of renal disease. AGE, advanced oxidation protein products (AOPP) and AGE- oxidized LDL form swiftly in oxidative stress and contribute to initial cell stress and signaling, all of which are elements theoretically linked to origination of renal disease. Once renal disease is initiated, the subsequent attraction of inflammatory cells to the nephron and their activation, results in release of pro-inflammatory RAGE ligands such as S100/calgranulins and HMGB1. The unimpeded feed-forward cycle oxidative stress and inflammatory signaling which generates more AGE leads to crosslinking of local ECM which leads to the formation of amyloid fibrils, further perturbing the kidney. There appears to be vicious positive feedback mechanisms in AGE accumulation through AGE-RAGE induced oxidative stress in the course of CKD progression. sRAGE or esRAGE could be potentially important in dampening this vicious cycle

disease similar to ESRD in humans. We lack a robust animal model of ESRD, although a transgenic mouse model that overexpresses megsin (a glomerular-specific serpin), RAGE and iNOSmay offer promise [93]. These triple transgenic mice developed massive albuminuria, glomerular hypertrophy, diffuse mesangial expansion, inflammatory cell infiltration and interstitial fibrosis [93].

14.4.5 Towards a Delineation of the Pathogenic Role of the AGE-RAGE Axis in Cardiovascular Complications of CKD

The findings discussed here implicate the pivotal role of AGEs-RAGE system in initiation and progression of cardiovascular disease in patients with CKD that we summarize in Fig. 14.4. There appears to be vicious positive feedback mechanisms in AGE accumulation through AGE-RAGE induced oxidative stress in the course of CKD progression. sRAGE or esRAGE could be potentially important in dampening this vicious cycle. Measuring tissue AGE accumulation could be useful to estimate how long and to what extent the patient have been exposed to oxidative stress. Moreover, plasma sRAGE or esRAGE could serve as novel biomarkers for estimation of the risk stratification of atherosclerotic disorders. Further examination of the molecular mechanisms underlying RAGE and esRAGE regulation will provide important insights into potential targets for the prevention and treatment of cardiovascular diseases associated with CKD.

14.5 Therapeutic Avenues to Target the AGE-RAGE Axis

The evidence presented in this review points to a key role of AGE accumulation and AGE engagement of RAGE in the progression and aggravation of CKD and ESRD. In this regard strategies to reduce the burden of both arms of the pathways are warranted as therapeutic avenues. Several approaches have been proposed that can be subdivided in strategies to lower the ligand burden (AGEs and other RAGE ligands) and strategies to dampen RAGE activation. We summarize these strategies in Fig. 14.5.

14.5.1 Strategies to Lower the Ligand Burden

14.5.1.1 Nutrition of Hemodialysis Patients

AGE intake correlates with circulating AGE levels in ESRD patients. As described earlier, AGEs in food are absorbed (albeit poorly), circulate and may be an important source of ligands for RAGE as well as directly damage tissue proteins. In CKD this burden may become more important due to the critical loss in elimination [20, 21]. Therefore, AGE intake is a modifiable risk factor amenable to be considered as a strategy. There is, however controversy on whether a reduction in tissue AGE accumulation (as measured by skin autofluorescence) can be achieved by reduction of AGE food intake in HD patients.

14.5.1.2 Low Advanced Glycation End-Products in Peritoneal Dialysis (PD) Solutions

PD solutions are sterilized by heat, which produces AGEs in most standard glucose-based PD solutions [4, 41]. The length of PD and the glucose exposure dose are independently associated with the level of SAF in PD patients. The difficulty with increased AGE accretion due to PD treatment may be fixed using neutral PD solutions with low AGE content [4, 41]. They have been available for more than 10 years on the market and only recently approved by the FDA in the USA.

14.5.1.3 Advanced Hemodialysis Techniques

The practice of diverse HD techniques to reduce AGE accumulation in HD patients has been studied [41, 94–97]. A comparison of the removal of free plasma AGE and AGE peptides by low, high, and super flux HD revealed that all of them efficiently remove free plasma AGE during a single HD session. Protein-bound AGE were not reduced during dialysis sessions with either modality. Super flux HD may be capable of reducing AGE peptides in the long-term [95].

Polysulfone membranes produce lower levels of plasma AGE than other membranes, such as modified or unmodified cellulose [95]. Vitamin E-coated HD membrane reduced plasma AGE levels. The use of ultrapure HD fluid results in lower AGE levels in HD patients. Consequently, reduction of AGE accumulation can be achieved by using emergent HD techniques that employ

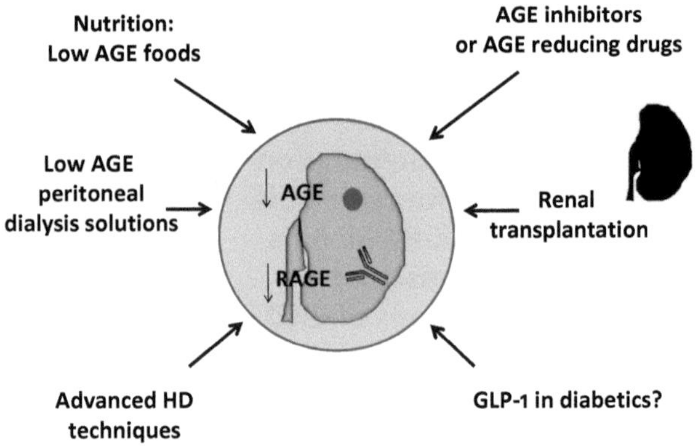

Fig. 14.5 Therapeutic avenues to target the AGE-RAGE axis. The diagram summarizes some of the approached to reduce ligand burden in CKD: nutrition, better dialysis solutions, advanced dialysis procedures, transplantation and some drugs

large pores membranes such as super flux, HDF or protein leaking HD since they filter protein-bound AGE. Employing biocompatible membranes and ultra-pure HD fluid can curtail AGE accumulation, by diminishing the level ofinflammation and immune response [41, 94–96, 98].

Daily or home dialysis could accomplish better removal of plasma AGE and thus lower AGE accumulation. Clinical data on the effects of this kind of alternative HD techniques on CVD events and mortality are lacking and deserve exploration.

14.5.1.4 Advanced Glycation End-Products Reducing Drugs

In vitro and animal experiments have shown that the specific AGE inhibitor aminoguanidine and the AGEs crosslink breaker alagebrium, as well as the B vitamins pyridoxamine, thiamine, and its synthetic derivative, benfotiamine [3, 39, 88, 99]. Several herbal medicines and experimental drugs can reduce of AGE accumulation as we and others have shown *in vitro* [100–108]. The conceivable clinical usefulness of these interventions, however, remains to be established. Clinical studies with the specific AGE inhibitor, aminoguanidine and the AGE breaker, alagebrium, unraveled problems with safety or efficacy [1, 7, 8, 38, 39, 109–113]. The clinical evidence on the potential AGE-inhibiting effects of B vitamins is still limited [1, 38, 39, 114, 115]. There is some evidence for AGE inhibition by treatments established for pathways unrelated to glycation [46, 99, 116]. Some examples are angiotensin receptor blockers and statins in PD patients [116]. However, due to a sizeable heterogeneity in the available studies design and techniques, there is no consensus [1, 80, 99, 116, 117]. Nonetheless, no long-term study has been conducted in CKD patients with either benfothiamine, piridoxamine or alagebrium [46, 99, 118–120]. Overall, the clinical evidence on interventions to inhibit AGE formation is presently weak and unpersuasive [46, 99, 118–120].

14.5.1.5 Renal Transplantation

Renal transplantation is the ultimate treatment of CKD and consequently should lead to total reversibility of AGE accumulation, at least in theory [4, 97]. Certainly, protein-bound serum pentosidine from renal transplant patients are similar to healthy subjects. Data on SAF however, indicates that patients with renal transplant maintain high tissue levels AGE [4, 97]. AGE accumulation in skin and other tissues with slow turnover is obviously not easily reversible, a concept already explained earlier in regards to the diabetic “glycemic memory” hypothesis.

14.5.2 Upcoming Therapeutic Strategies for CKD in Diabetes: The Role of the Mesangial Cell

Mesangial cells occupy a central anatomical position in the glomerulus, playing crucial roles in maintaining structure and function of glomerular capillary tufts. AGE can induce mesangial apoptosis and dysfunction which may contribute in part to glomerular hyperfiltration, an early renal dysfunction in diabetes [121, 122]. The incretin receptor GLP-1R is expressed in mesangial cells and proximal tubular cells and administration of GLP-1 reduces RAGE mRNA and protein levels and consequently inhibits the AGEs-induced ROS generation and MCP- 1 expression in human cultured mesangial cells [121]. These and other findings suggest that GLP-1 could inhibit the harmful effects of AGEs-RAGE axis on mesangial cells via GLP-1R-mediated cAMP elevation. AGEs induce RAGE and MCP-1 expression in mesangial cells via NADPH oxidase-mediated ROS generation. AGEs-RAGE-induced ROS generation augments RAGE expression in mesangial cells engaging in a vicious cycle [121]. So, the positive feedback loop between RAGE and NADPH oxidase-mediated ROS generation may be a molecular target of GLP-1- GLP-1R-cAMP axis in mesangial cells. Decreased production and/or bioavailability of NO play a role in vascular complications in diabetes. Serum levels of asymmetric dimethylarginine (ADMA), an endogenous NO synthase inhibitor, is increased in early diabetic nephropathy in type- 1 diabetes and associated with future cardiovascular events in these individuals. Serum levels of AGEs are positively associated with soluble form of RAGE and ADMA in patients with chronic kidney disease, thus suggesting the active involvement of the AGEs-RAGE system in the elevated levels of ADMA. Suppression of the AGEs-RAGE-induced ADMA generation by GLP-1 may be a novel therapeutic target for nephropathy and other vascular complications in diabetes [121].

References

1. Smit AJ, Gerrits EG. Skin autofluorescence as a measure of advanced glycation endproduct deposition: a novel risk marker in chronic kidney disease. Curr Opin Nephrol Hypertens. 2010;19:527–33.
2. Koyama H, Nishizawa Y. AGEs/RAGE in CKD: irreversible metabolic memory road toward CVD? Eur J Clin Invest. 2010;40:623–35.
3. Thallas-Bonke V, Coughlan MT, Tan AL, Harcourt BE, Morgan PE, Davies MJ, et al. Targeting the AGE-RAGE axis improves renal function in the context of a healthy diet low in advanced glycation end-product content. Nephrology (Carlton). 2013;18:47–56.
4. Arsov S, Graaff R, van Oeveren W, Stegmayr B, Sikole A, Rakhorst G, et al. Advanced glycation end-products and skin autofluorescence in end-stage renal disease: a review. Clin Chem Lab Med. 2013;4:1–10.
5. Yamagishi S, Matsui T. Advanced glycation end products, oxidative stress and diabetic nephropathy. Oxid Med Cell Longev. 2010;3:101–8.
6. Schmidt AM, Stern D. Atherosclerosis and diabetes: the RAGE connection. Curr Atheroscler Rep. 2000;2:430–6.
7. Jerums G, Panagiotopoulos S, Forbes J, Osicka T, Cooper M. Evolving concepts in advanced glycation, diabetic nephropathy, and diabetic vascular disease. Arch Biochem Biophys. 2003;419:55–62.
8. Heidland A, Sebekova K, Schinzel R. Advanced glycation end products and the progressive course of renal disease. Am J Kidney Dis. 2001;38(4 Suppl 1):S100–6.
9. Vlassara H. Protein glycation in the kidney: role in diabetes and aging. Kidney Int. 1996;49:1795–804.
10. Gugliucci A, Bendayan M. Renal fate of circulating advanced glycated end products (AGE): evidence for reabsorption and catabolism of AGE-peptides by renal proximal tubular cells. Diabetologia. 1996;39:149–60.
11. Vlassara H. Advanced glycation in diabetic renal and vascular disease. Kidney Int Suppl. 1995;51:S43–4.
12. Dyer DG, Blackledge JA, Katz BM, Hull CJ, Adkisson HD, Thorpe SR, et al. The Maillard reaction in vivo. Z Ernahrungswiss. 1991;30:29–45.
13. Mullarkey CJ, Edelstein D, Brownlee M. Free radical generation by early glycation products: a mechanism for accelerated atherogenesis in diabetes. Biochem Biophys Res Commun. 1990;173:932–9.
14. Watkins NG, Neglia-Fisher CI, Dyer DG, Thorpe SR, Baynes JW. Effect of phosphate on the kinetics and specificity of glycation of protein. J Biol Chem. 1987;262:7207–12.
15. Thornalley PJ. Advanced glycation end products in renal failure. J Ren Nutr. 2006;16:178–84.
16. Thornalley PJ. Glycation free adduct accumulation in renal disease: the new AGE. Pediatr Nephrol. 2005;20:1515–22.
17. Brouwers O, Niessen PM, Miyata T, Ostergaard JA, Flyvbjerg A, Peutz-Kootstra CJ, et al. Glyoxalase-1 overexpression reduces endothelial dysfunction and attenuates early renal impairment in a rat model of diabetes. Diabetologia. 2014;57:224–35.
18. Vlassara H, Torreggiani M, Post JB, Zheng F, Uribarri J, Striker GE. Role of oxidants/inflammation in declining renal function in chronic kidney disease and normal aging. Kidney Int Suppl. 2009;114:S3–11.
19. Linden E, Cai W, He JC, Xue C, Li Z, Winston J, et al. Endothelial dysfunction in patients with chronic kidney disease results from advanced glycation end products (AGE)-mediated inhibition of endothelial nitric oxide synthase through RAGE activation. Clin J Am Soc Nephrol. 2008;3:691–8.
20. Uribarri J, Peppa M, Cai W, Goldberg T, Lu M, He C, et al. Restriction of dietary glycotoxins reduces excessive advanced glycation end products in renal failure patients. J Am Soc Nephrol. 2003;14:728–31.
21. Uribarri J, Peppa M, Cai W, Goldberg T, Lu M, Baliga S, et al. Dietary glycotoxins correlate with circulating advanced glycation end product levels in renal failure patients. Am J Kidney Dis. 2003;42:532–8.
22. Gugliucci A, Bendayan M. Reaction of advanced glycation endproducts with renal tissue from normal and streptozotocin-induced diabetic rats: an ultrastructural study using colloidal gold cytochemistry. J Histochem Cytochem. 1995;43:591–600.
23. Popolo A, Autore G, Pinto A, Marzocco S. Oxidative stress in patients with cardiovascular disease and chronic renal failure. Free Radic Res. 2013;47:346–56.
24. Bowman MA, Schmidt AM. The next generation of RAGE modulators: implications for soluble RAGE therapies in vascular inflammation. J Mol Med (Berl). 2013;91:1329–31.
25. Ramasamy R, Yan SF, Schmidt AM. Advanced glycation endproducts: from precursors to RAGE: round and round we go. Amino Acids. 2012;42:1151–61.
26. Monnier VM, Sell DR, Nagaraj RH, Miyata S, Grandhee S, Odetti P, et al. Maillard reaction-mediated molecular damage to extracellular matrix

and other tissue proteins in diabetes, aging, and uremia. Diabetes. 1992;41 Suppl 2:36–41.
27. Dyer DG, Dunn JA, Thorpe SR, Bailie KE, Lyons TJ, McCance DR, et al. Accumulation of Maillard reaction products in skin collagen in diabetes and aging. J Clin Invest. 1993;91:2463–9.
28. Lee AT, Cerami A. Role of glycation in aging. Ann N Y Acad Sci. 1992;663:63–70.
29. Bucala R, Makita Z, Vega G, Grundy S, Koschinsky T, Cerami A, et al. Modification of low density lipoprotein by advanced glycation end products contributes to the dyslipidemia of diabetes and renal insufficiency. Proc Natl Acad Sci U S A. 1994; 91:9441–5.
30. Cerami C, Founds H, Nicholl I, Mitsuhashi T, Giordano D, Vanpatten S, et al. Tobacco smoke is a source of toxic reactive glycation products. Proc Natl Acad Sci U S A. 1997;94:13915–20.
31. Thornalley PJ, Rabbani N. Highlights and hotspots of protein glycation in end-stage renal disease. Semin Dial. 2009;22:400–4.
32. Grandhee SK, Monnier VM. Mechanism of formation of the Maillard protein cross-link pentosidine. Glucose, fructose, and ascorbate as pentosidine precursors. J Biol Chem. 1991;266:11649–53.
33. Sell DR, Nagaraj RH, Grandhee SK, Odetti P, Lapolla A, Fogarty J, et al. Pentosidine: a molecular marker for the cumulative damage to proteins in diabetes, aging, and uremia. Diabetes Metab Rev. 1991;7:239–51.
34. Hricik DE, Schulak JA, Sell DR, Fogarty JF, Monnier VM. Effects of kidney or kidney-pancreas transplantation on plasma pentosidine. Kidney Int. 1993;43:398–403.
35. Agalou S, Ahmed N, Babaei-Jadidi R, Dawnay A, Thornalley PJ. Profound mishandling of protein glycation degradation products in uremia and dialysis. J Am Soc Nephrol. 2005;16:1471–85.
36. Rabbani N, Sebekova K, Sebekova Jr K, Heidland A, Thornalley PJ. Accumulation of free adduct glycation, oxidation, and nitration products follows acute loss of renal function. Kidney Int. 2007;72:1113–21.
37. Rabbani N, Thornalley PJ. Quantitation of markers of protein damage by glycation, oxidation, and nitration in peritoneal dialysis. Perit Dial Int. 2009;29 Suppl 2:S51–6.
38. Karachalias N, Babaei-Jadidi R, Rabbani N, Thornalley PJ. Increased protein damage in renal glomeruli, retina, nerve, plasma and urine and its prevention by thiamine and benfotiamine therapy in a rat model of diabetes. Diabetologia. 2010; 53:1506–16.
39. Kihm LP, Muller-Krebs S, Klein J, Ehrlich G, Mertes L, Gross ML, et al. Benfotiamine protects against peritoneal and kidney damage in peritoneal dialysis. J Am Soc Nephrol. 2011;22:914–26.
40. Mallipattu SK, He JC, Uribarri J. Role of advanced glycation endproducts and potential therapeutic interventions in dialysis patients. Semin Dial. 2012;25:529–38.
41. Kim YL, Cho JH, Choi JY, Kim CD, Park SH. Systemic and local impact of glucose and glucose degradation products in peritoneal dialysis solution. J Ren Nutr. 2013;23:218–22.
42. Vlassara H. Serum advanced glycosylation end products: a new class of uremic toxins? Blood Purif. 1994;12:54–9.
43. Horl WH. Genesis of the uraemic syndrome: role of uraemic toxins. Wien Klin Wochenschr. 1998; 110:511–20.
44. Gugliucci A, Kinugasa E, Ogata H, Caccavello R, Kimura S. Activation of paraoxonase 1 after hemodialysis is associated with HDL remodeling and its increase in the HDL fraction and VLDL. Clin Chim Acta. 2013;430C:9–14.
45. Gugliucci A, Mehlhaff K, Kinugasa E, Ogata H, Hermo R, Schulze J, et al. Paraoxonase-1 concentrations in end-stage renal disease patients increase after hemodialysis: correlation with low molecular AGE adduct clearance. Clin Chim Acta. 2007; 377:213–20.
46. Wihler C, Schafer S, Schmid K, Deemer EK, Munch G, Bleich M, et al. Renal accumulation and clearance of advanced glycation end-products in type 2 diabetic nephropathy: effect of angiotensin-converting enzyme and vasopeptidase inhibition. Diabetologia. 2005;48:1645–53.
47. Nathan DM, Cleary PA, Backlund JY, Genuth SM, Lachin JM, Orchard TJ, et al. Intensive diabetes treatment and cardiovascular disease in patients with type 1 diabetes. N Engl J Med. 2005;353:2643–53.
48. Coughlan MT, Mibus AL, Forbes JM. Oxidative stress and advanced glycation in diabetic nephropathy. Ann N Y Acad Sci. 2008;1126:190–3.
49. Siems W, Quast S, Carluccio F, Wiswedel I, Hirsch D, Augustin W, et al. Oxidative stress in chronic renal failure as a cardiovascular risk factor. Clin Nephrol. 2002;58 Suppl 1:S12–9.
50. Massy ZA, Nguyen-Khoa T. Oxidative stress and chronic renal failure: markers and management. J Nephrol. 2002;15:336–41.
51. Miyata T, Hori O, Zhang J, Yan SD, Ferran L, Iida Y, et al. The receptor for advanced glycation end products (RAGE) is a central mediator of the interaction of AGE-beta2microglobulin with human mononuclear phagocytes via an oxidant-sensitive pathway. Implications for the pathogenesis of dialysis-related amyloidosis. J Clin Invest. 1996;98:1088–94.
52. Schmidt AM, Hori O, Cao R, Yan SD, Brett J, Wautier JL, et al. RAGE: a novel cellular receptor for advanced glycation end products. Diabetes. 1996;45 Suppl 3:S77–80.
53. Tanji N, Markowitz GS, Fu C, Kislinger T, Taguchi A, Pischetsrieder M, et al. Expression of advanced glycation end products and their cellular receptor RAGE in diabetic nephropathy and nondiabetic renal disease. J Am Soc Nephrol. 2000;11:1656–66.

54. Yamagishi S, Takeuchi M, Inagaki Y, Nakamura K, Imaizumi T. Role of advanced glycation end products (AGEs) and their receptor (RAGE) in the pathogenesis of diabetic microangiopathy. Int J Clin Pharmacol Res. 2003;23:129–34.
55. Rodriguez-Ayala E, Anderstam B, Suliman ME, Seeberger A, Heimburger O, Lindholm B, et al. Enhanced RAGE-mediated NFkappaB stimulation in inflamed hemodialysis patients. Atherosclerosis. 2005;180:333–40.
56. Coughlan MT, Cooper ME, Forbes JM. Renal microvascular injury in diabetes: RAGE and redox signaling. Antioxid Redox Signal. 2007;9:331–42.
57. Daffu G, del Pozo CH, O'Shea KM, Ananthakrishnan R, Ramasamy R, Schmidt AM. Radical roles for RAGE in the pathogenesis of oxidative stress in cardiovascular diseases and beyond. Int J Mol Sci. 2013;14:19891–910.
58. Zhou LL, Cao W, Xie C, Tian J, Zhou Z, Zhou Q, et al. The receptor of advanced glycation end products plays a central role in advanced oxidation protein products-induced podocyte apoptosis. Kidney Int. 2012;82:759–70.
59. Yamamoto Y, Yamamoto H. Interaction of receptor for advanced glycation end products with advanced oxidation protein products induces podocyte injury. Kidney Int. 2012;82:733–5.
60. Ramasamy R, Yan SF, Schmidt AM. Receptor for AGE (RAGE): signaling mechanisms in the pathogenesis of diabetes and its complications. Ann N Y Acad Sci. 2011;1243:88–102.
61. Yan SF, Ramasamy R, Schmidt AM. The RAGE axis: a fundamental mechanism signaling danger to the vulnerable vasculature. Circ Res. 2010;106:842–53.
62. Reiniger N, Lau K, McCalla D, Eby B, Cheng B, Lu Y, et al. Deletion of the receptor for advanced glycation end products reduces glomerulosclerosis and preserves renal function in the diabetic OVE26 mouse. Diabetes. 2010;59:2043–54.
63. D'Agati V, Yan SF, Ramasamy R, Schmidt AM. RAGE, glomerulosclerosis and proteinuria: roles in podocytes and endothelial cells. Trends Endocrinol Metab. 2010;21:50–6.
64. D'Agati V, Schmidt AM. RAGE and the pathogenesis of chronic kidney disease. Nat Rev Nephrol. 2010;6:352–60.
65. Cottone S, Lorito MC, Riccobene R, Nardi E, Mule G, Buscemi S, et al. Oxidative stress, inflammation and cardiovascular disease in chronic renal failure. J Nephrol. 2008;21:175–9.
66. Nakashima A, Carrero JJ, Qureshi AR, Miyamoto T, Anderstam B, Barany P, et al. Effect of circulating soluble receptor for advanced glycation end products (sRAGE) and the proinflammatory RAGE ligand (EN-RAGE, S100A12) on mortality in hemodialysis patients. Clin J Am Soc Nephrol. 2010;5:2213–9.
67. Kalousova M, Jachymova M, Mestek O, Hodkova M, Kazderova M, Tesar V, et al. Receptor for advanced glycation end products–soluble form and gene polymorphisms in chronic haemodialysis patients. Nephrol Dial Transplant. 2007;22:2020–6.
68. Forbes JM, Thorpe SR, Thallas-Bonke V, Pete J, Thomas MC, Deemer ER, et al. Modulation of soluble receptor for advanced glycation end products by angiotensin-converting enzyme-1 inhibition in diabetic nephropathy. J Am Soc Nephrol. 2005;16:2363–72.
69. Leonardis D, Basta G, Mallamaci F, Cutrupi S, Pizzini P, Tripepi R, et al. Circulating soluble receptor for advanced glycation end product (sRAGE) and left ventricular hypertrophy in patients with chronic kidney disease (CKD). Nutr Metab Cardiovasc Dis. 2012;22:748–55.
70. Yan SF, Ramasamy R, Schmidt AM. Soluble RAGE: therapy and biomarker in unraveling the RAGE axis in chronic disease and aging. Biochem Pharmacol. 2010;79:1379–86.
71. Raposeiras-Roubin S, Rodino-Janeiro BK, Grigorian-Shamagian L, Moure-Gonzalez M, Seoane-Blanco A, Varela-Roman A, et al. Soluble receptor of advanced glycation end products levels are related to ischaemic aetiology and extent of coronary disease in chronic heart failure patients, independent of advanced glycation end products levels: New Roles for Soluble RAGE. Eur J Heart Fail. 2010;12:1092–100.
72. Vazzana N, Santilli F, Cuccurullo C, Davi G. Soluble forms of RAGE in internal medicine. Intern Emerg Med. 2009;4:389–401.
73. Sung JY, Chung W, Kim AJ, Kim HS, Ro H, Chang JH, et al. Calcitriol treatment increases serum levels of the soluble receptor of advanced glycation end products in hemodialysis patients with secondary hyperparathyroidism. Tohoku J Exp Med. 2013;230:59–66.
74. Zakiyanov O, Kalousova M, Kriha V, Zima T, Tesar V. Serum S100A12 (EN-RAGE) levels in patients with decreased renal function and subclinical chronic inflammatory disease. Kidney Blood Press Res. 2011;34:457–64.
75. Menini T, Ikeda H, Kimura S, Gugliucci A. Circulating soluble RAGE increase after a cerebrovascular event. Clin Chem Lab Med. 2014;52:109–16.
76. Mahajan N, Mahmood S, Jain S, Dhawan V. Receptor for advanced glycation end products (RAGE), inflammatory ligand EN-RAGE and soluble RAGE (sRAGE) in subjects with Takayasu's arteritis. Int J Cardiol. 2013;168:532–4.
77. Fujisawa K, Katakami N, Kaneto H, Naka T, Takahara M, Sakamoto F, et al. Circulating soluble RAGE as a predictive biomarker of cardiovascular event risk in patients with type 2 diabetes. Atherosclerosis. 2013;227:425–8.
78. Falcone C, Bozzini S, Guasti L, D'Angelo A, Capettini AC, Paganini EM, et al. Soluble RAGE plasma levels in patients with coronary artery disease

and peripheral artery disease. ScientificWorldJournal. 2013;2013:584504.
79. Skrha Jr J, Kalousova M, Svarcova J, Muravska A, Kvasnicka J, Landova L, et al. Relationship of soluble RAGE and RAGE ligands HMGB1 and EN-RAGE to endothelial dysfunction in type 1 and type 2 diabetes mellitus. Exp Clin Endocrinol Diabetes. 2012;120:277–81.
80. Sourris KC, Morley AL, Koitka A, Samuel P, Coughlan MT, Penfold SA, et al. Receptor for AGEs (RAGE) blockade may exert its renoprotective effects in patients with diabetic nephropathy via induction of the angiotensin II type 2 (AT2) receptor. Diabetologia. 2010;53:2442–51.
81. Ishibashi Y, Yamagishi S, Matsui T, Ohta K, Tanoue R, Takeuchi M, et al. Pravastatin inhibits advanced glycation end products (AGEs)-induced proximal tubular cell apoptosis and injury by reducing receptor for AGEs (RAGE) level. Metabolism. 2012;61:1067–72.
82. Tang SC, Chan LY, Leung JC, Cheng AS, Lin M, Lan HY, et al. Differential effects of advanced glycation end-products on renal tubular cell inflammation. Nephrology (Carlton). 2011;16:417–25.
83. Yamagishi S, Nakamura K, Matsui T, Ueda S, Fukami K, Okuda S. Agents that block advanced glycation end product (AGE)-RAGE (receptor for AGEs)-oxidative stress system: a novel therapeutic strategy for diabetic vascular complications. Expert Opin Investig Drugs. 2008;17:983–96.
84. Gada E, Owens AW, Gore MO, See R, Abdullah SM, Ayers CR, et al. Discordant effects of rosiglitazone on novel inflammatory biomarkers. Am Heart J. 2013;165:609–14.
85. Lanati N, Emanuele E, Brondino N, Geroldi D. Soluble RAGE-modulating drugs: state-of-the-art and future perspectives for targeting vascular inflammation. Curr Vasc Pharmacol. 2010;8:86–92.
86. Tan KC, Chow WS, Tso AW, Xu A, Tse HF, Hoo RL, et al. Thiazolidinedione increases serum soluble receptor for advanced glycation end-products in type 2 diabetes. Diabetologia. 2007;50:1819–25.
87. Daroux M, Prevost G, Maillard-Lefebvre H, Gaxatte C, D'Agati VD, Schmidt AM, et al. Advanced glycation end-products: implications for diabetic and non-diabetic nephropathies. Diabetes Metab. 2010;36:1–10.
88. Harcourt BE, Sourris KC, Coughlan MT, Walker KZ, Dougherty SL, Andrikopoulos S, et al. Targeted reduction of advanced glycation improves renal function in obesity. Kidney Int. 2011;80:190–8.
89. Busch M, Franke S, Ruster C, Wolf G. Advanced glycation end-products and the kidney. Eur J Clin Invest. 2010;40:742–55.
90. Maillard-Lefebvre H, Boulanger E, Daroux M, Gaxatte C, Hudson BI, Lambert M. Soluble receptor for advanced glycation end products: a new biomarker in diagnosis and prognosis of chronic inflammatory diseases. Rheumatology (Oxford). 2009;48:1190–6.
91. Nishizawa Y, Koyama H. Endogenous secretory receptor for advanced glycation end-products and cardiovascular disease in end-stage renal disease. J Ren Nutr. 2008;18:76–82.
92. Kim JK, Park S, Lee MJ, Song YR, Han SH, Kim SG, et al. Plasma levels of soluble receptor for advanced glycation end products (sRAGE) and proinflammatory ligand for RAGE (EN-RAGE) are associated with carotid atherosclerosis in patients with peritoneal dialysis. Atherosclerosis. 2012;220:208–14.
93. Inagi R, Yamamoto Y, Nangaku M, Usuda N, Okamato H, Kurokawa K, et al. A severe diabetic nephropathy model with early development of nodule-like lesions induced by megsin overexpression in RAGE/iNOS transgenic mice. Diabetes. 2006;55:356–66.
94. Tanaka K, Nakayama M, Kanno M, Kimura H, Watanabe K, Tani Y, et al. Skin autofluorescence is associated with the progression of chronic kidney disease: a prospective observational study. PLoS One. 2013;8:e83799.
95. Susantitaphong P, Siribamrungwong M, Jaber BL. Convective therapies versus low-flux hemodialysis for chronic kidney failure: a meta-analysis of randomized controlled trials. Nephrol Dial Transplant. 2013;28:2859–74.
96. Mac-Way F, Couture V, Utescu MS, Ignace S, De Serres SA, Loignon RC, et al. Advanced glycation end products, aortic stiffness, and wave reflection in peritoneal dialysis as compared to hemodialysis. Int Urol Nephrol. 2014;46(4):817–24.
97. Crowley LE, Johnson CP, McIntyre N, Fluck RJ, McIntyre CW, Taal MW, et al. Tissue advanced glycation end product deposition after kidney transplantation. Nephron Clin Pract. 2013;124:54–9.
98. Makulska I, Szczepanska M, Drozdz D, Polak-Jonkisz D, Zwolinska D. Skin autofluorescence as a marker of cardiovascular risk in children with chronic kidney disease. Pediatr Nephrol. 2013;28:121–8.
99. Nagai R, Murray DB, Metz TO, Baynes JW. Chelation: a fundamental mechanism of action of AGE inhibitors, AGE breakers, and other inhibitors of diabetes complications. Diabetes. 2012; 61:549–59.
100. Lunceford N, Gugliucci A. Ilex paraguariensis extracts inhibit AGE formation more efficiently than green tea. Fitoterapia. 2005;76:419–27.
101. Gugliucci A, Bastos DH, Schulze J, Souza MF. Caffeic and chlorogenic acids in Ilex paraguariensis extracts are the main inhibitors of AGE generation by methylglyoxal in model proteins. Fitoterapia. 2009;80:339–44.
102. Gugliucci A, Menini T. The polyamines spermine and spermidine protect proteins from structural and functional damage by AGE precursors: a new role for old molecules? Life Sci. 2003;72:2603–16.
103. Gugliucci A, Menini T. The botanical extracts of Achyrocline satureoides and Ilex paraguariensis prevent methylglyoxal-induced inhibition of

plasminogen and antithrombin III. Life Sci. 2002; 72:279–92.
104. Sri Harsha PS, Gardana C, Simonetti P, Spigno G, Lavelli V. Characterization of phenolics, in vitro reducing capacity and anti-glycation activity of red grape skins recovered from winemaking by-products. Bioresour Technol. 2013;140:263–8.
105. Saraswat M, Reddy PY, Muthenna P, Reddy GB. Prevention of non-enzymic glycation of proteins by dietary agents: prospects for alleviating diabetic complications. Br J Nutr. 2009;101:1714–21.
106. Rasheed Z, Anbazhagan AN, Akhtar N, Ramamurthy S, Voss FR, Haqqi TM. Green tea polyphenol epigallocatechin-3-gallate inhibits advanced glycation end product-induced expression of tumor necrosis factor-alpha and matrix metalloproteinase-13 in human chondrocytes. Arthritis Res Ther. 2009;11:R71.
107. Babu PV, Sabitha KE, Shyamaladevi CS. Effect of green tea extract on advanced glycation and cross-linking of tail tendon collagen in streptozotocin induced diabetic rats. Food Chem Toxicol. 2008;46:280–5.
108. Ouyang P, Peng WL, Xu DL, Lai WY, Xu AL. Green tea polyphenols inhibit advanced glycation end product-induced rat vascular smooth muscle cell proliferation. Di Yi Jun Yi Da Xue Xue Bao. 2004;24:247–51.
109. Engelen L, Stehouwer CD, Schalkwijk CG. Current therapeutic interventions in the glycation pathway: evidence from clinical studies. Diabetes Obes Metab. 2013;15:677–89.
110. Desai K, Wu L. Methylglyoxal and advanced glycation endproducts: new therapeutic horizons? Recent Pat Cardiovasc Drug Discov. 2007;2:89–99.
111. Thomas MC, Baynes JW, Thorpe SR, Cooper ME. The role of AGEs and AGE inhibitors in diabetic cardiovascular disease. Curr Drug Targets. 2005;6:453–74.
112. Susic D. Cross-link breakers as a new therapeutic approach to cardiovascular disease. Biochem Soc Trans. 2007;35:853–6.
113. Coughlan MT, Forbes JM, Cooper ME. Role of the AGE crosslink breaker, alagebrium, as a renoprotective agent in diabetes. Kidney Int Suppl. 2007;106:S54–60.
114. Balakumar P, Rohilla A, Krishan P, Solairaj P, Thangathirupathi A. The multifaceted therapeutic potential of benfotiamine. Pharmacol Res. 2010;61:482–8.
115. Beltramo E, Berrone E, Tarallo S, Porta M. Effects of thiamine and benfotiamine on intracellular glucose metabolism and relevance in the prevention of diabetic complications. Acta Diabetol. 2008;45:131–41.
116. Matsui T, Yamagishi S, Takeuchi M, Ueda S, Fukami K, Okuda S. Irbesartan inhibits advanced glycation end product (AGE)-induced proximal tubular cell injury in vitro by suppressing receptor for AGEs (RAGE) expression. Pharmacol Res. 2010;61:34–9.
117. Yamagishi S, Nakamura K, Matsui T, Noda Y, Imaizumi T. Receptor for advanced glycation end products (RAGE): a novel therapeutic target for diabetic vascular complication. Curr Pharm Des. 2008;14:487–95.
118. Williams ME. New potential agents in treating diabetic kidney disease: the fourth act. Drugs. 2006;66:2287–98.
119. Miyata T, van Ypersele de Strihou C, Ueda Y, Ichimori K, Inagi R, Onogi H, et al. Angiotensin II receptor antagonists and angiotensin-converting enzyme inhibitors lower in vitro the formation of advanced glycation end products: biochemical mechanisms. J Am Soc Nephrol. 2002;13:2478–87.
120. Forbes JM, Cooper ME, Thallas V, Burns WC, Thomas MC, Brammar GC, et al. Reduction of the accumulation of advanced glycation end products by ACE inhibition in experimental diabetic nephropathy. Diabetes. 2002;51:3274–82.
121. Ishibashi Y, Nishino Y, Matsui T, Takeuchi M, Yamagishi S. Glucagon-like peptide-1 suppresses advanced glycation end product-induced monocyte chemoattractant protein-1 expression in mesangial cells by reducing advanced glycation end product receptor level. Metabolism. 2011;60:1271–7.
122. Lu C, He JC, Cai W, Liu H, Zhu L, Vlassara H. Advanced glycation endproduct (AGE) receptor 1 is a negative regulator of the inflammatory response to AGE in mesangial cells. Proc Natl Acad Sci U S A. 2004;101:11767–72.

The Chemokine (C-C Motif) Ligand 2 in Neuroinflammation and Neurodegeneration

15

José L.M. Madrigal and Javier R. Caso

Abstract

Among all the chemokines known so far, chemokine (C-C motif) ligand 2 (CCL2) is probably the best characterized. This is mainly due to the therapeutic potential attributed to its regulation. The suppression of CCL2 function may reduce the attraction of immune cells to the sites of inflammation and therefore slow down the progression of inflammation and the tissue damage that may be associated to it. While this has proven to be right in diverse conditions, it has also been described to have deleterious consequences such as a dual effect that is also frequently observed in other endogenous defense systems. This review discusses current knowledge about CCL2 involvement in different neurodegenerative diseases as well as its anti-inflammatory and neuro-protective actions.

Keywords

Alzheimer's disease • CCL2 • MCP-1 • Multiple sclerosis • Neurodegeneration • Neuroinflammation • Parkinson's disease • Stroke

J.L.M. Madrigal (✉)
Departamento de Farmacología, Facultad de Medicina, Universidad Complutense de Madrid (UCM), Centro de Investigación Biomédica en Red de Salud Mental (CIBERSAM), Instituto de Investigación Hospital 12 de Octubre (Imas12), Instituto de Investigación en Neuroquímica UCM, Avda. Complutense s/n, Madrid 28040, Spain
e-mail: jlmmadrigal@med.ucm.es

J.R. Caso
Departamento de Psiquiatría, Facultad de Medicina, Universidad Complutense de Madrid (UCM), Centro de Investigación Biomédica en Red de Salud Mental (CIBERSAM), Instituto de Investigación Hospital 12 de Octubre (Imas12), Instituto de Investigación en Neuroquímica UCM, Madrid, Spain
e-mail: jrcaso@med.ucm.es

15.1 Introduction

As it can be inferred from its common name (MCP-1 or monocyte chemoattractant protein-1), the chemokine (C-C motif) ligand 2 (CCL2) was initially categorized as an attractor for mononuclear cells [1]. Soon after its characterization, many other actions were attributed to this chemokine. Most of these actions were related to its role as a mediator of inflammation and immune function, but further research allowed to discover different actions of CCL2, not only related to cell migration, but seeming to work in the opposite direction allowing for tissue regeneration or even

J. Camps (ed.), *Oxidative Stress and Inflammation in Non-communicable Diseases – Molecular Mechanisms and Perspectives in Therapeutics*, Advances in Experimental Medicine and Biology 824, DOI 10.1007/978-3-319-07320-0_15, preventing damage and protecting against different types of injuries.

CCL2 and its main receptor (CCR2) are expressed by astrocytes, endothelial cells, microglia and neurons [2–8], in different brain regions [3, 4]. This gives us an idea of the relevance of its functions within the central nervous system (CNS). In fact, due to its role in the development of the inflammatory and immune responses, CCL2 is considered as an indicator that allows the detection and quantification of the progression of such process [9]. This gains special relevance in the CNS, an area where the initiation of inflammation can be more difficult to notice, and where its exaggerated progression can have worse consequences than in other areas.

15.2 CCL2 and Alzheimer's Disease

Alzheimer's disease (AD) is considered as one of the most important neurodegenerative diseases nowadays due to its prevalence which is expected to increase in the forthcoming years. In addition to this point, despite multiple efforts from different research perspectives and all the advances made towards the elucidation of the mechanisms regulating AD, there is no medication available that can really decelerate the disease progression. One interesting approach is the reduction of the neuroinflammation associated to AD and the subsequent tissue damage which is undoubtedly responsible for the neuronal degeneration characteristic of AD. It is at this point when chemokines and, particularly, CCL2 become relevant since they are key players in the development of the inflammatory response. However, while many studies propose CCL2 as a therapeutic target whose blockade would help reduce inflammation, others demonstrate that this strategy should be carefully considered because CCL2 correct functioning may be necessary to achieve recovery.

The elevation of CCL2 levels observed in brain tissue extracts [10], cerebrospinal fluid [11], and plasma [12] of AD patients as well as in cases of mild cognitive impairment [13], allows to propose it as an indicator that can help predict the development of AD. Even whether CCL2 by itself may not be a very reliable biomarker, its combination with the cerebrospinal fluid concentration of Tau, P-tau and Aβ42 [13] may altogether constitute a relatively reliable AD biomarker. On the other hand, the lack of CCL2 modifications observed by other authors [14] reduces the consistency of such suggestion. A possible explanation for this discrepancy could be that proposed by Galimberti et al. according to which the induction of CCL2 would take place during the first stages of the disease and be followed by a progressive reduction to pre-AD conditions [15].

This suggests that the role of CCL2 in AD would be more complex than the mere attraction of microglia and astrocytes towards damaged areas. The accumulation of CCL2 in the vicinity of Aβ plaques has been described by studies performed in human brains [16]. For this reason CCL2 has been regarded as a mediator that attracts cells which identify Aβ as a noxious agent to be isolated in order to prevent further damages [17]. However, overproduction of CCL2 (like many other inflammatory mediators) can become and added factor that expands the damage initially caused by Aβ [18]. In agreement with this hypothesis, overexpression of CCL2 in APP transgenic mice (Tg2576) has been described to result in enhanced Aβ deposition, oligomer formation and acceleration of memory impairments [19]. Aβ accumulation in this animal mode was proposed to be mediated through the increased expression of apolipoprotein E [20]. Based on these data, it seems reasonable to propose CCL2 as a therapeutic target since its inhibition could decrease Aβ production and accumulation reducing the damage associated to AD.

However, while the pro-inflammatory role of CCL2 is well known and its deleterious consequences have been abundantly described, its suppression by genetic alterations has proven to have the opposite effect to that expected. One study performed with APP transgenic mice (Tg2576) crossed with CCL2 KO concluded that the presence of both alterations results in accelerated β-amyloidosis, oligomer formation and memory impairment as well as reduced neurogenesis [21]. In agreement with these observations, two separate studies analyzed the consequences of reducing the activity of CCL2 main receptor (CCR2) in APP-AD mouse models by crossing APP and CCR2 KO mice. The first work observed an acceleration of Aβ accumulation together with an elevated rate of mortality [22]. Likewise, the second work also found that the resulting mice had a larger accumulation of Aβ and developed memory impairment and cognitive decline earlier that APP mice [23]. In a related study, the authors observed that the absence of CCL2 or CCR2 worsens the damage caused to photoreceptor cells by subretinal injections of Aβ [24]. According to the results presented in that paper, the attraction of activated microglial cells is regulated by CCL2 and these cells are responsible for the phagocytosis of Aβ and photoreceptor cells debris. These data suggest that some of the detrimental alterations characteristic of AD cannot be attributed to CCL2 and, more interestingly, that CCL2 may help to contain the progression of the disease. Indeed, AD is characterized by a reduction of brain noradrenaline levels [25] and the treatment of mice with the NA precursor l-threo-3,4-dihydroxyphenylserine induces astrocyte production of CCL2 [26] indicating that the loss of noradrenergic neurons may be associated to CCL2 deficits and this could be partly responsible for some AD symptoms.

While these conclusions are based on experimental models that reproduce AD only partially, they could be in agreement with the previously mentioned alterations of CCL2 found in AD patients. This lead us to analyze the mechanisms through which CCL2 may affect the neuronal damage caused by Aβ and the accumulation of plaques. One explanation for this is the attraction of microglia and astrocytes that can phagocyte Aβ. Supporting this concept, it has been proven that CCL2 functions as a signal that stimulates astrocytes migration towards Aβ plaques, and once in contact with the plaque, astrocytes stop their movement and proceed to degrade the deposit [27]. In parallel, similar effects have been described for microglia [28].

Regardless of the possible indirect effects that CCL2 may have on neurons through the attraction of astrocytes or microglia, it may also affect neuronal viability through a direct interaction with these cells. *In vitro* analysis have shown that the treatment with Bindarit, an inhibitor of CCL2 synthesis, reduces neuronal damage in primary mixed neuronal cultures [18]. While CCL2 may be toxic to neurons in culture, previous studies showed that CCL2 is neuroprotective against excitotoxic injuries [29, 30]. This is not directly opposing to the results obtained with the inhibition of CCL2 synthesis, but the possible involvement of glutamate and glutamate receptors in Aβ-induced neuronal alterations [31, 32] makes difficult to segregate both observations.

15.3 CCL2 and Parkinson's Disease

Like AD, Parkinson's disease (PD) is a degenerative disorder characterized by a progressive loss of neurons, being the dopaminergic ones the main type of neurons affected in PD. In addition, AD and PD have some other features in common. Interestingly from this review's perspective, one of these common features could also be the production and expression of CCL2 which is increased in the peripheral mononuclear cells obtained from PD patients [33]. Additionally, the production of CCL2 caused by lipopolysaccharide treatment in cells from PD patients is higher than in the cells obtained from healthy controls. When the analysis of CCL2 levels was performed in cerebrospinal fluid, a correlation was found between the concentration of CCL2 (and other inflammatory markers) and the degree of certain symptoms characteristic of PD such as depression, anxiety, fatigue or cognition [34]. In agreement with this, animal studies in which 1-methyl-4-phenyl-1,2,3,6-tetrahydropyridine (MPTP) administration was used as a model to reproduce PD symptoms, observed that this treatment causes an increase in CCL2 mRNA levels in mice striatum [35]. CCL2 seems to play a key role in the development of PD, however, a direct cause-effect connection between CCL2 and dopaminergic neuronal loss or disease progression could not be established. Furthermore, inspired by this possibility, different authors analyzed the effects of CCL2 suppression on MPTP toxicity, and contrarily to what could be expected, no differences were found between CCL2/CCR2 KO and WT animals. In both types of mice, MPTP caused a similar degree of toxicity for dopaminergic neurons [36] and dopamine levels were similar [37].

According to these data, we can conclude that CCL2 production is induced in the brain of PD patients but this chemokine may not be responsible for the loss of dopamine and dopaminergic neurons characteristic of this disease. Indeed, it seems to have the opposite effect; Guyon et al. described how CCL2 intranigral injections in rats increase the excitability of dopaminergic neurons, dopamine release and locomotor activity [38]. In a parallel process, CCL2 could also help recover the population of dopaminergic neurons since it can promote the differentiation of precursor cells into dopaminergic neurons and neuritogenesis in midbrain dopaminergic neuron cultures [39].

In conclusion, similarly to its potential effects in AD, the levels of CCL2 in PD could be elevated as a result of a compensatory mechanism that in this case may help restore dopamine concentrations.

15.4 CCL2 and Stroke

Cerebrovascular accidents represent one of the most common causes of permanent disability in adults of the developed world, being acute ischemic stroke the third leading cause of death in these regions. A stroke is caused by the suppression of blood supply to the brain; it can be produced by an obstruction of blood vessels or by the loss of their structural integrity, which would result in hemorrhage. Obviously, this type of lesion results in numerous alterations of brain conditions, and depending of the magnitude of the lesion and the time until blood flow is restored, the degree of the damage can range between a minor lesion without functional alterations to the loss of tissue viability and consequently death. One of the main mediators responsible for the cell damage characteristic of this kind of injuries is the excitatory amino acid glutamate which, due to the decreased production of ATP, accumulates in the synaptic space and over activates N-methyl-D-aspartate (NMDA) receptors causing an excessive influx of calcium into the cells [40]. This excitotoxic process is associated to an inflammatory response that seems to be induced by the release of reactive oxygen species and cytokines resulting from the death of neurons and other cells [41]. Chemokines are part of this inflammatory response as they attract leukocytes to the injured area [42]. Experimental studies confirm the induction of CCL2 in ischemic areas [43] and propose the inhibition of CCL2 function as a therapeutic strategy [44, 45]. Accordingly, studies performed on mice using the middle cerebral artery occlusion model of stroke indicate that the genetic suppression of CCL2 [46] or CCR2 [47] results in smaller infarct volumes.

The analysis of cerebrospinal fluid samples shows an elevation of CCL2 in patients with ischemic stroke [48]. This strengthens the therapeutic potential of those strategies aimed at lowering CCL2 function in stroke. However, the increased concentration of CCL2, while attracting immune-related cells, may not be directly responsible for the neuronal damage. In fact, the use of transgenic bone marrow chimeras allowed Schilling et al. to prove that CCL2 is necessary for recruiting blood-borne cells to the injury site but it does not affect microglia activation and migration [49].

The chemoattractant effects of CCL2 seem to contribute to the positioning of immune cells in the vicinity of the injured area and help to contain the progression of the damage, remove debris or attack harmful cells. It is well known that this process can have deleterious consequences in the CNS when the by-products of such reaction become toxic for the properly-functioning nervous cells. In addition to neuronal regeneration and migration of progenitor cells, another function was discovered for CCL2: This chemokine can also attract neuronal precursors and direct them towards injured areas [50].

This effect becomes particularly relevant in pathologies where the brain damage is not widely distributed (as is in AD) but is rather precisely located, such as certain types of trauma or stroke. In ischemic lesions is relatively easy to detect if progenitors migrate to the lesion site. For example, it was shown that CCL2 attracts neural progenitor cells from the subventricular zone towards the injured area created by middle cerebral artery occlusion in rodent brains, and also stimulates their differentiation into neurons [51]. Similar results were found by Yan et al. using CCL2 infusion into the striatum to demonstrate that this chemokine induces the migration of neuroblasts [52]. Besides attracting cells from different brain areas, CCL2 has also been described to promote transendothelial recruitment of neural stem cells [53]. This process could be mediated by the expression of CCL2 from endothelial cells which seems to be necessary for the translocation of monocytes through the blood brain barrier [54].

In addition to attracting progenitor cells, CCL2 has shown to reduce infarct volumes by cooperating in the development of hypoxic preconditioning-induced resistance to ischemic injuries [55]. Interestingly, the authors of this study point to neurons as the main source of CCL2 caused by hypoxic pre-conditioning. While CCL2 expression has been observed for different brain cell types [3], astrocytes seem to be the main source within the CNS [56, 57]. Given the particular conditions leading to the production CCL2 in this model, this chemokine could be considered to act here as a messenger that propagates an alarm signal and helps to prepare the cells for a new aggression of the same nature but of greater intensity. This reminds of another chemokine for which such an effect has been largely described: CX3CL1, also called fractalkine. This chemokine is known to be released by neurons and to activate microglial CX3CR1 [58] receptors inhibiting the activation of these cells and therefore indirectly protecting neurons against an excessive inflammatory response [59]. Thereby, neuronal CCL2 may have an effect similar to that described for CX3CL1.

The main agents responsible for neuronal death during ischemic conditions are considered to be oxygen and glucose deprivation (OGD). Based on this, an *in vitro* model consisting of temporary removal of nutrients and oxygen displacement from the culture medium is commonly used to reproduce ischemic conditions in cell cultures. Thereby, we could observe how astrocytic CCL2 can prevent neuronal death caused by OGD when CCL2 was induced by noradrenaline treatment of astrocytes [60]. However, a similar effect of neuronal CCL2 cannot be discarded.

15.5 CCL2 and Multiple Sclerosis

Multiple sclerosis (MS) is an autoimmune inflammatory disease characterized by the destruction of myelin sheaths and the loss of axons which results in neurological impairment. This damage is mainly attributed to the infiltration of autoreactive T cells. In order for this infiltration to be produced, the blood brain barrier integrity has to be altered and a transendothelial passage of leukocytes must take place. Since the migration of immune cells can be directed by chemokines, CCL2 is considered a key factor in the development of MS. In support of this hypothesis, CCL2 serum levels have been found to be elevated in MS patients, being their concentration directly related to the severity of the disease [61]. Deeper examinations performed on brain samples obtained from human MS patients confirm the involvement of CCL2 in MS lesions. In these studies, immunohistochemical and *in situ* hybridization techniques allowed to localize CCL2 in the injured areas [62]. Similar studies performed by Simpson et al. confirm the presence of CCL2 within MS lesions and indicate that this chemokine is produced by reactive astrocytes in the parenchyma surrounding the injured area [63]. This research group also described the presence of foamy macrophages, infiltrating lymphocytes and activated microglia expressing CCR2 in the vicinity of MS lesions [64].

These data are reinforced by animal studies using a MS model such as experimental autoimmune encephalomyelitis (EAE). In this model, the correct functioning of CCR2 receptor [65, 66] is necessary for the induction of EAE.

In agreement with this, the use of bindarit, an inhibitor of CCL2 synthesis, has been shown to reduce the incidence and onset of the disease, and even produce some signs of reversal [67]. However, another study performed on mice with a genetic modification that caused a sustained expression of CCL2 in astrocytes, indicates that this alteration prevents the development of EAE [68].

15.6 CCL2 in Other CNS Pathologies

Probably due to the large number of situations where chemoattraction is involved, CCL2 has been related to many other CNS-related pathologies besides from those mentioned above. One of them is amyotrophic lateral sclerosis (ALS), a neurodegenerative disease characterized by a degeneration of motor neurons that causes progressive muscle atrophy and weakness. As for other neurodegenerative diseases, elevated concentrations of CCL2 have been found in serum and cerebrospinal fluid samples obtained from ALS patients [69] as well as in peripheral blood mononuclear cells [70]. These data suggest that CCL2 function may contribute to the progression of ALS-associated neurodegeneration.

The ability of human immunodeficiency virus (HIV) to cross the blood-brain barrier can be the cause of alterations in brain homeostasis. These alterations result in neurological complications comprised in the medical term HIV-associated neurocognitive disorders (HAND). This disorder, depending on the degree and type of lesion present, may have different manifestations, being dementia the most characteristic, and it is commonly referred to as HIV-associated dementia (HAD). Before the use of antiretroviral therapies, between 40 and 60 % of patients infected with HIV developed HAD; fortunately, new medications helped to reduce the prevalence of this pathology, but still there is a large proportion of HIV patients who develop HAD, particularly at advanced stages of the disease. HAD is characterized by the activation of glial cells and the death of neurons throughout the brain [71]. This death could be due to the direct interaction of neurons with viral proteins released from infected monocyte-derived cells [72], it could also be the result of the inflammatory response mounted by non-neuronal cells [73, 74].

Regardless of the mechanism responsible for neuronal degeneration, in order to exert a direct effect on brain cells, HIV must first cross the blood-brain barrier. This step is in part regulated by chemokines since they are responsible for the attraction of perivascular macrophages and $CD4^+$ T cells that serve as vehicles for HIV and allow its entrance into the CNS. CCL2 concentrations have been found to be elevated in cerebrospinal fluid samples obtained from HAD patients [75, 76]. The elevation of CCL2 can be induced by Tat protein, the transactivator of HIV-1 long-terminal repeat necessary for viral replication. Direct induction of CCL2 by Tat has been described in rat hippocampal slice cultures [77], and in the vascular endothelium [78].

The increase of CCL2 found in HAD samples, the direct induction of CCL2 caused by Tat and the ability of CCL2 to attract glial cells (which could be responsible for the neuronal damage associated to HAD), all suggest that CCL2 may contribute to the progression of HAD. However, the analysis of CCL2 effect on cultured human neurons and astrocytes revealed a protective effect for this chemokine against Tat-induced apoptosis. This neuroprotection seems to be related to the excitotoxic pathway of glutamate because CCL2 treatment reduced the increase of extracellular glutamate and the expression of NMDA receptor 1 induced by Tat in this *in vitro* model of HAD [30]. Further analysis of this effect allowed confirming it and discovering that CCL2 requires viable transient receptor potential canonical channels to suppress Tat toxicity [79].

While the involvement of CCL2 in other neurological disorders has not been so thoroughly examined as for those here mentioned so far, CCL2 elevations have also been found in schizophrenia [80], traumatic brain injury [81] and epilepsy [82] among others. These data confirm the clinical relevance of CCL2, but more information is still needed in order to be able to decide whether CCL2 blockade would be helpful from a therapeutic point of view or on the contrary, a potentiation of its actions is necessary to help our defense mechanisms cope with these pathologies.

15.7 Neuroprotection by CCL2

Some of the best known neuroprotective actions of CCL2 have already been mentioned above, particularly the prevention of glutamate toxicity observed in stroke and HAD models. This anti-excitotoxic potential of CCL2 was previously described by Bruno et al. in a study analyzing thr protective actions of several chemokines against NMDA and β amyloid peptide [29]. CCL2 may share it anti-excitotoxic effects with other chemokines; for example, CXCL16 can prevent the degeneration of neurons caused by exposure to elevated amounts of glutamate or to OGD. More interestingly, this effect seems to be mediated by CCL2 too, because CXCL16 stimulates astrocytes to release CCL2 and the neuroprotection provided by CXCL16 is reduced in the presence of a CCL2 neutralizing antibody [83].

Besides glutamate and NMDA receptor activation, CCL2 has also shown to reduce neuronal damage caused by methyl mercury (MeHg), a pollutant known to affect brain development and neuronal activity that can lead to the destruction of these cells. The protection provided by CCL2 in this instance is proposed by the authors of the study to be mediated by the upregulation of antioxidant systems [84].

CCL2 has also proven to provide protection against another potentially neurotoxic agent such as ethanol [85]. In this case, a genetic alteration causing chronically elevated levels of CCL2 in the hippocampus was shown to prevent the development of long term depression produced by ethanol exposure.

The ability of CCL2 to help reduce neuronal damage in such different scenarios suggests its regulation of additional anti-inflammatory mechanisms other than just the interference with NMDA-related excito-toxic pathways. Semple et al. analyzed this possibility using astrocyte cultures obtained from wild type and CCL2-deficient mice. They observed that CCL2 suppression results in an exacerbation of the production of inflammatory cytokines, while pretreatment with CCL2 not only did not induce the production of inflammatory cytokines, but reduced IL-6 protein and gene expression induced by IL-1β treatment [86]. Another inflammatory cytokine such as IL-12 has also been described to be regulated by CCL2 in human monocytes, avoiding the induction caused by the inflammatory agents Staphylococcus aureus Cowan strain 1 (SAC) and IFN-gamma [87].

The lack of inflammatory alterations or the suppression of cytokine release found in cell cultures treated with CCL2 may not accurately reproduce CCL2 actions *in vivo* where those cells are interacting with many other cell types and substances present in their environment. This way, while protecting neurons against glutamate or reducing the production of inflammatory cytokines from astrocytes, CCL2 may attract and activate microglia, the CNS main inflammatory cells, a process that could result to be toxic for neurons. In an attempt to explore this possibility, we treated primary rat microglia cultures with different concentrations of CCL2 and could not find any signs of activation or the production of pro-inflammatory mediators. Additionally, CCL2-treated microglia cells were cultured on top of primary neurons while separated by a membrane that prevented cell migration but allowed for the passage of smaller molecules. This conditions did not cause the death of neurons, suggesting that CCL2 does not activate microglia or cause any change on them that leads to the production of neurotoxic substances [88].

Moreover, in this study we also considered the possible inhibition of microglial trophic factors by CCL2 as a way to alter neuronal viability. None of the factors analyzed was found to be inhibited by CCL2 but, surprisingly, insulin like growth factor 1 (IGF) was induced. IGF is known to protect neurons and to promote the proliferation of neuronal progenitors [89].

15.8 Conclusions

The review of all available information regarding CCL2 involvement in neuroinflammation and neurodegeneration confirms the relevance of this chemokine in CNS activity. CCL2 was first known for its ability to attract leukocytes to injured areas. Therefore, it was commonly thought that the suppression of its actions would help to slowdown the progression of inflammation and in this way reduce the associated damage. However, later findings attributed more complex roles to CCL2 and warned about the risk of CCL2 blockade which, not only may not reduce the progression of inflammation, but can also develop new unexpected alterations. Having this point in mind, it does not seem unreasonable to propose that CCL2 actions may depend on its concentration, having a protective/homeostasis-maintenance function at constitutive levels and being toxic at higher concentrations. This phenomenon has been observed in retinal ganglion cells [90] and is in agreement with our recent observations [91]. In this study we observed that while in control conditions a stimulus such as noradrenaline, induces CCL2 expression and synthesis by astrocytes, in the presence of an inflammatory agent such as lipopolysaccharide that causes the production of large amounts of CCL2, the same stimulus has the opposite effect lowering CCL2 levels. This suggests the existence of mechanisms that control CCL2 synthesis preventing an excessive release while maintaining the levels necessary to support homeostasis. The elucidation of these mechanisms will probably provide new information that will help understand CCL2 nature and provide the basis for the development of new therapeutic strategies for neurodegenerative diseases.

Acknowledgements This work was supported by a grant from the Spanish Ministry of Science and Innovation (SAF2010-21948) to JLMM.

References

1. Yoshimura T, Robinson EA, Tanaka S, Appella E, Kuratsu J, Leonard EJ. Purification and amino acid analysis of two human glioma-derived monocyte chemoattractants. J Exp Med. 1989;169(4):1449–59.
2. Andjelkovic AV, Kerkovich D, Shanley J, Pulliam L, Pachter JS. Expression of binding sites for beta chemokines on human astrocytes. Glia. 1999;28(3):225–35.
3. Banisadr G, Gosselin RD, Mechighel P, Kitabgi P, Rostene W, Parsadaniantz SM. Highly regionalized neuronal expression of monocyte chemoattractant protein-1 (MCP-1/CCL2) in rat brain: evidence for its colocalization with neurotransmitters and neuropeptides. J Comp Neurol. 2005;489(3):275–92.
4. Banisadr G, Gosselin RD, Mechighel P, Rostene W, Kitabgi P, Melik PS. Constitutive neuronal expression of CCR2 chemokine receptor and its colocalization with neurotransmitters in normal rat brain: functional effect of MCP-1/CCL2 on calcium mobilization in primary cultured neurons. J Comp Neurol. 2005;492(2):178–92.
5. Boddeke EW, Meigel I, Frentzel S, Gourmala NG, Harrison JK, Buttini M, et al. Cultured rat microglia express functional beta-chemokine receptors. J Neuroimmunol. 1999;98(2):176–84.
6. Coughlan CM, McManus CM, Sharron M, Gao Z, Murphy D, Jaffer S, et al. Expression of multiple functional chemokine receptors and monocyte chemoattractant protein-1 in human neurons. Neuroscience. 2000;97(3):591–600.
7. Dorf ME, Berman MA, Tanabe S, Heesen M, Luo Y. Astrocytes express functional chemokine receptors. J Neuroimmunol. 2000;111(1–2):109–21.
8. Gosselin RD, Varela C, Banisadr G, Mechighel P, Rostene W, Kitabgi P, et al. Constitutive expression of CCR2 chemokine receptor and inhibition by MCP-1/CCL2 of GABA-induced currents in spinal cord neurones. J Neurochem. 2005;95(4):1023–34.
9. Daly C, Rollins BJ. Monocyte chemoattractant protein-1 (CCL2) in inflammatory disease and adaptive immunity: therapeutic opportunities and controversies. Microcirculation. 2003;10(3–4):247–57.
10. Sokolova A, Hill MD, Rahimi F, Warden LA, Halliday GM, Shepherd CE. Monocyte chemoattractant protein-1 plays a dominant role in the chronic inflammation observed in Alzheimer's disease. Brain Pathol. 2009;19(3):392–8.
11. Correa JD, Starling D, Teixeira AL, Caramelli P, Silva TA. Chemokines in CSF of Alzheimer's disease patients. Arq Neuropsiquiatr. 2011;69(3):455–9.
12. Zhang R, Miller RG, Madison C, Jin X, Honrada R, Harris W, et al. Systemic immune system alterations in early stages of Alzheimer's disease. J Neuroimmunol. 2013;256(1–2):38–42.
13. Westin K, Buchhave P, Nielsen H, Minthon L, Janciauskiene S, Hansson O. CCL2 is associated with a faster rate of cognitive decline during early stages of Alzheimer's disease. PLoS One. 2012;7(1):e30525.
14. Mattsson N, Tabatabaei S, Johansson P, Hansson O, Andreasson U, Mansson JE, et al. Cerebrospinal fluid microglial markers in Alzheimer's disease: elevated chitotriosidase activity but lack of diagnostic utility. Neuromolecular Med. 2011;13(2):151–9.
15. Galimberti D, Fenoglio C, Lovati C, Venturelli E, Guidi I, Corra B, et al. Serum MCP-1 levels are increased in

mild cognitive impairment and mild Alzheimer's disease. Neurobiol Aging. 2006;27(12):1763–8.
16. Ishizuka K, Kimura T, Igata-Yi R, Katsuragi S, Takamatsu J, Miyakawa T. Identification of monocyte chemoattractant protein-1 in senile plaques and reactive microglia of Alzheimer's disease. Psychiatry Clin Neurosci. 1997;51(3):135–8.
17. Nagele RG, Wegiel J, Venkataraman V, Imaki H, Wang KC, Wegiel J. Contribution of glial cells to the development of amyloid plaques in Alzheimer's disease. Neurobiol Aging. 2004;25(5):663–74.
18. Severini C, Passeri PP, Ciotti M, Florenzano F, Possenti R, Zona C, et al. Bindarit, inhibitor of CCL2 synthesis, protects neurons against amyloid-beta-induced toxicity. J Alzheimers Dis. 2014;38(2):281–93.
19. Kiyota T, Yamamoto M, Xiong H, Lambert MP, Klein WL, Gendelman HE, et al. CCL2 accelerates microglia-mediated Abeta oligomer formation and progression of neurocognitive dysfunction. PLoS One. 2009;4(7):e6197.
20. Yamamoto M, Horiba M, Buescher JL, Huang D, Gendelman HE, Ransohoff RM, et al. Overexpression of monocyte chemotactic protein-1/CCL2 in beta-amyloid precursor protein transgenic mice show accelerated diffuse beta-amyloid deposition. Am J Pathol. 2005;166(5):1475–85.
21. Kiyota T, Gendelman HE, Weir RA, Higgins EE, Zhang G, Jain M. CCL2 affects beta-amyloidosis and progressive neurocognitive dysfunction in a mouse model of Alzheimer's disease. Neurobiol Aging. 2013;34(4):1060–8.
22. El KJ, Toft M, Hickman SE, Means TK, Terada K, Geula C, et al. Ccr2 deficiency impairs microglial accumulation and accelerates progression of Alzheimer-like disease. Nat Med. 2007;13(4):432–8.
23. Naert G, Rivest S. CC chemokine receptor 2 deficiency aggravates cognitive impairments and amyloid pathology in a transgenic mouse model of Alzheimer's disease. J Neurosci. 2011;31(16):6208–20.
24. Bruban J, Maoui A, Chalour N, An N, Jonet L, Feumi C, et al. CCR2/CCL2-mediated inflammation protects photoreceptor cells from amyloid-beta-induced apoptosis. Neurobiol Dis. 2011;42(1):55–72.
25. Bondareff W, Mountjoy CQ, Roth M. Selective loss of neurones of origin of adrenergic projection to cerebral cortex (nucleus locus coeruleus) in senile dementia. Lancet. 1981;1(8223):783–4.
26. Madrigal JL, Garcia-Bueno B, Hinojosa AE, Polak P, Feinstein DL, Leza JC. Regulation of MCP-1 production in brain by stress and noradrenaline-modulating drugs. J Neurochem. 2010;113(2):543–51.
27. Wyss-Coray T, Loike JD, Brionne TC, Lu E, Anankov R, Yan F, et al. Adult mouse astrocytes degrade amyloid-beta in vitro and in situ. Nat Med. 2003;9(4):453–7.
28. Lee CY, Landreth GE. The role of microglia in amyloid clearance from the AD brain. J Neural Transm. 2010;117(8):949–60.
29. Bruno V, Copani A, Besong G, Scoto G, Nicoletti F. Neuroprotective activity of chemokines against N-methyl-D-aspartate or beta-amyloid-induced toxicity in culture. Eur J Pharmacol. 2000;399(2–3):117–21.
30. Eugenin EA, D'Aversa TG, Lopez L, Calderon TM, Berman JW. MCP-1 (CCL2) protects human neurons and astrocytes from NMDA or HIV-tat-induced apoptosis. J Neurochem. 2003;85(5):1299–311.
31. Harkany T, Abraham I, Timmerman W, Laskay G, Toth B, Sasvari M, et al. beta-amyloid neurotoxicity is mediated by a glutamate-triggered excitotoxic cascade in rat nucleus basalis. Eur J Neurosci. 2000;12(8):2735–45.
32. Kessels HW, Nabavi S, Malinow R. Metabotropic NMDA receptor function is required for beta-amyloid-induced synaptic depression. Proc Natl Acad Sci U S A. 2013;110(10):4033–8.
33. Reale M, Iarlori C, Thomas A, Gambi D, Perfetti B, Di NM, et al. Peripheral cytokines profile in Parkinson's disease. Brain Behav Immun. 2009; 23(1):55–63.
34. Lindqvist D, Hall S, Surova Y, Nielsen HM, Janelidze S, Brundin L, et al. Cerebrospinal fluid inflammatory markers in Parkinson's disease – associations with depression, fatigue, and cognitive impairment. Brain Behav Immun. 2013;33:183–9.
35. Pattarini R, Smeyne RJ, Morgan JI. Temporal mRNA profiles of inflammatory mediators in the murine 1-methyl-4-phenyl-1,2,3,6-tetrahydropyrimidine model of Parkinson's disease. Neuroscience. 2007;145(2):654–68.
36. Sriram K, Miller DB, O'Callaghan JP. Minocycline attenuates microglial activation but fails to mitigate striatal dopaminergic neurotoxicity: role of tumor necrosis factor-alpha. J Neurochem. 2006;96(3):706–18.
37. Kalkonde YV, Morgan WW, Sigala J, Maffi SK, Condello C, Kuziel W, et al. Chemokines in the MPTP model of Parkinson's disease: absence of CCL2 and its receptor CCR2 does not protect against striatal neurodegeneration. Brain Res. 2007;1128(1):1–11.
38. Guyon A, Skrzydelski D, De Giry I, Rovere C, Conductier G, Trocello JM, et al. Long term exposure to the chemokine CCL2 activates the nigrostriatal dopamine system: a novel mechanism for the control of dopamine release. Neuroscience. 2009;162(4):1072–80.
39. Edman LC, Mira H, Arenas E. The beta-chemokines CCL2 and CCL7 are two novel differentiation factors for midbrain dopaminergic precursors and neurons. Exp Cell Res. 2008;314(10):2123–30.
40. Grewer C, Gameiro A, Zhang Z, Tao Z, Braams S, Rauen T. Glutamate forward and reverse transport: from molecular mechanism to transporter-mediated release after ischemia. IUBMB Life. 2008;60(9): 609–19.
41. Lakhan SE, Kirchgessner A, Hofer M. Inflammatory mechanisms in ischemic stroke: therapeutic approaches. J Transl Med. 2009;7:97.
42. Frangogiannis NG. Chemokines in ischemia and reperfusion. Thromb Haemost. 2007;97(5):738–47.
43. Kim JS, Gautam SC, Chopp M, Zaloga C, Jones ML, Ward PA, et al. Expression of monocyte chemoattractant protein-1 and macrophage inflammatory protein-1 after focal cerebral ischemia in the rat. J Neuroimmunol. 1995;56(2):127–34.

44. Minami M, Satoh M. Chemokines and their receptors in the brain: pathophysiological roles in ischemic brain injury. Life Sci. 2003;74(2–3):321–7.
45. Tsukuda K, Mogi M, Iwanami J, Min LJ, Jing F, Oshima K, et al. Irbesartan attenuates ischemic brain damage by inhibition of MCP-1/CCR2 signaling pathway beyond AT(1) receptor blockade. Biochem Biophys Res Commun. 2011;409(2):275–9.
46. Hughes PM, Allegrini PR, Rudin M, Perry VH, Mir AK, Wiessner C. Monocyte chemoattractant protein-1 deficiency is protective in a murine stroke model. J Cereb Blood Flow Metab. 2002;22(3):308–17.
47. Dimitrijevic OB, Stamatovic SM, Keep RF, Andjelkovic AV. Absence of the chemokine receptor CCR2 protects against cerebral ischemia/reperfusion injury in mice. Stroke. 2007;38(4):1345–53.
48. Losy J, Zaremba J. Monocyte chemoattractant protein-1 is increased in the cerebrospinal fluid of patients with ischemic stroke. Stroke. 2001;32(11):2695–6.
49. Schilling M, Strecker JK, Schabitz WR, Ringelstein EB, Kiefer R. Effects of monocyte chemoattractant protein 1 on blood-borne cell recruitment after transient focal cerebral ischemia in mice. Neuroscience. 2009;161(3):806–12.
50. Belmadani A, Tran PB, Ren D, Miller RJ. Chemokines regulate the migration of neural progenitors to sites of neuroinflammation. J Neurosci. 2006;26(12):3182–91.
51. Liu XS, Zhang ZG, Zhang RL, Gregg SR, Wang L, Yier T, et al. Chemokine ligand 2 (CCL2) induces migration and differentiation of subventricular zone cells after stroke. J Neurosci Res. 2007;85(10):2120–5.
52. Yan YP, Sailor KA, Lang BT, Park SW, Vemuganti R, Dempsey RJ. Monocyte chemoattractant protein-1 plays a critical role in neuroblast migration after focal cerebral ischemia. J Cereb Blood Flow Metab. 2007;27(6):1213–24.
53. Andres RH, Choi R, Pendharkar AV, Gaeta X, Wang N, Nathan JK, et al. The CCR2/CCL2 interaction mediates the transendothelial recruitment of intravascularly delivered neural stem cells to the ischemic brain. Stroke. 2011;42(10):2923–31.
54. Tei N, Tanaka J, Sugimoto K, Nishihara T, Nishioka R, Takahashi H, et al. Expression of MCP-1 and fractalkine on endothelial cells and astrocytes may contribute to the invasion and migration of brain macrophages in ischemic rat brain lesions. J Neurosci Res. 2013;91(5):681–93.
55. Stowe AM, Wacker BK, Cravens PD, Perfater JL, Li MK, Hu R, et al. CCL2 upregulation triggers hypoxic preconditioning-induced protection from stroke. J Neuroinflammation. 2012;9:33.
56. Berman JW, Guida MP, Warren J, Amat J, Brosnan CF. Localization of monocyte chemoattractant peptide-1 expression in the central nervous system in experimental autoimmune encephalomyelitis and trauma in the rat. J Immunol. 1996;156(8):3017–23.
57. Glabinski AR, Balasingam V, Tani M, Kunkel SL, Strieter RM, Yong VW, et al. Chemokine monocyte chemoattractant protein-1 is expressed by astrocytes after mechanical injury to the brain. J Immunol. 1996;156(11):4363–8.
58. Harrison JK, Jiang Y, Chen S, Xia Y, Maciejewski D, McNamara RK, et al. Role for neuronally derived fractalkine in mediating interactions between neurons and CX3CR1-expressing microglia. Proc Natl Acad Sci U S A. 1998;95(18):10896–901.
59. Cardona AE, Pioro EP, Sasse ME, Kostenko V, Cardona SM, Dijkstra IM, et al. Control of microglial neurotoxicity by the fractalkine receptor. Nat Neurosci. 2006;9(7):917–24.
60. Madrigal JL, Leza JC, Polak P, Kalinin S, Feinstein DL. Astrocyte-derived MCP-1 mediates neuroprotective effects of noradrenaline. J Neurosci. 2009;29(1):263–7.
61. Comini-Frota ER, Teixeira AL, Angelo JP, Andrade MV, Brum DG, Kaimen-Maciel DR, et al. Evaluation of serum levels of chemokines during interferon-beta treatment in multiple sclerosis patients: a 1-year, observational cohort study. CNS Drugs. 2011;25(11):971–81.
62. McManus C, Berman JW, Brett FM, Staunton H, Farrell M, Brosnan CF. MCP-1, MCP-2 and MCP-3 expression in multiple sclerosis lesions: an immunohistochemical and in situ hybridization study. J Neuroimmunol. 1998;86(1):20–9.
63. Simpson JE, Newcombe J, Cuzner ML, Woodroofe MN. Expression of monocyte chemoattractant protein-1 and other beta-chemokines by resident glia and inflammatory cells in multiple sclerosis lesions. J Neuroimmunol. 1998;84(2):238–49.
64. Simpson J, Rezaie P, Newcombe J, Cuzner ML, Male D, Woodroofe MN. Expression of the beta-chemokine receptors CCR2, CCR3 and CCR5 in multiple sclerosis central nervous system tissue. J Neuroimmunol. 2000;108(1–2):192–200.
65. Fife BT, Huffnagle GB, Kuziel WA, Karpus WJ. CC chemokine receptor 2 is critical for induction of experimental autoimmune encephalomyelitis. J Exp Med. 2000;192(6):899–905.
66. Izikson L, Klein RS, Charo IF, Weiner HL, Luster AD. Resistance to experimental autoimmune encephalomyelitis in mice lacking the CC chemokine receptor (CCR)2. J Exp Med. 2000;192(7):1075–80.
67. Ge S, Shrestha B, Paul D, Keating C, Cone R, Guglielmotti A, et al. The CCL2 synthesis inhibitor bindarit targets cells of the neurovascular unit, and suppresses experimental autoimmune encephalomyelitis. J Neuroinflammation. 2012;9:171.
68. Elhofy A, Wang J, Tani M, Fife BT, Kennedy KJ, Bennett J, et al. Transgenic expression of CCL2 in the central nervous system prevents experimental autoimmune encephalomyelitis. J Leukoc Biol. 2005;77(2):229–37.
69. Baron P, Bussini S, Cardin V, Corbo M, Conti G, Galimberti D, et al. Production of monocyte chemoattractant protein-1 in amyotrophic lateral sclerosis. Muscle Nerve. 2005;32(4):541–4.

70. Gupta PK, Prabhakar S, Abburi C, Sharma NK, Anand A. Vascular endothelial growth factor-A and chemokine ligand (CCL2) genes are upregulated in peripheral blood mononuclear cells in Indian amyotrophic lateral sclerosis patients. J Neuroinflammation. 2011;8:114.
71. Lindl KA, Marks DR, Kolson DL, Jordan-Sciutto KL. HIV-associated neurocognitive disorder: pathogenesis and therapeutic opportunities. J Neuroimmune Pharmacol. 2010;5(3):294–309.
72. Liu Y, Jones M, Hingtgen CM, Bu G, Laribee N, Tanzi RE, et al. Uptake of HIV-1 tat protein mediated by low-density lipoprotein receptor-related protein disrupts the neuronal metabolic balance of the receptor ligands. Nat Med. 2000;6(12):1380–7.
73. Giulian D, Vaca K, Noonan CA. Secretion of neurotoxins by mononuclear phagocytes infected with HIV-1. Science. 1990;250(4987):1593–6.
74. Pulliam L, Herndier BG, Tang NM, McGrath MS. Human immunodeficiency virus-infected macrophages produce soluble factors that cause histological and neurochemical alterations in cultured human brains. J Clin Invest. 1991;87(2):503–12.
75. Cinque P, Vago L, Mengozzi M, Torri V, Ceresa D, Vicenzi E, et al. Elevated cerebrospinal fluid levels of monocyte chemotactic protein-1 correlate with HIV-1 encephalitis and local viral replication. AIDS. 1998;12(11):1327–32.
76. Kelder W, McArthur JC, Nance-Sproson T, McClernon D, Griffin DE. Beta-chemokines MCP-1 and RANTES are selectively increased in cerebrospinal fluid of patients with human immunodeficiency virus-associated dementia. Ann Neurol. 1998;44(5):831–5.
77. Lee EO, Kim SE, Park HK, Kang JL, Chong YH. Extracellular HIV-1 Tat upregulates TNF-alpha dependent MCP-1/CCL2 production via activation of ERK1/2 pathway in rat hippocampal slice cultures: inhibition by resveratrol, a polyphenolic phytostilbene. Exp Neurol. 2011;229(2):399–408.
78. Toborek M, Lee YW, Pu H, Malecki A, Flora G, Garrido R, et al. HIV-Tat protein induces oxidative and inflammatory pathways in brain endothelium. J Neurochem. 2003;84(1):169–79.
79. Yao H, Peng F, Dhillon N, Callen S, Bokhari S, Stehno-Bittel L, et al. Involvement of TRPC channels in CCL2-mediated neuroprotection against tat toxicity. J Neurosci. 2009;29(6):1657–69.
80. Drexhage RC, Padmos RC, de Wit H, Versnel MA, Hooijkaas H, van der Lely AJ, et al. Patients with schizophrenia show raised serum levels of the pro-inflammatory chemokine CCL2: association with the metabolic syndrome in patients? Schizophr Res. 2008;102(1–3):352–5.
81. Semple BD, Bye N, Rancan M, Ziebell JM, Morganti-Kossmann MC. Role of CCL2 (MCP-1) in traumatic brain injury (TBI): evidence from severe TBI patients and CCL2–/– mice. J Cereb Blood Flow Metab. 2010;30(4):769–82.
82. Wu Y, Wang X, Mo X, Xi Z, Xiao F, Li J, et al. Expression of monocyte chemoattractant protein-1 in brain tissue of patients with intractable epilepsy. Clin Neuropathol. 2008;27(2):55–63.
83. Rosito M, Deflorio C, Limatola C, Trettel F. CXCL16 orchestrates adenosine A3 receptor and MCP-1/CCL2 activity to protect neurons from excitotoxic cell death in the CNS. J Neurosci. 2012;32(9):3154–63.
84. Godefroy D, Gosselin RD, Yasutake A, Fujimura M, Combadiere C, Maury-Brachet R, et al. The chemokine CCL2 protects against methylmercury neurotoxicity. Toxicol Sci. 2012;125(1):209–18.
85. Bray JG, Reyes KC, Roberts AJ, Ransohoff RM, Gruol DL. Synaptic plasticity in the hippocampus shows resistance to acute ethanol exposure in transgenic mice with astrocyte-targeted enhanced CCL2 expression. Neuropharmacology. 2013;67:115–25.
86. Semple BD, Frugier T, Morganti-Kossmann MC. CCL2 modulates cytokine production in cultured mouse astrocytes. J Neuroinflammation. 2010;7:67.
87. Braun MC, Lahey E, Kelsall BL. Selective suppression of IL-12 production by chemoattractants. J Immunol. 2000;164(6):3009–17.
88. Hinojosa AE, Garcia-Bueno B, Leza JC, Madrigal JL. CCL2/MCP-1 modulation of microglial activation and proliferation. J Neuroinflammation. 2011;8:77.
89. Aberg ND, Brywe KG, Isgaard J. Aspects of growth hormone and insulin-like growth factor-I related to neuroprotection, regeneration, and functional plasticity in the adult brain. ScientificWorldJournal. 2006;6:53–80.
90. Chiu K, Yeung SC, So KF, Chang RC. Modulation of morphological changes of microglia and neuroprotection by monocyte chemoattractant protein-1 in experimental glaucoma. Cell Mol Immunol. 2010;7(1):61–8.
91. Hinojosa AE, Caso JR, Garcia-Bueno B, Leza JC, Madrigal JL. Dual effects of noradrenaline on astroglial production of chemokines and pro-inflammatory mediators. J Neuroinflammation. 2013;10(1):81.

Index

J. Camps (ed.), *Oxidative Stress and Inflammation in Non-communicable Diseases – Molecular Mechanisms and Perspectives in Therapeutics*, Advances in Experimental Medicine and Biology 824, DOI 10.1007/978-3-319-07320-0, © Springer International Publishing Switzerland 2014

MIX
Papier aus verantwortungsvollen Quellen
Paper from responsible sources
FSC® C105338

If you have any concerns about our products,
you can contact us on
ProductSafety@springernature.com

In case Publisher is established outside the EU,
the EU authorized representative is:
Springer Nature Customer Service Center GmbH
Europaplatz 3, 69115 Heidelberg, Germany

Printed by Libri Plureos GmbH
in Hamburg, Germany